GEOGRAPHY OF TRANSPORT DEVELOPMENT IN INDIA

About the Editor

Balkrishan C. Vaidya (born at Sirasgaon, Dist. Wardha in Maharashtra) holds M.A. (Geography), M.A. (Sociology), M.Ed., M. Phil., Dip. T. English and Ph. D. He possesses brilliant academic carrier and has awarded Robertson and Korea Darbar Gold Medals for his outstanding performance. Dr. Vaidya is actively engaged in research work of applied nature. His interest is in the field of transport, population, tribal development, agricultural innovation and writing since his college days. He has contributed more than two dozen articles in scholarly reputed Journals and is a member of many academic societies. Dr. Vaidya is a author of two books, namely, *Agricultural Landuse in India,* and *Readings in Transport Geography: A Regional Perspective*. The authro has successfully completed IRDP project on 'A Study of Soil Types and Cropping Pattern in Yadhoda Basin.' Dr. Vaidya is currently working on project entitled "Socio-economic Status of Tribal Population in Maharashtra—A Geographical Perspective". Dr. Vaidya is presently working as Reader in Department of Geography, University of Pune.

GEOGRAPHY OF TRANSPORT DEVELOPMENT IN INDIA

Edited by
B.C. VAIDYA

CONCEPT PUBLISHING COMPANY PVT. LTD.
NEW DELHI-110059

ISBN: 978-81-7022-957-5

First Published 2003

Published and Printed by

Concept
Since 1974

Concept Publishing Company Pvt. Ltd.
A/15-16, Commercial Block, Mohan Garden,
New Delhi-110059 (India)
T : 91-11-41101460
E : publishing@conceptpub.com, *W*: www.conceptpub.com

Dedicated

To

Shri Sanghpriya Gautam

who enlightened the downtrodden people

through his modest speeches and

made men with some mission

FOREWORD

The development index of the country is measured through the transport network and linkages. After independence, India has developed one of the largest network of roads, railways and air transport. Amongst these, the roads and rail transport development is highly dependent on geographical terrains. Availability of such transport facilities allows the smooth movement of commodities and people and accelerates the development. The produce can reach at relatively low cost and in a short time to those who require, with the existence of well-developed transport network. India is a large country with various types of geographical terrains. The highly developed transport systems not only helps in transporting perishable goods from production site to the users in good condition, bust also helps in bringing in cultural unity. India is a country of diverse religions, practices and cultures and can be united through better development of transport network.

The road technology developed by the Indian researchers cannot be implemented due to lack of resources. At the same time, in a developing country like India, goods and people are transported using various mades such as bullock-carts, motorized vehicles of various dimensions and other transport system including railways. The wear and tear of the roads is, therefore, much faster. There is thus, urgent need to invest heavily on improving Indian road quality as we have entered in the second phase of development.

The book edited by Dr. B.C. Vaidya entitled "Geography of Transport Development" becomes thus highly important. I have scanned through the various articles and chapters in this book and I am confident that this book will be very useful not only the students of Economic Geography but also to those who are in transport industry and transport planning area.

I have thus great pleasure while writing the foreward for this important book that will help the students of Economics, Economic Geography and Transport Industry. The book will be useful to understand Indian transport development and will give the researchers an idea about the current status of Transport Geography.

DR. A.S. KOLASKAR

Vice-Chancellor
University of Pune

PREFACE

The transportation is considered as an important aspect to be a pace setter in human life and it is an urgent needs to gear up the avaialble resources and population for better exploitation. The present project tries to open new direction on transport aspect at national, regional and and at the level of metropolitan cities to understand their intra and inter-linkages with other nodes, junctions, port and airports. And, thus, the exploring and movement of primary resources, it is very much required road and railway access to interior places wherever the potential of such resources are found in the region.

The present book studies the various facets of raod and rail transport in India with giving geographical significance in the region. An attempt was also made to provide recent statistics, if possible, in the text that would certainly support to understand transport development in India. From this point this volume carries immense value and provides the base for understanding transport development and would be sound enough to provide background of transport to the Indian transport in general, and teachers and students who are pursuing researchers in this branch.

This treatise has an outcome of help received from many friends and well wishers. The few names can be enlisted here. During completion of this volume Shri Sanghpriya Gautam, New Delhi was instrumental to encourage me. I an greatful to our present Vice-Chancellor Dr. Ashok Kolaskar for well-wishing and ever encouraging attitude in this academic presuit. Faculty member of Geography Department, Pune University and G.D. Walvekar and Shashank Waghmare for their kind help. Shri Ramdas Jagtap and many individuals are stimulus in guiding this project. My beloved wife Chhaya, despite her heavy housework cheered me in every stage of completion of this work. Thanks to Shri Sadashiv Surve for his kind enough in timely giving typing work and Shri M.D. Pathak for proof reading of this manuscript.

A special thanks to Shri Ashok Kumar who was undertaken this task for printing and publishing with personal touch and maintain high quality.

Pune

Dr. B.C. Vaidya
Department of Geography,
University of Pune, Pune-411007

SECTION-I

INDIAN TRANSPORT SYSTEM

SECTION-II

METROPOLITAN TRANSPORT SYSTEM

SECTION-III

REGIONAL TRANSPORT SYSTEM

LIST OF CONTRIBUTORS

G.S. Badigar
Department of Geography
G.V. M.S. College,
Ponda-403 401(Goa)

K.R. Badigar
Department of Geography
Govt. College of Arts, Science & Commerce
Sanquelim-403 401 (Goa)

A.P. Mahanta
Department of Geography
J.B. College, Johrat
Johrat-785 001 (Assam)

Anup Saikia
Department of Geography
Arunachal University
Itanagar-791 112
(Arunachal Pradesh)

Y.V. Ramanaiah
Department of Geography
S.K. University
Anantapur-515 003
(Andhra Pradesh)

K. Krishnaiah
Department of Geography
S.K. University, Anantapur-515003
(Andhra Pradesh)

B. Sreenivasa Reddy
(Research Scholar)
Department of Geography
S.K. University, Anantapur-515003
(Andhra Pradesh)

Salil Latay
(Reearch Scholar)
M.S. University, Vadodhara
Vadodhara-390 002 (Gujarat)

B.C. Vaidya
Department of Geography
University of Pune
Pune-411 007 (Maharashtra)

Tapati Mukhopadhyay
Department of Geography
Siddarth College of Arts, Science & Commerce,
Mumbai-400 001 (Maharashtra)

T.A. Kant
Department of Geography
University of Kashmir,
Srinagar-190006 (Jammu & Kashmir)

M.Y. Bhat
Department of Geography
University of Kashmir
Srinagar-190006 (Jammu & Kashmir)

R.C. Tiwari
Department of Geography
University of Allahabad,
Allahabad-211 004 (Uttar Pradesh)

Shankeshwar Tripathi
Department of Geography
University of Allahabad
Allahabad-211004 (Uttar Pradesh)

H. S. Yadav
Department of Geography
University of Allahabad
Allahabad-211004 (Uttar Pradesh)

Kuntala Lahiri Dutta
Department of Geography
University of Burdwan
Burdwan-713 104 (West Bengal)

Gopa Samant
Department of Geography
University of Burdwan
Burdwan-713 104 (West Bengal)

K. Laxmi
Department of Geography
Madurai Kamaraj University
Madurai-625 021

N. Krishnan
Department of Geography
Madurai Kamaraj University
Madurai-625 021

Mondira Dutta Zutshi
Centre for Asian Division
School of International Studies
Jawaharlal Nehru University
New Delhi-110 067

Bupinder Zutshi
C.S.R.D.
Jawaharlal Nehru University
New Delhi-110 067

G. Marthandan
(Research Scholar)
Department of Geography
University of Chennai
Chennai-600 005
(Tamil Nadu)

S. Subhiah
Department of Geography
University of Chennai
Chennai-600 005
(Tamil Nadu)

Sudesh Nangia
C.S.R.D.
Jawaharlal Nehru University
New Delhi-110 067

L.C. Mahajan
Ministry of Railway
Jawaharlal Nehru University
New Delhi-110 067

Vijaya Bhole
Department of Geography
Osmania University
Hyderabad-500 007
(Andhra Pradesh)

Amresh Deshpande
10, Dahanukar Colony
Kothrud, Pune-411 029
(Maharashtra)

M.V. Bagade
383, Sukrawarpeth,
Opp. Phadgate, Pune-411 002
(Maharashtra)

M. Parida
Transport Engineering
Department of Civil Engineering
University of Roorkee,
Roorkee-247 667 (Uttar Pradesh)

S.S. Jain
Centre for Transport Engineering
& Transport Engineering Science
Dept. of Civil Engineering
University of Roorkee,
Roorkee-247 667 (Uttar Pradesh)

Vidyadhar Deshpande
Dy. Director of Town Planning (Traffic) Traffic & Transportation Department,
S.N. 74/2, Sahakarnagar,
Pune-411 009 (Maharashtra)

Patrali Chatterjee
Department of Geography
Utkal University,
Bhubaneswar-751 004 (Orissa)

V. L. Singh
Department of Geography,
Utkal University,
Bhubaneswar-751 004 (Orissa)

Arup Khan
Central Institure of Road Transport
Nashik Road, Pune-411 003
(Maharashtra)

Sukla Bhaduri
Department of Geography
Calcutta University,
Calcutta-700 019

B. R. Meena
Department of Geography
Govt. P. G. College
Dausa-303 303 (Rajasthan)

G.D. Singhal
Department of Geography
Magadh University,
Bodh-Gaya (Bihar)

Awadh Tiwary
Department of Geography
Gaya College, Gaya (Bihar)

Abha Pathak
Department of Geography
Gaya College, Gaya (Bihar)

S.C. Sharma
Department of Geography
M.L.K. (P.G.) College,
Balrampur, Gonda,
(Uttar Pradesh)

D.N. Tripathi
Department of Geography
M.L.K. (P.G.) College,
Balrampur, Gonda
(Uttar Pradesh)

O.P. Mishra
Department of Geography
M.L.K. (P.G.) College,
Balrampur, Gonda
(Uttar Pradesh)

INTRODUCTION

Transport geography is a branch of economic geography dealing with patterns and modes of transportaion, quantitative studies of the movement of goods and people, and the relationship between transport and other geographical factors. Transport is considered as mojor plank of infrastructure for growth and development of any region as it provides basically linkages of nodes and centres with means and modes. The meaning of transport has been understood by economists, planners, engineers and geographers with different sense and widening the horizon of text. To geographers, transport geography is studied in term of spatial expression examining the geographical pattern of transport. Broadly speaking, the subject matter of transport geography can be categorized as : (i) Geographical pattern of transport network, (ii) Study of junctions and nodal points or terminals i.e. port and air-port. (iii) Study of commodity flow analysis, (iv) Study of people movement at the local regional, national and intenational levels, and (v) The entire system of hinterlands and hierarchical realationship associated with transport network.

Transport network in transport geography plays a pivotal role in reducing the spatial disparities and bringing about a balance and integrated development. This could be basic intra-structure helping in proper exploitation of the regional resources of the region. Transport network is thus, a necessary element of spatial expression. The linkages and flow between centres, their nature and size, function and accessibility are major consideration in structural aspects in transport geography.

The present project on "Geography of Transport Development in India" consists of 29 chapters. These chapters are divided into three sections.

(i) Indian Transport System

(ii) Metropolitan Transport System, and

(iii) Regional Transport System.

The first part evaluates **"Indian Transport System"** on aspects, namely, road transport, railway transport, road network and social interface, railway freight rate and emerging express highway. The emerging of expressways for linking big cities was rightly explained in fifth chapter.

The second section is devoted to assess the **"Metropolitan Transport System"**. This part explained transport of seven metropolitan cities of India, namely, Delhi, Mumbai, Kolkata, Madurai, Chennai, Hyderabad, and Pune. This part also explains the manifold problems of these city and suggestions are made to improve transport facility by giving transport plans.

Third section was fully dealt with to make exhaustive study of **"Regional Transport System"**, where authors have attempted to evaluate transport at state level, district level, and taluka level. Authors have fully justified in their treatise by preparing scientific plan of transport network of concerned area or region. Moreover, the study on "Konkan Railway - A Case Study of Regional Development" has explored the various centres/nodes to be developed in order to extract resource, manufacturing them and transported by Konkan Railway within the study region and outside too.

Dr. (Mrs.) Tapati Mukhopadhay has written paper on "Railway Transport Network in India". She made an in-depth study of railway development since its inception to India and assessed the railway network through Five Year Plans. Beside this, author examined the fast growth of people and commodity transported both by public trains and commodity trains from 1970 to 1998. She also stressed that new lines of railway be constructed in mineral resources area and backward region in order to bring these area forward. The paper entitled "Road Transport Network Development in India" by Dr. Arup Khan. Dr. Khan has systematically attempted to assess the road development in India since Independence. The historical study of state highway and national highway through Five Year Plans have been thoroughly examined. According to author, expressway is must for fast development as it provides manifold benefit to users and Government and finally, author seeks to explore alternative sources for funds to construct the Zilla Parishad, roads, village roads, national highways and state highways.

Dr. Sudesh Nangia and L.C. Mahajan have written paper on "Road Transport Network and Social Interface in India". Authors examined decadal vehicle population in India from 1951 to 1999 and focussed on transport development in relation to infrastructure. This article is also given thought to rural agricultural production, price fluctuation, non-agricultural employment, transport pollution as well as depletion of natural resources, education, health, polluton and trade. The present paper attempted and examined the influence of involvement of voluntary, resettlement, landuse and management, landuse in road sector on one hand and roadside encroachment, roaduser education, traffic enforcement and traffic management on other hand, and finally suggested to exist transport means to cope up with demand and other social infrastructures. Mondira Dutta and Bupinder Zutshi have written paper on "A Study of Railway Freight Rate and Regional Development in India". The main objective of this paper was to analyse the major features of the railway freight. Authors have examined freight rate policy of railway transport and changes occurred from 1970 to 1998 by giving stastistical facts and figures for both light and heavy commodities.

Shri Amresh Deshpande has examined the present situation of the road network of India and highlighted the policy formulated by Government of India regarding expressway in our country in his paper entitled "A Study of Road Network with Special Reference to Expressway - Some Possibilities and Alternatives". Author has clearly identified some practical difficulties in construction of expressway and attempted to give certain suggestions in order to fast implementation of expressway construction scheme in India. G. Marthandan and S. Subhiah have written an article on "Development of Transport in Chennai Metropolitan Area". Authors have highlighted road and rail network of Chennai. With increasing industrial, educational, trade and commercial activities.The transport has to play a basic role to meet the all types of demand of the city. Moreover, authors examined the growth of personalized transport means in Chennai Metropolitan area and the growth of vehicled, bus fleet and persons trip in the study area. Authors discussed four railwaylines that handled the suburban traffic besides water and air transport in Chennai Metropolitan area.

In an article of Dr. (Mrs.) Tapati Mukhopadhyay on "Geography of Transport Network - A Case Study of Mumbai Metropolitan Region" has thoroughly studies the transport situation in study area and brought out various interesting features related to road and railway transport. The efforts were made to identify the gap regions within Mumbai Metropolitan Region and found out major nodal points/centres on periphery area of the study region. Hence, author has suggested further plan for study region based on nodes/centres for transport network to Mumbai Metropolitan Region. M. Parida and S.S. Jain in their article have made modest attempt to sudy the "MRTS Services of Delhi Metropolitan Area". This article studies the growth of vehicle population and its increasing problems in study region for the period from 1968 to 1999. Moreover, authors looked at new users difficulties of ring railway of Delhi Metropolitan Region.

Dr. Sukla Bhaduri has written an article on "Analytical Study of Mass Transport in Kolkata Metropolitan area", wherein she attempted to examine past and present mass transport service of study region with giving due emphasis on performance and perspectives of both state bus services and private bus services, their decadal growth and route composition. Moreover, she computed traffic density and fare structure in Kolkata metropolitan area.

The paper entitled "A study of Road Transport and Resource Development in Madurai Metropolitan Area" written by Dr. (Mrs.) K. Laxmi. She chosen 14 variables of resources and 18 variables for transport connectivity and has tried to evaluate functional inter-dependence and complementary character of resoures and transport by applying multiple correlation and regression analysis with a "Computer Aided Programe". This study has clearly brought out spatial changes in transport network and resource in the development of Madurai Metropolitan city. Moreover, this study has identified urban influence zones in rural surrounding area in relation to the availability of resources and road transport and finally suggested transport plan at micro level to avoid transport problems in future. Dr. Vijaya Bhole in her paper entitled "Dynamics of Urban Travel and its Impact on Urban Environment - A Case Study of Metropolitan Hyderabad" wherein author examined the urban travel pattern and evaluates the impact of urban environment. Trip generation analysis was computed out to determine traffic generating capacity of existing landuse. The efforts were made to identify traffic zones by taking four per cent household sample of Hyderabad City.

Dr. Bagade in his paper on "Transport Development in Pune Metropolitan Region", has studied the development of road transport in Pune city. He emphasised on changing industrial scenario around Pune city and rapid growth of population resulting fast growth of various types of means. Author has proposed various measures to improve public transport personalized transport in Pune Metropolitan city. Dr. Meena has written an article on "Railway and Road Transport in Rajasthan State". He evaluated the stages of development of rail and road transport network in 18th, 19th and 20th centuries in study region. After examining types of roads, and railway density and finally, suggestions are made to improve both railway and road transport in Rajasthan State.

Dr. (Mrs.) L.V. Singh has written an article on "A Study of Spatial Variations in Road Transport in Orissa State". She attempted to highlight the analytical framework of road network by applying beta, gama, kohning indices in present study for the study region. Moreover, author computed road accessibility and connectivity for intra-district settlements and has identified level of transport development in study area. Dr. B.C.Vaidya has written an article on "Road and Railway Transport Development in Maharashtra State". Author has made modest attempt to examine trend of road and railway development from 1951 to 1999 in Maharashtra State. The road and railway density was computed for per 100 km and per 100,00 persons by applying cartographic method. Moreover, author studies historical evolution of state highway, Zilla Parishad and villages roads in Maharashtra. Though, railways are well-connected with major cities and big market centres, it was suggested to make rail web in backward regions in Maharashtra state.

An article written by Dr. G. J. Badigar and Mr. K. R. Badigar on "A Study of Impact of Transport Development on

Industrialization in Gao state" examined the historical development of transport means and modes in study region. The structural characteristics of transport was studied by applying Alpha, Beta, Gamma and Cyclomatic number as well as nodal accessibility was found in urban centres within Gao state. In addition to this, relationship was computed between transport availability and industrial development for years 1970, 1980, 1990 and 1998. Shri Salil Latay has presented his thought in an article on "Transport Development in Gujarat state: A Case Study of Vadodara Railway Station".

He attempted to examine the development of Vadodra railway for the period 16 years since 1983. He found out noteworth changes occurred to Vadodra railway station and its surrounding. Vadodra is linked directly with Mumbai, Delhi and other prominent places in India. Vadodra is cosmopolitan in nature hence announcements are made in Marathi, Bengali besides Gujarathi at Vadodra railway station.

Dr. Anup Saikia on his paper on "Transport Development in the Northeastern region in India". He evaluated historical development railway and road transport in study region. He further attempted to computer the density of road in relation to population for year 1979 to 1999. It is identified that still rate of road and railway route is needed in this region in order to exploit more resources like jute, tea and petroleum. Dr. B.C. Vaidya has written an article on "Konkan Railway Transport — A Study in Regional Development". Author has studies the newly constructed Konkan Railway of 720 km from Roha to Mangalore along the West coast in India. He highlighted the latest technology used while constructing minor and major bridge and digging out the tunnels. Moreover, Dr. Vaidya explained that how Konkan Railway be benefited to local people for movement and commodity transport to this region. He identified certain resources that may be moved to increase industrial production by Konkan Railway. Dr. Vidyadhar Deshpande has written an article on "Urban Transport in Maharashtra". He examined transport of seven urban centers in Maharashtra. He studies the urban transport problems in detail. Beside this, author evaluated the transport efforts made in the state.

Dr. Y. V. Ramanaiah and others have written jointly paper

on "Structural Characteristics and Development of Road Network of Drought Prone Area in Anantapur District". The authors have deliberately chosen backward and drought prone area for study to suggest better strategy for the development of road transport in study region. After examining the history of road development, authors computed degree of accessibility and connectivity for the district. The spatial structure of road network is analyzed in term of topological and geometric component (graph-theoretic measures) by applying alpha, beta, gama, pie, eta, theta indices as well as knowing number and shimbel index for the study area. Moreover, bus traffic has been thoroughly studies in present treatise and measures were suggested to bring up backward district. The study on "Impact of Transportation on Toursits Development in Gulmarg and Pahelgaon in Kashmir State" was thoroughly justified by Dr. Kanth and Dr. Bhat. The movement pattern, trip generation and tourist flow have been examined for ten years at Gulmarg and Pahelgaon. Moreover, impact of flow concentration on environment was rightly evaluated by way of detecting land use changes and working out the recreational carrying capacity by computing users density at both places namely, Gulmarg and Pahelgaon.

Dr. Singhal, Dr. Awadh Tiwary and Dr. Abha Pathak have studied the road transport scenario of Magadh plain in Bihar in paper on "Road Transport in Magadh Plain, Bihar State". Author evaluated the historical development of transport network. This study identified nodal points/centers for transport and further volumes, bus trips services were computed in 'order to get traffic density flow. Moreover, new roads have been proposed in Magadh Plain by author in order to meet the increasing tourist of Gaya, Bodhgaya and other tourist places in study region and suggested better plan for future transport network in Magadh plain of Bihar State.

An article on "Analysis and Planning for Transport Network of Transport System in Gorakhpur District, Uttar Pradesh" was written by Dr. R.C. Tiwari and Dr. S. Tripathi. In this article authors evaluated the existing network of rural transport in Gorakhpur district. Accessibility traffic flow, connectivity matrix and gamma, beta indices were employed to find out the nodal points/centres. Moreover, transport problems have been rightly pinpointed and suggested a suitable guideline for its scientific planning. The paper entitled on "Spatial characteristics of Transport in Allahabad district, Uttar Pradesh", examined the multi-functional role of transport in rural area in Allahabad district. Rail road accessibility was computed and proposed new roads schematic linkage plan to rural area in Allahabad district.

Dr. (Mrs.) P. Chatterjee in her paper on "A Study of Transport Network in Ganjam District; Orissa State". She calculated accessibility and dispersion index to understand the scenario of the study region. Author has also provided further guideline for viabal plan of transport network to initiate the development at micro level and regional economy to get healthy growth of living environment.

Ms Gopa Samant and Kuntala Lahiri-Dutta have written on "Transport Network and Rural Development in Burdwan District, West Bengal". Authors examined the influence of urban market centers for promoting rural urban fringe linkages resulting improvement of transport network in study area. This paper written by Dr. A. P. Mahanta on "Evolution and Commodity Transacting Behaviour of Railway Transport in Brahmaputra Valley in Assam". Author evaluated the historical evolution of transportation in study area. The V shape railway network connecting tea, timber, coal and petroleum producing region. Moreover, he examined the trend of commodity transport of various types from 1951 to 1998. It is found that Brahmaputra valley and its environments has upward trends for tea, jute, timber, plywood, coal and petroleum production and transport. Dr. S.S. Sharma, Dr. D.N. Tripathi and O.P. Mishra have written an article on "The Geographical Study of Transport Network in Amethi Taluka of Uttar Pradesh". Authors computed accessibility of road this taluka. Traffic density was assessed, and, finally found out area on low density and high density of transport network. The density of roads and connectivity are higher in the block Amethi and Sanagrampur than those of Bhadarand Bhatna. It is found that links or feeder roads are inadequate in the rural areas.

B.C. VAIDYA

SECTION -I

INDIAN TRANSPORT SYSTEM

1

RAILWAY TRANSPORT NETWORK IN INDIA

Tapati Mukhopadhyay

Introduction

Transport is considered as an index of economic, social and commercial progress of a country. Railway transport is one of modes of transport for movement of people and commodities in India. Better railway transport always makes possible to farmers to put their less accessible land to more productive use. Transport of railway route connects both developed and underdeveloped regions. The vast Indian territory spreaded its railway transport throughout the country. The present treatise is an humble attempt to examine the historical facts of railway and to assess the development of railway transport in India.

The idea of Railway in India first occurred to Mr. George Clark, the Chief Engineer of Bombay Government in 1853 to connect Mumbai with Thane-Kalyan and with the Thal and Bhol Ghat, inclines successively Lord Dalhousie dispatched his famous minute in July 1850, emphasising the great social, political and commercial benefits which he expected from the Railways in India. Dalhousie's minute highlighted three objectives : (i) to attain highest degree of efficiency for military and civil administration, (ii) to collect the raw material to help British industry. He stated that England was calling aloud for raw material like cotton, and (iii) uniform dispersion of food and finished product to increase consumption of English goods.

The first Railway built in India was laid and worked as private companies incorporated in England with the Board of Directors in

England and with British capital on the guarantee of a fixed rate of exchange and a specified return on the capital invested by them. The first train in India began in middle of 19th century on 16th April, 1853 from Mumbai to Thane of about 32 km length. The policy of Guarantee System continued till 1869. During this period (1844-69) 6,847 km of railway routes were opened. Attempt was made in 1862 to secure the construction of Railway on terms more favourable to the Government by private investors. The term failed to attract private capital. Guarantee System was discussed very seriously in 1867 and 1869 and finally Secretary of State replied that state may construct political lines leaving the commercial lines open for private companies. Vigorous programmes of railway construction then followed with satisfactory results as far as cost were concerned. Among the lines constructed by the State Agency and from State Capital between 1969 and 1980 were Indus valley. Punjab Northern, Rajputana Malwa, Northern Bengal, Rangoon and Irravaly Valley and the Tirhut Railway. Of the total railway at the end of 1879 the state and the rest had directly constructed mileage of 8,998 at the end of year 1879, 2,175 miles had been directly constructed by State and rest by the companies.

Development of Railway

Railway developed rapidly as a result of a new and much more vigorous policy of national development affecting almost every branch of economic life. In 1905 a railway branch of public works department was abolished and railway board was established at the head of the railway system under the Department of Commerce and Industry. A fresh impetus was given in 1908 where the Mackey Committee on railway finance recommended an annual programme of pound 12,500,000 for capital expenditure on railway subject to periodical revision. Railway mileage increased from 24,725 in 1900 to 34,656 in 1914 and capital outlay from 32,953 crores to 49,509 crores rupees. During 1914-1920 only 2,333 km of new lines were opened. In 1920-1930 the annual increase of new rail routes was 803 km and in 1930 the total route km stood at 67,146 km.

As on March 31,1940 the Indian Railway route was 66,231 km. However at the dawn of Independence the country had to face the unfortunate partition of nation that had resulted into the partition of Railway mileage and assests also. In 1950-51 railway mileage became 53,596 km and 1,26,000 workers opted for India. These numbers of workers had to be absorbed by the Indian railway. Table 1.1 shows the growth of Indian railway route since 1950. At the time of Independence there were forty-two railway which were eventually nationalized and brought under central management. The entire system was regrouped into eight major zonal Railway with a view to make them more effective, economic and uniform.

Table 1.1 : *Post Independence growth of Railway in India*

Category	Years		
	1950-51	1997-98	Growth
Routes (000 km.)	53.6	63.0	9.4
Electrified (000 km.)	0.4	13.9	13.5
Revenue carring Freight (tonnes)	73.0	42.9	356.0
Passenger traffic (million)	1284.0	4342.0	3058.0
Passenger fare	1.5	18.5	17.0

Source : Ministry of Railway, New Delhi, 1999

Indian Railways at present have 63,000 km. of route length that had added 9,400 in 48 years. Which means an average in a year it had added 195 km. But in case passengers it had added 3,058 million passengers there is a phenomenal increase of freight and passengers but not exactly thė asset to Railway. Undoubtedly there is a phenomenal growth in the movement of freight and passenger. In 1950-51 the freight revenue was 139.30 crores rupees which was 52.9 per cent of the gross revenue earned by the railway that freight in 1990-91 reached 8247.0 crores rupees which contributed 66.2 per cent of the gross revenue and thus there is 15 per cent increase in revenue earning in last 40 years. There is rise in passengers revenue in absolute term (1980-90) but in relative term it is fluctuated between 1970-80 and 1980-90. Table 1.2 shows the passengers growth in Indian Railways.

The total number of passengers were 1,284 millions in 1950-51 which reached 3,858 million in 1990-91. Thus, there is 198.13 per cent increase in number of passengers in last 40 years. But in case of average rate per passenger in 1950-51 was 1.48 paise per passenger per km. In 1990-91 the average rate increased from

Table 1.2 : *Passenger Growth in Indian Railway*

Years	Passengers (Millions)	Increase/Decrease	Per cent of Increase/ Decrease
1950-51	1284	—	—
1955-56	1275	-9	0.70
1960-61	1594	310	24.14
1965-66	2082	798	62.14
1970-71	2431	1147	89.33
1975-76	2945	1661	129.36
1980-81	3613	2329	181.38
1985-86	3433	2149	167.36
1990-91	3858	2544	198.13
1997-98	4342	3059	238.23

Source : Annual Reports, Ministry of Railway, New Delhi.

1.48 paise to 10.60 paise. The actual increase 9.123 as per passenger km over 40 years of time. In fact, there is a small increase in passenger fare in since 1965-1980. But between 1980-85 there is an increase of 3.18 paise and further 1990-91 there is further increase 3.45 and in 1997-98 7.90 paise. It is apparent from the railway statistics that in last 20 years the Government of India has shifted the original policy of subsidizing the public utility services . On one hand, there is continuous increase of freight charge which is indirectly shared by the people on the other form 1980-97 there is 30 times increase of passenger fare.

Suburbanisation and the Impact on Railway

Prior to the independence major passengers for Indian Railways were long distance passengers as the railway were connecting between the raw material producing centres and the markets. In 1951 there were 872 million non-suburban passengers while suburban passengers were less than half the numbers (412 millions).

The Table 1.3 presents the suburban and non-suburban passengers in India from 1951 to 1991. This structure gets completely inverted in 1990-91 where the suburban passengers superseded the non-suburban passengers. In 1990-91 the suburban passengers were 2.259 million while non-suburban passengers 1599. Taking into consideration of percentage increase in passengers

Table 1.3 : *Suburban and non Suburban Passengers in India (1951-1991)*

Passengers (Million)	Years 1950-51	1990-91	Increase/ Decrease	Per cent of Increase/ Decrease
1. Passengers (million)	1284	3858	2574	200.4
i) Suburban	412	2,259	1847	448.0
ii) Non-suburban	872	1599	727	83.4
2. Passengers Travel (km)	66517	295644	229127	344.4
i) Suburban	6551	59578	53027	809.4
ii)Non-Suburban	59966	236066	17600	293.6
3. Passengers Earning (Crores Rupees)	98.2	314.0	3046.8	310.2
i) Suburban	6.58	357.0	5325.5	5325.5
ii) Non-suburban	91.6	2788.0	2696	2943.7

Source : Ministry of Railways, New Delhi, 1999.

over the 40 years. In case of non-suburban passengers there are 83.4 per cent increase while suburban passengers there is 448 per cent increase. The trend continues in case of passengers km. also the suburban passengers km increased by 809 per cent while non-suburban 293.6 per cent. Regarding the railway earnings it shows that during 1950-51 earning from suburban passengers were 65.80 million which became 3569.80 million in 1990-91. Thus, there is an absolute increase of 3,504 million. In case of earning from non-suburban passengers which was 916.6 million rupees in 1950-51 increased upto 314.0 million. These are absolute increase of 3102 per cent.

Thus suburban increase is closely associated with the expansion of urban area especially around the metropolitan cities like Kolkata, Mumbai, etc. On one hand there is an expansion of urban areas and on the other there is complete centralization of jobs within the old city core. That resulted into huge rush of population from the suburban as well as peri-urban areas towards the city centre and that had contributed a new face in Indian Railways. The importance of the suburban railway can be seen from the increase of suburban passengers as well as earning as shown in Table 1.3.

Planning and its Impact on Indian Railway

The process of planning with reference to Indian Railways began

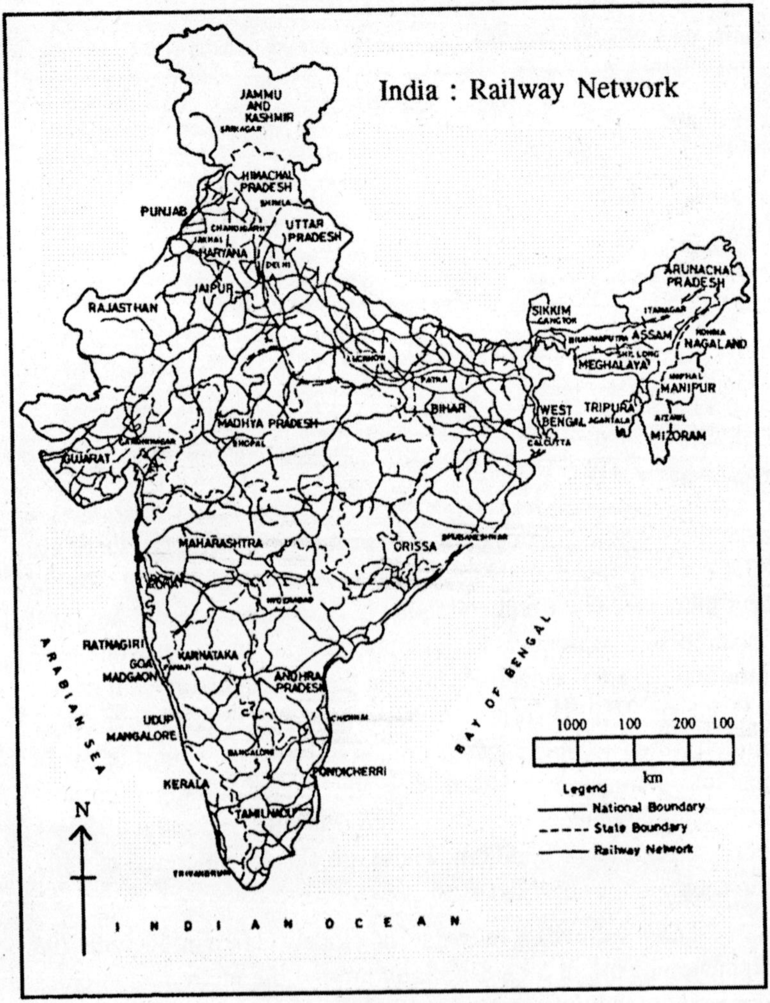

Fig. 1.1 : Indian Railway Network

with Independence. At the time of independence there were forty-two railway system consisting of 33 class-I railway, 10 class-II railway and 19 class-III railway. This also includes 32 lines owned by the ex-Indian states. The Government of India has regrouped the railway system and put into workable group of six zones and at present sixteen zones have emerged (1999) (Table 1.4 and Fig.1.1).

Table 1.4 : *Railway Zones in India*

Sr. No.	Railway Zones	Headquarters	Distance(Kms)
1.	Eastern Railway	Mumbai CST terminus	705
2.	Northern Railway	Kolkata	4228
3.	Northern Railway	New Delhi	11026
4.	North Eastern Railway	Gorakhpur	5011
5.	Southern Railway	Chennai	7010
6.	Western Railway	Mumbai (Churchgate	10013
7.	North Eastern Railway	Maligaon	3749
8.	South Central Railway	Secunderabad	
9.	South Eastern Railway	Kolkata	7270
10.	East Coast Railway	Bhuvaneshwar	Not yet demarcated
11.	East Central Railway	Hajipur	"
12.	North Central Railway	Allahabad	"
13.	North Western Railway	Jaipur	"
14.	South Central Railway	Bangalore	"
15.	West Central Railway	Jabalpur	"
16.	Bilaspur Railway	Bilaspur	"

Source : Annual Report, Minister of Railways, New Delhi,1999.

The main objective of planning of Indian Railway in the post Independence era has been to develop a rail system to provide supporting system for the planning growth of the nation. An indepth analysis can further be carried with Railway allotment from 1950-51 to 1985-90.

It is clear that there had been 48 times increase in the allotment. In 1950-51 it was 257 crores rupees which increased to 12,334 crores rupees in 1985-90. This apparent huge rise of Railway allotment does not necessarily bring the pattern of relative distribution of public finance. Table 1.5 shows in 1951-55 during the First Five-Year Plan a sum of 257 crores rupees was allotted for railway which contributed 11 per cent of the total public outlays which was a substantial increase in railway allotment does not necessarily bring the pattern of relative distribution of public finance. Table 1.5 shows in 1951-52 during the First Five Years Plan a sum of 257 crores rupees was alloted for railway that contributed 11 per cent of the total public outlays which was 2,378 crores during that period. However, during the, Second plan period there was a substantial increase in railway allotment from 257 crores rupees to 896 crores rupees that had contributed 19.17 per cent of the total plan public expenditure. There was a net rise of 8.37 per cent of the earlier plan period. This booster of Second plan period did not sustain for longer. There was fall in railway allotment both in absolute and relative terms in the next plan period.

Table 1.5 : *Railways and Five Year Plan in India*

Plan period	Total allotment (in crores rupees)	Allotment for Railways (Crores rupees)	Per cent of total public outlay for Railways	Increase/Decrease
1st Plan (1951-55)	2372	257	10.80	—
2nd Plan (1956-61)	4672	896	19.17	8.37
3rd Plan (1961-66)	8577	890	10.37	-8.08
4th Plan (1969-74)	15779	1050	6.65	-3.72
5th Plan (1975-79)	39426	2200	5.58	-1.07
6th Plan (1980-85)	110467	5100	4.61	-0.97
7th Plan (1985-90)	221436	12334	5.57	0.96

Source : Annual Report, Ministry of Railways, New Delhi.

The Second Plan has 896 crores rupees for railway allotment which became 890 crores rupees in the Third Plan period which shows a decline of 6 crores and there was a simultaneous decline in the share of railway allotment from the total plan public expenditure.The railway ratio which was 10.37 per cent of the plan public expenditure during the Third Plan period which precisely indicates decline of 8.8 per cent from the earlier plan period. This trend of decline has continued in the successive plan period also. In case of the Fourth Plan period from 1969-74 the allotment for the railway increased from 890 crores to 1050 crores rupees. There was an absolute increase of 160 crores rupees, but the relative share from the plan public expenditure which was earlier period 10.37 per cent became 6.65 per cent, thus between Second and Third plan period the decline was 8.8 per cent and between third and Fourth Plan that was from 1956-69 actual decline for the railway expenditure can be estimated at 12.52 per cent. These rates of decline became less in the next two plan periods though the relative decline of railway allotment from the plan public expenditure continued even for the next plan. In the fifth plan the railway allotment was 2200 crores rupees and that had contributed 5.50 per cent of the public expenditure which was again less than the earlier plan period a decline of 1.07 per cent.

In the Sixth Plan period the railway allotment was 5,100 crores rupees, there was an absolute rise of 2,900 crores rupees from the earlier plan, but taking into consideration the total plan expenditure it constituted 4.61 per cent and that further indicated a decline of 0.97 per cent from the earlier plan period. But the situation has changed from the Seventh Plan period. The Railway allotment during that period became 12,334 crores rupees that showed an absolute increase of 7,234 crores. It further showed the share of railway in relation to total plan public expenditure also increased which became 5.5 per cent. Thus there is an increase of 0.96 per cent from the earlier plan period. The Second Plan period may be taken as a benchmark year for the development of the railway system where railway allotment was 19.17 per cent of plan public expenditure. In the Seventh Plan though it has turned towards the positive side but still it remains 13 per cent less in relative terms from the Second Plan period. It is necessary to mention here that there is a structural change in the expenditure pattern of the Government. During the Second, Third, and Fourth Plan more than 60 per cent of the overall expenditure allowed for the public sector but from Fifth Plan onward there is a considerable decline of public sector expenditure and the rise of private sectors. In fact the Fifth and Sixth Plan 59 per cent of the overall became 49 per cent of the total plan outlay. Thus from the Third Plan period there is 18 per cent decline in the public sector expenditure as railway is an important component of public sector the decline in the relative expenditure the further growth and development of railway system of the nation. With the advancement of railway during the British period Indian economy had undergone a thorough change. A new spatial order of development emerged from the pre-colonial period which was mainly concentrated in the interior part of the country with local administrator gradually shifted towards the coastal area around the newly developed port based towns like Kolkata, Mumbai or Chennai. The Railway System acted as the major force for this coastal development.

Railway provides transportation facilities for various commodities from the reach hinterland was possible that further facilitated for the development of large wholesale markets adjacent to the city centre followed by the warehouse became the common metropolitan landuse features. The local and regional markets due to its connection to the railway system ultimately transform into national market system. Predominating commodities where the agricultural products which used to be pulled from the various producing centres to the market centres undoubtedly that had brought a revolutionary change in the pattern of distribution of the commodities. And also added a new dimension in the agricultural scenario of the nation, and that was commercial agriculture. The cultivation of tea, coffee, rubber and cotton were undertaken at the remote places like Assam, Darjeeling, Nilgiris, Maharashtra, etc. In July 1928 the Agricultural Commissions in their report recommended a close correlation between the railway authority and the agriculture department and that was the era of expansion of agro-based industry in India like Jute, cotton, sugar. The predominance of the agricultural commodity in the railway freight began during the first half of this century had continued even during the post-independence period. The railway board statistics for the year 1957-58 showed the importance of agricultural product in the freight. This can be further substained by taking into consideration the wagon loaded by various commodities. It is seen that the highest number of wagon were for grain and pulses followed by sugarcane, oilseeds, raw cotton and tea.

This typical freight composition of Indian railway consisting of mainly agricultural commodities has continued even 20 year after the Independence. But gradually there has been a change in composition of commodities that can be seen from Table 1.6 and 1.7. Table 1.6 indicates the decline in traffic for major commodities (between 1975-90) and Table 1.7 indicates increase in traffic in certain commodities of the same period. In case of decline the highest decline has taken place from raw Jute (-97.29) followed by Jaggary (-90.33) and organic manure (-79.24), rubber (-78.14), bricks and tiles (-74.39), bidi leaves (-73.4), oil seeds (-68.48), Fodder (-54.66), Paper (-46.27). These are the mainly agricultural products followed by some primary resources like Bidi leaves or sand. The decline of these commodities from the railway freight can be adjusted by the increase of road km.

Table 1.6 : *Decline in Major Commodities (1975 to 1990)*

(Production in 000 tons)

Commodity	Years		Increase/Decrease	
	1975	1990		
Fruits & Vegetable	713	71	642	-90.04
Bidi Leaves	173	46	127	-73.4
Organic Manure	265	55	21	79.24
Raw Jute	962	26	936	97.29
Tea	118	48	70	-59.32
Rubber	151	33	118	-78.14
Row Cotton	437	185	252	57.66
Jaggery	766	74	692	-90.33
Edible Oil	419	294	125	57.66
Spices	515	202	313	-60.77
Bricks and Tiles	582	149	433	-74.39
Sand	645	295	650	-57.29
Psper	899	483	416	-46.27
Oilseeds	1168	368	800	68.49
Food	1013	359	654	-64.56

Source : Ministry of Railway, New Delhi.

Table 1.7 : *Increase on Traffic for Major Commodities (1975 to 1990)*

Commodity	Years		Total Increase	Increase in Per cent
	1975	1990		
Coal	64309	135162	70853	1101.1
Iron Ore	25596	34481	8885	35.05
Mineral Ore	11663	24988	13325	114.25
Cement	11597	28870	17273	148.94
Iron and Steel	10758	12480	1722	16.00
Chemical	7169	18358	11189	156.07
Gypseem	979	1071	92	9.39
Lime Stone	8349	12480	4131	49.47
Soda Ore	442	560	118	26.69
Food Grains	16184	25348	9164	56.62

Source : Ministry of Railway, New Delhi.

The Table 1.8 shows increase road length and number of trucks on road between 1971 to 1995 there is 261 per cent increase of road km, and it is successively followed automobile revolution from 1971 to 1995, there was an increase of 1,25,6000 number trucks which shows in relative terms 366 per cent increase. This has taken away sufficient load from railway. Further there is an expansion of domestic market during the post-independence period, that has brought considerable changes in the merchant wholesaling character, and that has created a new mechanism of connectivity by the surface roadways. As a sequel to the decline of traffic in certain commodities there is also increase in some commodities they are mainly raw materials and industrial product (Table 1.7). The highest amount of increase has taken place in case of chemical manure (156.07) followed by cement (148.94), then minerals oil (114.25) and coal (110.1). Rests of the item are mainly the raw material for basic industrial production like limestone (49.47). Iron ore (35.07). It is necessary to mention here, though there is decline in some agricultural product but there is increase in foodgrain movement. The movement of the industrial raw material mainly started with the industrial and infrastructure development that country had undertaken during the Second Plan period and continued even after. These were development plan under public sector where railways plays an important role in commodity movement. The foodgrain movements are mainly due to the initiative of the Food Corporation of India. This is the major Government organ for collection and distribution of foodgrain in the nation and railway provides the cheapest mode of transportation.

Table 1.8 : *Road Length and Road Vehicles in India (1951-1999)*

Year	Road Km (000)	Trucks on Road
1951	400	—
1971	918	343
1981	1491	542
1991	2017	1411
1995-99	3320	1599

A close look at the pattern of distribution of railway in India as its stands today distinctly brought to the notice the spatial characteristics of concentration. Undoubtedly, railway expansion is closely associated with the physiographic set-up of the country followed by economic growth. In case of India, there was a third force that the desire and advantage of British rulers. The present railway distribution pattern is the outcome of combined forces. A high concentration of railway is seen in north Indian plan, which is covered by three zones eastern, northeastern and northern zones. The Kolkata is a headquarters for both Eastern and South-eastern railway. A broad guage electrified double line is connecting Kolkata city with coal field in Damodar valley in south Bihar. This was the original railway line constructed during the British period to connect the industrial belt of Kolkata with the raw material like coal.

Though the coal field belt is connected both with northern part of India through eastern Railway and southern part through South-eastern Railway. Eastern railway also connects north Bengal and Brahmaputra valley in Assam with the Kolkata city and port. Eastern Railway merges with northeastern Railway at Katihar then follows the north Indian plain along the river Ganga. This is the major trunk railway line in the country following the river and also simultaneously dividing the north Indian plain between north and southern part. A well coordinated railway system has expanded over the plain land consisting of two important Railways Zones northeastern and northern railways. The major part of Northern railway at Varanasi in Uttar Pradesh a well knit railway system spreaded between the river Ganga and Ghagra doab, Ghagra Gandak doab. Several urban centres developed, namely, Aunriha Zafarabad, Sultanpur, Baphamer also Got with bit metropolitan centre like kanpur, Lucknow. Beyond, Uttar Pradesh northern railway countinue and covers Punjab and Haryana. However, large parts of the track are till single line and non-electrified.

The intensity of the railway system that had developed in the north India plain specially north of the Ganga river gradually declined towards south. Though the most important trunk railway line of north India passes throuh the Ganga-Yamuna in the southern part of Bihar as well as in northern part of Madhya Pradesh, there is a sharp decline of railway intensity. Though, Damodar Valley Coal field had been connected with Kolkata city and port by the British as one of the initial initiative but that endeavour did not reach at Chattishgarh plateau presently located in eastern. Madhya Pradesh.

The area, in spite of, being rich in minerals is not reach connectivity. Poor connectivity continued in the whole of eastern Madhya Pradesh barring Bilaspur and Raipur there are no important railway junction. This situation has extended in western

Orissa also engulfed a part of eastern Maharashtra. The major concentration of railway in eastern coast along the coastal plain covering three states covering Orissa, Andhra Pradesh and Tamil Nadu. The Southeastern Railway passes through the coastal area Orissa and merges with the South Central Railway near the Vishakhapattanam in Andhra Pradesh which further merges with southern railway at Renigunta at the South Eastern border of Andhra Pradesh and continued down south. Till the emergence of Konkan Railway whole western coast was a railway deprive zone. But the establishment of Konkan railway has brought a new scenario with reference to the railway connectivity of the coastal districts belong to Maharashtra, Karnataka and Kerala. A new expectation for the development of western coastal districts may be materialised in the next millennium. The vast potentiality which the Konkan region have with reference to the horticulture industry and tourism remain unfolded even after fifty-two years of Independence that was mainly due to the lack of connectivity. Today a new area is opened that can be seen from the investment only in Maharashtra and up to October 1998, 56119.64 crores and 419 projects approved by the Government. Moreover, a rosy global trade figure of various horticulture products also are emerging.

Concluding Remarks

This exercise has undertaken a systematic analysis of the Indian railways in a temporal frame that has brought to the notice of several hidden contradictions that the railway system had acquired over one hundred and fifty years of its growth and expansion. These are the matters for concern of the railway management in the next millennium. Today, the Indian railway system have a route km of 62,976 km (1997-98) interlinked with 7,084 railway stations. It had 16 lacks regular employees carrying 4342 million passengers and 492 millions of originating freight with the help of 9,161 locomotives, 30,474 passenger coaches and 3,46,884 wagons. Undoubtedly, this is one of the largest public sector organized railway system in the world has its own centralized mechanism acquired through the trial adjustment for its administration and management.

Taking into consideration railway expansion, it is amazing to note that the major expansion in India that had taken place in the last century from 34 km in 1853 to 43,000 km in 1903 but in the next 50 years up to 1950 it had added only 10,000 km and in the post-independence period till the golden jubilee it had further added 10,000 km. Though it had added only 9.4 thousand km in the post-independence period but from this of passenger traffic it had added 3,058 million passengers and 356 million tons of freight. This non-expansion of route km further associated with non-renewable of the existing infrastructure with a slow pace of modernization. This tremendous pressure virtually brought the railway system in to collapse. A security of the planned expenditure on Indian Railway further highlighted that there is a systematic withdrawal of the Government from the railway expenditure. It was the pragmatic Second Five-Year plan that had made an allotment of 19.127 per cent of the total public expenditure on railway and today on the Seventh Plan it has been reduced to 5.57 per cent of the public expenditure. The actual fall from Second to Seventh Plan is 13.60 per cent. But the absolute rises of passengers traffic of 3,067 million and freight 346 million. This is the genesis of railway management today.

The public utility service like railways of a vast democratic nation like India having a passenger traffic of about 5 million cannot service without Government subsidy. This is the obligation of to provide the minimum services to the people of the nation. The railway cannot be converted into a profit making organization by shading their social responsibility. This is the time to the restructuring the Government allotment toward the railway and also to strengthen the technological support for skillfull management of railway. In spite of 63,000 km of railway it is much away from the uniform distribution of the railway system. Railways deprived area still exist in central part of the nation. Spatial spread of railway still maintain a colonial structure, heavily, located on the coastal plain and concentrated toward the administrative centre. There is no remarkable development expect the development of Konkan railway in the post independent period. The deviation from the original model is waiting the next century. There is a huge expansion of the class-III and class-IV type of towns in the last three decades and these are not yet connected by railway system. The connectivity of these newly emerging towns mainly dependent on roadways which is supported by the automobile revolution. This is the high time railway deprived area should be given prime importance for national interest.

To conclude the one hundred and fifty years old railway system of India today awaits for a facelift in all spheres to increase its productivity and to improve customer satisfaction. The system which has developed through public financing, needs to maintain its public utility stance. A steady decline in the Plan outlay in real terms is a very demoralising situation. With continuous decline in budgetary support, the Indian railways has to move increasing number of passengers and increasing volume of freight with nothing much left for the overall development of the railway system and network. This should change. The expansion, modernization and streamlining of the top- heavy administration should be the priority in the coming decade. It cannot merely be a commercial proposition the Indian railway should retain its age-old public utility characteristics with total commitment of the Government. This is the challenge of the future.

References

Amba Prasad (1968) : *Indian Railways*, Asia Publishing House, New Delhi.

Bhatnagar, N.L. (1961) : *Transport in India and Abroad*, Jaiprakash, Meerut.

Chakraborty, P.K.(1998) : *Transport and Economic Development*.

Govt. of India (1998) : Statistical Article of India (1998-99), Tata Service Ltd, New Delhi.

Govt. of India (1958) : *A Reveiw of Performance of Indian Railway*, Ministry of Railway, New Delhi.

Lord Dalhousie (1858) : *Minutes of Lord Dalhousie*,(4th July, 1850).

Ovasdi, J.M. (1973) : *Railway Administration*, Deep and Deep Publication, New Delhi.

Sahni, J.N.: *One Hundred Years of Indian Railway*, Ministry of Railways, New Delhi.

2

ROAD TRANSPORT NETWORK DEVELOPMENT IN INDIA

Arup Khan

Introduction

Roads are key for transportation of men and materials. In a developing country like India, with the rapid increase in intensity of traffic and introduction of heavier vehicles, the road pavements of better specialization and quality have become very necessary. There had been phenomenal increase in the vehicle population from 3 lakhs in 1951 to 275 lakhs in 1995-96 (MOST, 1997). In the same period, passenger traffic had increased by 65 times, road freight 60 times (Planning Commission, 1995-96). Whereas the road length was increased only 5 times (MOST, 1997). The road network in the country presents a vast spectrum ranging from rural roads to 4 lane highways/expressways. On the other hand, the country's population is projected to increase to 986 million by the year 2000 with the urban population increasing by 338 million (UNESCAP,1993). The growth in urban population will be mainly in the metropolitan and other large cities. The significant structural changes are taking place in the national economy with the highest growth in manufacturing followed by electricity, telecommunication, automobiles and gas sectors. Export trade is taking a big leap forward and is projected to touch 76 billion US dollors in 2000 from 25 billion US dollors in 1994-1995. Similarly imports will also approximately triple by 2000 (World Bank,1996). These activities are directly exerting pressure on transport as well as road system of India. There has also mismatched between resource distribution and the road system. The badly congested national highway system that accounts for about 40 per cent of the national traffic received about 20 per cent of the Eighth Five-Year Plan investment allocation (World Bank,1995). Less congested state highways and major district roads which account for about 60 per cent of the traffic received 57 per cent of the resources, lightly trafficked rural roads under the Minimum Needs Programme which account for less then 1 per cent of traffic received 23 per cent of the Eighth Plan Allocation (World Bank, 1995). The bulk of the transportation infrastructure in India is publicly owned and directly or indirectly financed by the Government. According to a recent study conducted by the World Bank explains that the Indian Transport System is predominantly important to serve fairly to rural-based economy.

ROAD DEVELOPMENT–A HISTORICAL PRESPECTIVE

Historically, the road development aspect was neglected in India in pre-independence period. The roads were built principally from the administrative and strategic view points. Yet, it must be admitted that the pace of road construction activity was accelerated with the advent of the British rule in India. The East-India Company, mainly a commercial corporation did not evince much interest in road making. Lord William Bentinck revived the idea of constructing roads by connecting Peshawar, Delhi and Kolkata. In his time, Military Boards used to look after their maintenance. It was only during the regime of Lord Dalhousie that a Central Public Works Department was created. In 1855 such departments were created in provinces also, eliminating military boards. The Report of the Royal Commission on Agriculture in India (1928) mentioned that "Along with railway construction, road building was necessary to feed the railways leading to a demand which remains today far from being completely satisfied for bridges and metalled roads at right angles to the railways and giving access to them in all the seasons of the year". The progressive policies of Lord May and Lord Rippon acted as a stimulus to road development in India as local affairs came under the direct control of local boards. This commission made an important remark in regard with transport network. "Transportation is an integral part of marketing and commercial development tends everywhere to enhance the value and importance of good road communication". "During the Second World War, road construction activities increased. It was considered urgent to repair and build new roads in the frontiers for movement of military and materials". The road development policy adopted by the British Government in India has become the subject of strong criticism. While Great Britain spent $500 millions on road development between 1921 and 1931 over and over and above the local expenditure, India was busy converting her roads in tracks, while United States sent a part a sum equal to rupees 6250 for the maintenance of one km of road. India followed since 1930 the policy of curtailment of road expenditure and was content with patch repairs only. During the same period, road expenditure of India was cut down from rupees 625 to 375 and later, to rupees 125 a kilometer. Therefore, it can be concluded that the whole system of road development has been somewhat unbalanced in pre-independence era.

Nagpur Plan

The Second World War period witnessed rapid increase in the road transportation which caused serious damages to extending roads. A coordinated and scientific plan was required to remedy the situation. In 1943, a committee was set up consisting of Chief Engineers of the various state known as Nagpur plan to devise methods of construction of roads and fix targets for next 20 years. The Nagpur Plan attempted to take an overall and comprehensive view of road network planning at the country level for providing connectivity linkages and communications by recommending short- and long-term measures to restore the damaged roads and also take future actions for stable transportation system. The Committee's broad recommendation were that the Central Government should take action for balanced and coordinated road development. For this purpose, the roads were divided into four classes viz. National Highways to traverse all the states sand form a principle network for overall commercial and strategic transportation. The State Highways to serve as main roads in the states, district roads to take the traffic from the main roads to the interior of the districts further classified as major district roads and other district roads depending upon their importance and village roads to provide links between the villages and other roads. National Highways should be regarded as main framework and these should be controlled, maintained by the Central Government. The roads should be designed, constructed and maintained by the highway departments of the states. All National and State Highways and major district road should be provided with the sufficient land and durable pavements and should have adequate bridges and crossways. Other roads should have proper earth and/or gravel surface, macadam tracks etc. as per traffic necessity. By the end of the Second Five Year Plan the targets of Nagpur Plan were more or less achieved. But in view of large scale industrialisation and agricultural developments these roads were not enough. Therefore, the Indian Parliament took a decision to devise a programme of road development which should be adequate for the next 20-30 years. With these objectives, a Committee of Chief Engineers of the States met under the Chairmanship of the Director General (Road Development), Government of India. This Committee suggested a 20 years (1961-1981) Road Development Programme known as 'Bombay Plan'.

Bombay Plan

The objective of this plan was to provide better linkages and communication in rural areas to check the population migration to the urban areas to create better connectivity with semi-developed and undeveloped areas, state headquarters, pilgrimage centres, ports important railway junctions, commercial and industrial centres and the areas of strategic importance. The road length was recommended to be increased to give a density of 32 km per 100 sq km and all the villages were to have a connectivity distance of 6.5 km to 13 km with metal roads and 2.4 km to 20 km with any other road depending upon degree of the development of the village. The Bombay Plan (1961-81) suggested some positive steps for developed, semi-developed , backward areas, agricultural areas, agricultural centres, industrial centres, social and administrative centres to be connected with major in the country. Maximum emphases was laid on connecting most of the villages with all, weather roads. Under this plan the road target was kept at 10,67,330 km. The most interesting feature of this plan was the road density which was 40 km/250sq km in Nagpur plan has been doubled i.e. 80 km/250 sq km. It was also decided that of the total road network

40 per cent road will be surfaced and 60 per cent unsurfaced roads.

Lucknow Plan

The Road Development Programme (1981-2000) known as Lucknow plan was launched during IRC Session at Lucknow in 1985. The Lucknow Plan also considered the criteria of connectivity and accessibility of the village with population over 5,000 by the year 2,000 in underdeveloped areas to attract industrialisation and to provide employment. It also recommended the increase in National Highways length to forms 100 km side grid. Construction of expressways in high density corridors linked with the improvement places of commercial tourists, major industrial, agricultural centres and the aspects of environment, proper maintenance and road safety etc.

Limitations of 20 Years Plans

The Nagpur plan and subsequent twenty year plans are considered as landmark in the road network planning at macro level. These plans classified the roads and envisaged a balanced connectivity of the places according to their importance and development of the country. These also recommended the standards and specifications for various types of the roads. The traffic demand during the last three decades has however, undergone a significant quantitative jump due to higher industrial and agricultural growth, population explosion, shift to urban areas thereby increasing pressure on cities and the concentration of development around and along existing developed centres. The twenty Year Plans did not take into account the total passenger and freight demands the origination and destination and the share of various modes like railways, coastal shipping etc. and the need for coordinated development. The planning of all surface transport modes has been so far done in isolation. This has resulted in extreme congestion on certain routes and all modes are being developed as independent competitive facilities rather than complimentary and supporting sub- system. However, the present plan of making express highways is one of the bold step by the Government of India coordinating each mode. The most of the expressways will run parallel to existing national highways.

During the past four decades the growth of traffic on Indian

Table 2.1 : *Road Length and Category-wise Growth Indices in India (1981 to 1996)*

Year	National Highways	Index Highways	State	Index PWD Rd.	Other	Index Raj Roads	Panchayat	Index Roads	Urban	Index Roads	Project	Index Roads	JRY	Index	Total	Index
1951	19811	100.0	@		173727	100.0	206408	100.0	–	–	–	–	–	–	399942	100.0
1961	23798	120.1	@		257125	148.0	197194	95.5	46361	100.0	–	–	–	–	524478	131.1
1971	23838	120.3	56765	100.0	276833	159.4	354530	171.8	72120	155.6	130893	100.0	–	–	914979	228.8
1981	31671	159.9	94359	166.2	421895	242.9	628864	304.7	123120	265.6	185511	141.7	–	–	1485421	371.4
1982	31777	160.4	95654	168.5	436882	251.5	651975	315.9	126095	272.0	187882	143.5	–	–	1530265	382.6
1983	31787	160.5	98969	174.3	455174	262.0	675392	327.2	129801	280.0	190003	145.2	–	–	1581126	395.3
1984	31817	160.6	97041	171.0	463210	266.6	69377	336.1	132416	285.6	202260	154.5	–	–	1620504	405.2
1985	31852	160.8	99262	174.9	489807	281.9	727077	352.3	135299	291.8	203651	155.6	–	–	1686948	421.8
1986	32088	162.0	100461	177.0	500869	288.3	747710	362.2	138647	299.1	206276	157.6	–	–	1726051	431.6
1987	32305	163.1	100970	177.9	525264	302.4	773421	374.7	141618	305.5	207014	158.2	–	–	1780592	445.2
1988	32446	163.8	116989	204.3	484052	278.6	864414	418.8	143537	309.6	207332	158.4	–	–	1847770	462.0
1989	32886	166.0	123131	216.9	488512	281.2	907556	439.7	157193	339.1	210607	160.9	–	–	1919885	480.0
1990	33479	169.0	124266	218.9	497476	286.4	941410	456.1	181171	390.8	206065	157.4	–	–	1983867	496.0
1991	33650	169.9	127311	224.3	509435	293.2	931288	451.2	186799	402.9	209737	160.2	329142	100.0	2327362	581.9
1992	33666	169.9	128622	226.6	521028	299.9	936288	453.6	189352	408.4	212485	162.3	440848	133.9	2462289	615.7

(Contd...)

Table 2.1 (*Contd.*)

1993	33916	171.2	129876	228.8	495438	285.4	1037663	502.7	195150	420.9	222057	169.6	564143	171.4	2678643	669.8
1994	34249	172.2	132401	233.2	500615	288.2	1048907	508.2	202013	435.7	224606	171.6	812440	246.8	2955231	738.9
1995	34262	172.9	134085	236.2	511046	294.2	1063790	515.4	202671	437.2	225481	172.3	886070	269.2	3057411	764.5
1996	34508	174.2	135187	238.2	763959	439.8	1021807	495.0	210680	454.4	237513	181.5	915990	278.3	3319644	830.0

Notes

1. Other PWD Roads are major district Roads
2. Panchayat Raj Roads include Zilla Parishad Roads, Village Roads and Panchayat Roads
3. Urban Roads include Municipal Roads, MES Rodas, Railway Roads and Port Roads
4. Project Roads include roads under forest, irrigation and electricity departments as well as Sugarcane Authority, Coal Mines Authority and Steel Authority Roads.

   @ Included in Other PWD Roads
5. Compiled from "Basic Road Statistics of India-1995-96" published by Transport Research Wing of Ministry of Surface Transport.

roads has increased from 10 to 12 per cent per annum and the road network has increased 4.5 per cent per annum from 4 lakhs km in 1950 to 33 lakhs km in 1996 (Padam, S. 1999). Number of trucks and vehicles has increased 10 to 40 times during this period, with consequent increase in share of passenger (26 to 80 per cent) and goods (11 to 60 per cent) traffic carried by roads. Goods traffic carried by roads has increased 35 folds since 1951 and passenger traffic 40 fold and this demonstrates the importance of roads in national economy. The road transport is expected to grow another 4 to 5 folds, over the present levels by the turn of the century (World Bank, 1995). The total road length of 3.3 million km (46 per cent are surfaced) comprises only 34,508 km (1.03 per cent) of National Highway, 135,187 km (4.07 per cent) of State Highway and 31,49,449 km (94.99 per cent) of district and rural roads as per 1996 statistics published by the Government of India. As mentioned earlier India's road network is seemingly very large (3.3 millions km), however the network cannot meet the requirements in term of accessibility and mobility for the economic growth and development of the country. The figure of total roads in kilometre has swelled recently because of the inclusion, first time village roads constructed under various employment generation programme. The all India figure of 72.4 km per 100 sq of area comes out as fairly large in international comparisons. But the picture varies widely across the country. Many areas like north-east and Jammu & Kashmir are inadequately served by the road network. Similarly, only half of India's 6 lakhs villages are connected by all weather roads with nearby market towns, while many habitations remain to be connected although successive plans, have focused on the problem of rural connectivity. Provides the detailed statistics about the road network in India and their growth 1981-1996). There has been a rapid expansion of the network in the 5 years between 1991 and 1996 increasing is of 67 per cent. This statistics disguises a major weakness the National and State Highway Road Systems which are the only roads capable of carrying heavy traffic, have been virtually stagnant. During 1991 to 1996 the National Highways grown by 3 per cent and State Highway by 9 per cent. The growth of roads under JRY was 178 per cent.

National Highways

National Highways is a central subject in terms of "Entry 23" of the Union list under the seventh schedule. The entire expenditure on the development, construction, improvement and maintenance of National Highways is met by the Government of India. The National Highway Act, 1956 (48 of 1956) governs the matters related to the National Highway System. The works are however, executed by the respective State Governments and Border Road Development Board in respect of certain highways on "Agency" basis through annual plan/Five Year Plans. Declaration of Roads as National Highways depends on several factors such as the criteria laid down by the Government for deciding new National Highways recommendations of the National Transport Policy Committee (NTPC, 1981) and inter-State priority of a road on all India basis including proximity to other National Highways in the region and overall availability of funds. These national highways are laid down and designed to – (i) connecting roads to adjacent countries, (ii) connecting roads to state capitals, (iii) connecting roads to major ports and important industrial or tourist centres, (iv) meeting roads to a very important strategic requirements, (v) connecting roads with carrying high density of traffic over an adequate length; and (vi) connecting roads which will enable sizeable reduction interval distance and achievement of substantial economy thereby.

The national highways are obviously most traffic-laden roads in India. The majority of this network is single carriage width only. just 1.04 per cent of the total network, they carry 40 per cent of the total road traffic. During the Eighth Plan Period (1992-1997) only 609 km of roads were added to the national highway system. The investment in national highways has also declined and its share came down from 1.4 of the public expenditure in the First Five Year Plan (1951-56) to 0.6 per cent in the Eighth Plan period (1992-97). Merely 1669 km (5 per cent) of roads on National Highways are multi-laned and 21,521 km are double laned. Shows the detailed analysis of distribution of National Highways and its growth. In India from 1986 to 1996. The National Highways has grown only 8 per cent. More than 10 per cent growth has been observed. Andhra

Table 2.2 : *Growth of National Highways in India (1986-96)*

Sr. No. States	Years 1986 (Length in km)	Share	Year 1992 (Length in km)	Share	Year 1996 (Length in km)	Share
All India	32088	100.00	33650	100.00	34508	100.00
States						
1. Andhra Pradesh	2352	7.3	2587	7.7	2949	8.5
2 Arunachal Pradesh	368	1.1	368	1.1	368	1.1
3. Assam	2227	6.9	2227	6.9	2254	6.5
4. Bihar	2118	6.6	2118	6.6	2218	6.4
5. Goa	224	0.7	225	0.7	225	0.7
6. Gujarat	1421	4.4	1572	4.7	1570	4.5
7. Haryana	656	2.0	655	1.9	656	1.9
8. Himachal Pradesh	589	1.8	722	2.1	232	2.1
9. Jammu & Kashmir	648	2.0	648	1.9	648	1.9
10. Karnataka	1968	6.1	1997	5.9	1997	5.8
11. Kerala	821	2.6	817	2.4	1011	2.9
12. Madhya Pradesh	2756	8.6	2976	8.8	2976	8.6
13. Maharashtra	2937	9.2	2969	8.8	2958	8.6
14. Manipur	434	1.4	434	1.4	438	1.3
15. Maghalaya	460	1.4	462	1.4	454	1.3
16. Mizoram	240	0.7	571	1.7	566	1.6
17. Nagaland	113	0.4	113	0.3	113	0.3
18. Orrisa	1624	5.1	124	5.8	1625	4.7
19. Punjab	977	3.0	988	4.8	988	2.9
20. Rajasthan	2521	7.9	2846	8.5	2846	8.5
21. Sikkim	62	0.2	41	0.1	53	0.2
22. Tamilnadu	1883	5.9	2002	5.9	2002	5.8
23. Tripura	198	0.6	195	0.6	198	0.6
24. Uttar Pradesh	2739	8.5	2754	8.1	2862	8.3
25. West Bangel	1630	5.1	1651	4.9	1651	4.8
Union Territories						
26. A & N Island	–	–	–	–	–	–
27. Chandigarh	15	0.1	34	0.1	34	0.1
28. Dadra & N. Haweli	–	–	–	–	–	–
29. Daman & Diu	–	–	–	–	–	–
30. Delhi	84	0.2	75	0.2	75	0.2
31. Lakshadweep	–	–	–	–	–	–
32. Pondicherry	23	0.1	25	–	–	–

Source : Basis Road Statistics of India, 1995-96. Transport Research Wing, Ministry of Surface Transport, Government of India, New Delhi, 1998.

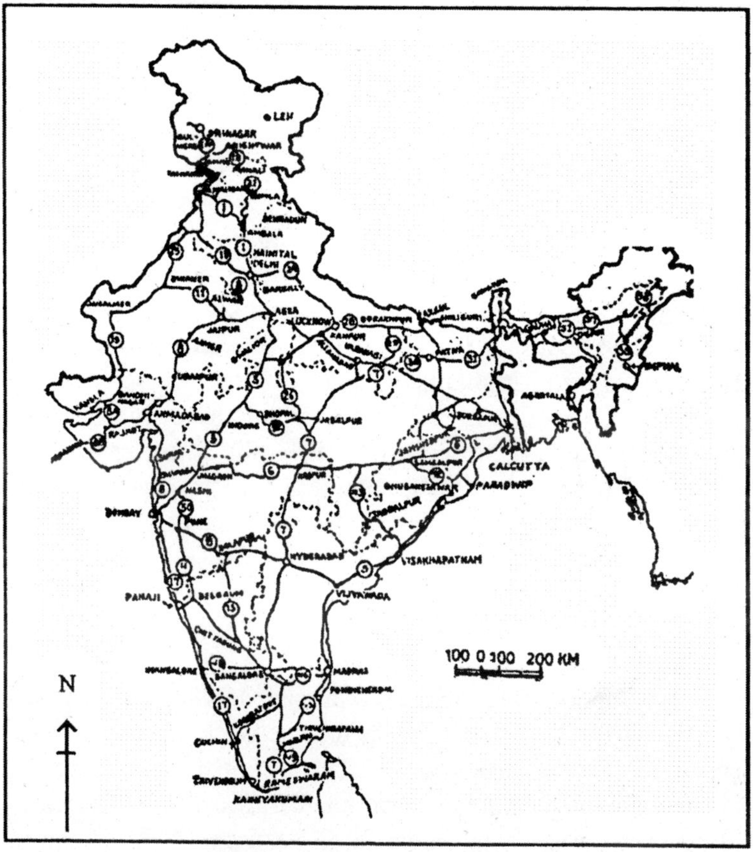

Fig. 2.1 : *India : National Highway*

Pradesh (25 per cent), Gujarat (11 per cent). Himachal Pradesh (24 per cent), Kerala (23 per cent) and Rajasthan (13 per cent). Rest of the States it is below 10 per cent. As per the State share of national highways is concerned from 1986 to 1996, none the States are having more than 10 per cent share. In 1996 following States are having the share of more than 5 per cent Andhra Pradesh (8.5 per cent), Assam (6.5 per cent), Bihar (6.4 per cent), Karnataka (5.8 per cent), Madhya Pradesh 8.6 per cent and Maharashtra (18.6 per cent), Rajasthan (6.2 per cent) and Tamil Nadu (5.0 per cent) (Table 2.2).

State Highways

The state highways carry bulk of the intra-state traffic. They constitute approximately 4.07 per cent of the total road network. Sixty per cent of the state highways are single lane, a mere 1.1 per cent are multi-lane highways. Table 2.3 presents the state-wise distribution of state highways and its growth from 1986 to 1996. It is interesting to mention here that from 1986 to 1996 the growth of state highway sector in India is only 35 per cent. The state-wise share of state highways is showing very interesting result. These are two states i.e. Maharashtra and Gujarat, where the road development is maximum. Of the country's total road in the state sector, the percentage share of Maharashtra is 24 per cent, Uttar Pradesh 7.7 per cent and Andhra Pradesh 6.49 per cent. As far as the growth is connected the highest growth of state highways from 1986 to 1996 has taken place in Gujarat (109 per cent). In Karnataka the growth of road development during the same decade was 44 per cent. In Maharashtra, it was 67 per cent. In Uttar Pradesh it was 26 per cent.

It is important to note here that the national and state highways together constitute 8 per cent of the total network and carry 80 per cent of the traffic volume. The differential growth rates in capacity augmentation on the arterial network and traffic demand has led to the present situation in which large sections of the National Highways and some sections of the State Highways are used well over 100 per cent of their intended economic capacity. The result of this mis-match between traffic growth and capacity is slow speeds, high vehicle operating costs, environmental pollution and high incidence of accidents. Further, the roads asset deteriorate fast because of increased traffic and heavier loads. Poor maintenance of asset also resulting much larger proportion of the network becoming unserviceable. It has been estimated that on 20 per cent of the paved roads are in goods condition in India compared to 70 per cent in South Korea and 50 per cent in Thailand.

PWD and Other Roads

The growth of PWD roads during the decade (1986-1996) is not showing very encouraging results. During the period between 1986 and 1996, it has only grown by 57 per cent. Andhra Pradesh (40 per cent), Karnataka (41 per cent), Gujarat (48 per cent), Maharashtra (62 per cent), Tamil Nadu (48 per cent). Zilla Parishad roads are basically the reorganised road of PWD roads. Hence no comparative analysis can be done. However, it can be mentioned here that the states wherever democratic decentralisation process of Governance has taken place those States are showing better result. The states like West Bengal, Maharashtra and Gujarat are the examples. The growth of Panchayat roads in India during this decade is around 72 per cent. In Arunachal Pradesh it is 40.4 per cent and in Orissa it is 42 per cent. During this decade maximum growth of community development and panchayat roads have been taken place in northeastern states, Manipur (22.4 per cent), Meghalaya (48 per cent), Nagaland (31.7 per cent). Orissa, Tamil Nadu and Rajasthan also registered higher growth. The urban roads includs municipal roads, railway roads and road built under port trust. Though there is no direct correlation between urbanisation and the distribution of urban roads statewise, however, it is interesting to see that the state like Andhra Pradesh and Himachal Pradesh, the growth in the road sector is astronomical. The urbanisation in Andhra Pradesh and Himachal Pradesh is higher that the national average. The growth in the road sector is 22.3 per cent and 38.0 per cent respectively. The state share of urban road is showing intersting results. The states having more than 5 per cent share of urban roads are: namely, Andhra Pradesh 8.1 per cent, Gujarat (5.9 per cent), Tamil Nadu (6.3 per cent), Uttar Pradesh (17.1 per cent) and West Bengal (9.1 per cent).

Project Roads

The project roads include road under forest, irrigation, electricity, sugarcane development authority, coal mines department and steel authority. Table 2.4 shows the state-wise distribution and growth of project roads from 1986 to 1996 the project roads has grown 15 per cent in the decade of 1986 to 1996. In 1996, Madhya Pradesh and Maharashtra are having 31 per cent of total project roads in India with a break-up of 20.0 per cent and 10.0 per cent respectively. The miximum growth of project roads during the decade has observed by Meghalaya 139 per cent followed by Karnataka 89

Table 2.3 : *State highways in India (1986-96)*

Sr. No. States	Years 1986 (Length in km)	Share	Year 1992 (Length in km)	Share	Year 1992 (Length in km)	Share
All India	100461	100.00	128622	100.00	135187	100.00
States						
1. Andhra Pradesh	8604	8.6	8851	6.88	8777	6.5
2 Arunachal Pradesh	–	–				
3. Assam	1873	1.9	1895	1.47	1811	1.34
4. Bihar	41.92	4.2	4192	3.26	4092	3.03
5. Goa	231	0.2	231	0.18	232	0.17
6. Gujarat	9442	9.4	19390	15.10	19717	14.60
7. Haryana	3136	3.1	3136	2.42	3135	2.32
8. Himachal Pradesh	3539	3.5	3680	2.86	3672	2.72
9. Jammu & Kashmir	688	0.7	688	0.53	688	0.51
10. Karnataka	7912	7.9	11288	8.78	11395	8.43
11. Kerala	2020	2.0	2048	1.59	2361	1.72
12. Madhya Pradesh	11637	11.6	11735	9.12	11789	8.70
13. Maharashtra	19260	19.2	31290	24.30	32249	23.90
14. Manipur	537	0.5	880	0.60	1437	1.06
15. Meghalaya	917	0.9	984	0.70	988	0.73
16. Mizoram	–	–	40	0.03	240	0.18
17. Nagaland	398	0.4	398	0.31	398	0.29
18. Orissa	2928	2.9	2928	2.28	4360	3.23
19. Punjab	1963	2.0	2178	1.69	2166	1.60
20. Rajasthan	7457	7.4	7136	5.55	10006	7.40
21. Sikkim	217	0.2	225	0.17	186	0.14
22. Tamilnadu	1864	1.9	1920	1.49	1934	1.43
23. Tripura	136	0.1	136	0.11	136	0.10
24. Uttar Pradesh	7769	7.7	9782	7.61	9647	7.14
25. West Bangel	3455	3.4	3455	2.65	3388	2.50
Union Territories						
26. A & N Island	250	0.2	273	0.21	273	0.2
27. Chandigarh	–	–	–	–		
28. Dadra & N. Haweli	–	–	–	–	42	0.03
29. Daman & Diu	–	–	–	–	–	–
30. Delhi	–	–	–	–	–	–
31. Lakshadweep	–	–	–	–	–	–
32. Pondicherry	36	0.04	63	0.05	68	0.05

Source : Basis Road Statistics of India, 1995-96. Transport Research Wing, Ministry of Surface Transport, Government of India, New Delhi, 1998.

Table 2.4: *Growth of project in India (1986-96)*

Sr. No. States	Years 1986 (Length	Share in km)	Year 1992 (Length	Share in km)	Year 1992 (Length	Share in km)
All India	206276	100.00	212485	100.00	237513	100.00
States						
1. Andhra Pradesh	7097	3.4	77742	3.6	7436	3.1
2 Arunachal Pradesh	987	0.5	1038	0.5	1055	0.4
3. Assam	5291	2.6	6098	2.9	6900	2.9
4. Bihar	14210	6.9	14944	7.0	13193	5.6
5. Goa	239	0.1	238	0.1	84	0.0
6. Gujarat	2408	1.2	2747	1.3	3167	1.3
7. Haryana	781	0.4	823	0.4	685	0.3
8. Himachal Pradesh	2229	1.1	2359	1.2	1591	0.7
9. Jammu & Kashmir	1352	0.7	1403	0.7	155536	0.6
10. Karnataka	8847	44.3	12039	5.7	16701	7.0
11. Kerala	3327	1.6	5164	2.4	5013	2.1
12. Madhya Pradesh	36821	17.9	44119	20.8	48379	20.4
13. Maharashtra	25250	12.2	20541	9.7	24990	10.5
14. Manipur	363	0.2	417	0.7	2417	0.2
15. Meghalaya	147	0.4	781	0.4	998	0.4
16. Mizoram	–	–	–	–	–	–
17. Nagaland	415	0.2	552	0.3	579	0.2
18. Orissa	9673	4.7	9682	4.6	13805	5.8
19. Punjab	9720	4.7	10087	4.7	10087	4.7
20. Rajsthan	11642	5.6	13508	6.4	13598	5.7
21. Sikkim	59	0.0	59	0.0	65	0.0
22. Tamilnadu	8201	4.0	6617	3.1	6508	2.7
23. Tripura	1417	0.7	1420	0.7	1420	0.6
24. Uttar Pradesh	51852	25.1	43895	20.0	52438	22.1
25. West Bangel	5537	2.7	5824	2.7	6593	2.8
Union Territories						
26. A & N Island	08	0.0	55	0.0	13	0.0
27. Pondicherry	133	0.1	153	0.1	153	0.1

Source : Basis Road Statistics of India, 1995-96. Transport Research Wing, Ministry of Surface Transport, Government of India, New Delhi, 1998.

per cent, Kerala 51 per cent, Orissa 43 per cent, Gujarat 32 per cent and Madhya Pradesh 31 per cent.

ROAD DEVELOPMENT THROUGH FIVE-YEAR PLAN

As regards the contribution of roads to the economy, there is direct contribution to the exchequer in the form of various types of taxes such as vehicle tax, road tax, licence fee and other taxes on fuel, tyres, spares etc. which generate revenue to the tune of 8,500 crore rupees annually (AITD, 1998). Whereas, the indirect contribution of roads to the economy comes through saving in transportation costs, such as Vehicle Operating Cost (VOC) which includes fuel, tyres, spares and crew etc. As per the details given by the Indian Roads Congress in their memorandum submitted to the Committee (Sub-Committee of India Infrastructure Report, 1996) that the annual vehicle operating cost on roads in the country has been estimated nearly 100,000 crore rupees of which 40,000 crore rupees is accounted for by national highways alone. According to Indian Roads Congress, extensive road user cost studies have established that by bringing improvement in roads it is possible to effect a saving upto 15 per cent VOC. Thus, this would mean an annual saving of 15,000 crore rupees consisting of 6,000 crores on national highways and 9,000 crore rupees on state roads in VOC. As the component of fuel constitutes about 15 per cent of VOC, the resultant savings in fuel consumption will be of the order of 2,250 crore rupees per annum including 900 crore rupees on national highways. This, in nutshell quantities the cost, bad roads, inflict on the economy due to increased vehicle operating cost. In addition, improved roads will cause travel time saving which will give a tremendous boost to economy by increasing the efficiency of transport system.

In several other countries the expenditure on roads was 80 to 100 per cent of even more of the total amount of road user taxation. A country like Japan spends on the roads as much as 130 per cent of the revenue collected from road and U.S.A. 97 per cent, whereas India spends barely 40 per cent. Thus, roads sector in India heavily cross subsidizes other sectors of the economy while itself remains starved of funds (World Bank, 1995). An analysis of the data of financial expenditure by Planning Commission through various Five Year Plans expenditure on transport sector as percentage to public sector is showing decreasing trend. In the First Five-Year Plan the percentage was 22.1 and it reduced to 12.6 per cent in the Eight Plans. As far as the allocation of the road sector as percentage to total transport sector is concerned. It has reduced to almost half of the First Five-Year Plan allocation. In the First Five-Year Plan the road sector received 6.5 per cent to the total transport sector allocation and it reduced to 3.0 per cent at the end of the Eighth Five-Year Plan. The picture on the National Highway sector is also showing the same trend. In the First Five-Year Plan the National Highways received 1.3 per cent of the road sector allocation and it reduced to 0.56 per cent at the end of the Eighth Five-Year Plan.

First Five-Year Plan

In 1950 is was realized by the Planning Commission that taking into consideration the area and population of the country the total kilometerage of the roads was less than the required level. It is therefore, suggested that all the states must prioritize its investment for the development of the road. The First Five Year Plan had originally provided for a sum of 97.6 crore rupees (later revised to 131.3 crores) for road development programmes to be undertaken by the centres and various states. The Government of India had assumed with effect from April 1,1947 financial responsibility for the construction and maintenance of a system of national highways upto a length of about 21,440 km. In 1947, when Five-Year state road plans were launched, 2560 km of roads and 120 bridges over big rivers have not been taken into consideration. During that period of 5 years 256 km of new roads were constructed and 2104 km of existing roads were improved. In addition 17 large bridges and numerous small ones were constructed. When the First Five-Year plan was launched,18 large bridges and 512 km of new roads were under construction. The First Five-Year plan provided for the completion of the work already undertaken and for the construction of 720 km of new roads and 48 very large bridges. Further, about 3420 km of roads were to be improved out of which about two-thirds were expected to be completed by 1955-56 and the rest would be in progress under the Second Five Year Plan. For the construction of national highways the following priorities were fixed (i) completion of all capital works which had been sanctioned and were in progress; (ii) provision of new bridges and missing road links along arterial highways; and (iii) replacement or strengthening of dangerous bridges and improvement of these sections of roads which were dangerous to traffic and were un-economical to maintain these bridges.

The First Plan provided for 27 crore rupees for the development of national highways and 4.24 crore rupees for the development of certain selected roads. In addition, a sum of rupees 21.15 lacks was provided for the Central Road Research Institute. By the end of the First Five-Year Plan, 1,194 km of missing links were constructed on the national highways; and 6,400 km of one way and 640 km of two-way modern roads were built. During the plan period, only 200 km of new roads were constructed; and about 800 km of existing roads improved. Taken as a whole, about 16,000 km of new surfaced roads and about 32,000 km of unmetalled roads were constructed; and about 16,000 km of existing roads were improved.

Second Five-Year Plan

The Second Five-Year Plan primarily aimed at the development of the industrial resources of the country which calls for a rapid development of transport sector. It was also realized that the roads has to play a major role. Accordingly, the Second Five-Year Plan provided for 246 crore rupees for the implementation of Central and State road plans. In addition a sum of 25 crore rupees was made available from the Central Road Fund. The road development programme under the Second Five-Year Plan mainly comprised the construction of missing links, major bridges and the improvement of existing roads. The total outlay on the works which were started in 1956 was estimated at 87.5 crores rupees. Three important roads were included for construction. These are, namely, the Passi-Badarpur Road, the West Cost Road connection between Pathankot and Udhampur. It was expected that by the end of the Second Plan about 240 km of new roads would be constructed more than 800 km would be upgraded.

During the period of the Second Five-Year Plan road

development programmes achieved a considerable measures of success. In 1956, just after the completion of First Five-Year Plan the total surface and unsurface roads were 1,95,200 and 3,16,800 km and it increased to 2,30,400 and over 4,00,000 km respectively on the eve of the Second Five-Year Plan in 1961. In so far as national highways are concerned, during the same period 1024 km of missing links and 40 major bridges were constructed. In addition 5,600 km of existing roads were improved and 3,680 km of carriage ways were widened.

Third Five Year Plan

The Second Five-Year Plan gave special consideration to the areas which were comparatively backward in the means of communication like those of Tripura, Assam, Manipur, Himachal Pradesh, Rajasthan, Jammu & Kashmir etc. The Central Government financially assisted the road development programmes to these areas. The road development programmes for the Third Plan were formulated in accordance with the broad objectives laid down in the 20 years road development plan (1961-81). The new road plan proposed targets of 4,03,200 km for surfaced and 6,48,000 km.for unsurfaced roads. The expenditure on road development in the Second Plan was estimated as 222 crore rupees. The cost of road development programmes included in the Third Plan amounted to 324 crore rupees of which the programmes in the state sector were to account for 244 crore rupees. In the Third Five-Year Plan against the original programme of the central for the construction of 60 new bridges and improvement of 3360 kilometres of roads the programme provided for 85 new brides and improvement of 4,500 kilometres in addition to the construction of 480 kilometres of missing links.

Fourth Five-Year Plan

By the end of the Third Five-Year Plan, around 1200 crore rupees had been spent on road development programmes. More than 5,00,000 km had been added to the road network. Even then the road system had large deficiencies and a great deal still remained to be done to equip the country with roadways of the length, breadth and weight-bearing capacity which is necessary to meet the needs of the country. It was estimated that by the end of the Fourth Five-Year Plan the country would require not only a much longer network of roads but also surfaces capable of withstanding heavier roads. At the same time, the need for faster movement would require an improvements in the quality of roads. A sum of 876 crores rupees was allocated to the road sector in the Fourth Five-Year Plan.Out of this, a provision of 418 crores rupees was made for road development programmes in the central sector. To bring about improvement in the existing network of national highways it was proposed to complete all missing road links and to improve all low-grade sections. Of the 17 missing bridges on the national highway system, 16 were to be completed and work on the remaining one would be in progress by the end of the plan.

As far as the road development in States and Union Territories is concerned, the plan provided for 458 crore rupees, priority being given to the removal of deficiencies in the existing road system such as the completion of missing links, improvement of low grade sections and reconstruction of weak bridges. The length of surfaced roads was expected to increase from about 3,25,000 kilometres at the end of 1968-69 to about 3,85,000 kilometres at the end of the Fourth Plan. The Fourth Five-Year Plan has achieved a new dimension by way of introducing a new organization for constructed roads along the border of India The Border Road Development Board (BRDB) was constituted in 1960. The BRDB is charged with the task of accelerating the economic development of border areas by making them accessible through the development of arterial routes. The work of building roads in these areas has been assigned to the General Reserve Engineering Force. The construction of 6,900 km of new roads and improvement of 4,700 km of existing roads are included in the board's immediate programme.

Fifth Five-Year Plan

The Fifth Plan laid emphasis on the completion of spill over works of the Fourth Plan which included and number of missing links and bridges. Provision was also made for certain new schemes of essential character particularly those relating to the safety of traffic. The important aspect of this plan was so activate the programme of BRDB. In this plan the construction of about 7,931 km of new roads, improvement of about 7,507 km of existing roads and surfacing of about 12,135 km of roads. Of these 7,470 km of new roads have been constructed, 490 km of existing roads improved and 8,913 km of roads surfaced. The BRDB has constructed the world's highest road from Manali (Himachal Pradesh) to Leh (Kashmir) the average altitude of the terrain being 4270 metres.

Sixth Five-Year Plan

It was realized at the last phase of Fifth Five-Year Plan that the quantum of traffic on the road system, both for passenger and freight has been showing a steady upward trend. The brunt of additional traffic is borne by the national highways. The total surfaced road length in the country rose from 2,62,700 kms in 1961 to 6,23,402 km in 1979. In the same year the length of national highways stood at about 29,340 km which was approximately 5 of the entire surfaced road length, but was estimated to carry about 25 to 30 per cent of the total road transportation load. It clearly meant that the development of national highways must receive priority which is deserved. Keeping in view the deficiencies existing on the national highways the Sixth Plan proposed to do the tasks, namely, double laning including strengthening, double laning without strengthening, widening to four lanes, construction of bye-passes. Besides the development of national highways the plan also laid emphasis on the development of strategic roads, centrally aided state roads and road communication in the sensitive border areas.

In the state sector, the most important programme included the construction of rural roads under the minimum needs programme, which envisages the provision of all weather link roads for all villages with a population of 1500 and above and for 50 of villages with a population of 1000-1500 within a period of ten years. It was expected to provide link roads to about 20,000 villages during the Sixth Plan period. One of the major policy thrusts during the period 1980-85 was the removal of deficiencies in the national highway system and upgrading some selected stretches of roads taking into account the projected traffic growth. The plan emphasis on the needs of the north-eastern region. The Sixth Plan provided an outlay of 630 crores rupees for roads under the central sector.

Seventh Five-Year Plan

Though an outlay of 830 crore rupees was provided in the Sixth Plan for the development of central sector roads, actual expenditure, however amounted to about 760 crore rupees the shortfall occurring largely as a result of cutbacks to allocations. During the last plan, 2,687 km of roads were upgraded as national highways and 5.77 lakhs km of different types of roads were added to the road network. In addition, 18,000 villages were connected with roads under the minimum needs programme 166 km of missing links were constructed, 4,224 km were widened to two lanes, 90 km were widened to four lanes and 50 by-passes, 7 missing major bridges and 467 minor bridges were constructed The objectives for road development in the Seventh Plan were – (i) Progressive removal of the existing deficiencies in the national highways state highways and major district road systems, (ii) Energy conservation, (iii) Environmental quality of high ways, (iv) Reduction of road accident rates, (v) Improvement in the road system and (vi) Ganeration of employment. The plan has provided 1019.75 crore rupees for Central sector roads. In the state sector, an outlay of 4180.29 crore rupees has been provided for road development, out of which an amount of 1,729 crore rupees has been earmarked for roads under the minimum needs programme.

Eighth Five Year Plan

The Eighth Five-Year Plan has laid emphasis on removing existing deficiencies in national highways through four laning and two laning of various sections, construction of bridges and by-passes etc. The first priority has been given to complete the ongoing work. Rural roads are essential for achieving the objective of integrated rural development. The priority of rural road development in the Eighth Plan were – (i) Linking of all villages with a population of 1,000 and above, and (ii) Special efforts to accelerate village connectivity in respect of backward regions as tribal areas the total outlay of Eighth Five-Year Plan in the road sector was 2600 crores rupees. In the state sector, an outlay of 10,610 crore rupees was provided for roads and bridges including 3,066.10 crore rupees for rural roads under the Minimum Needs Programme.

Ninth Five Year Plan

The general approach of the Eighth Five-Year Plan was to speed up the completion of the ongoing works with a special emphasis on externally aided projects in order to contain the cost and time over runs and their spill over to the next plan. The overall limitation of funds consequently meant insufficient provision for other highway works only about 600 km were added to the national highway network during the Eighth Plan. Similar constraints affected most of the state level programmes also. The rural roads programme all fall short of the target of 100 per cent connectivity for villages with population of more than one thousand. The focus of the road development programme in the Ninth Plan are as under (i) Strengthening and improving the crucial sections of the highway network through phased removal of deficiencies and multi-laning of high density corridors. (ii) Improving the road communication in remote areas such as the northeast. (iii) Providing all weather connectivity to remaining villages and promoting energy conservation, safety and environment protection.

An integrated policy for the transport sector would inform the road development programme with accent on gradual modification of the modal mix through a suitable inter-modal coordination mechanism. While the road network will be expanded and strengthened where traffic density warrants it, sectoral solutions like express highways will not be encouraged because of the high social costs of such projects and the impetus that such policies may provide to a further deterioration of the modal split. While private enterprise will be encouraged in funding the constructing and maintenance of roads and bridges, the role of budgetary support will remain predominant in view of the crucial role that basic infrastructure like roads play in catalyzing development. In mobilizing resources for the road sector the instrumentalities of user changes ande dedicated levies will be used to the extent possible for carrying out expansion and improvement work. According to the Planning Commission the requirement is around 62,000 crores. The external assistance of the order to 1,212 crores rupees was available from international donor agencies for a number of National Highways/ bridge works during the Eighth Plan. Despite higher allocation for

Central/State road projects during the Ninth Plan this source of funding will continue to receive priority. The estimated availability of external assistance for the National Highways is of the order of 4,000 crore rupees during the Ninth Plan. Similarly some State Highways projects including the State roads development packages in northeastern region and some rural road development network packages are expected to be posed from foreign funding. The larger external assistance would, however need to be balanced by higher domestic resources so that counter part fund requirements of the externally aided project do not affect local road development programme adversely. Over the years the maintenance of roads had received inadequate attention due to lack of funds. Availability of funds for maintenance ranged between 40 to 60 per cent of the normal requirements both at the central and the state levels. The Tenth Finance Commission had considered the maintenance issue and evolved a formula based on the norms devised by the Ministry of Surface Transport. The norms were applied to the total road length in each states/union territories to assess the requirements during the forecast period. Since the amount worked out was very large, the commission limited the provision to double the level provided by the Ninth Finance Commission.

The demand for transport is basically a derived demand. The tempo of agricultural and industrial development, trade and commerce reflects itself in the demand for movement of freight. Passenger traffic demand owes to a variety of factors including travel for business purposes, employment, social and cultural activities etc. Out of all the modes in transport sector railways and roadways account for a major share in meeting the traffic and transportation requirements of both freights and passengers. Due to historical reasons as mentioned earlier, rail was the predominant mode in the earlier years. However, road transport has steadily expanded its share in meeting the growing transport demand. This has been due to its versatility there is a hierarchical network of roads, ranging from national and state highways to mojor district roads, other district roads, village roads etc. permitting movement of a variety of mechanized and non-mechanized traffic modes, besides offering convenience. The road transport sector is characterised by the inherent heterogeneity of vehicles in organized and unorganised sectors in comparison with rail sector. Under this situation the movement on road transport is likely to grow much faster done ever. This would definitely put more pressure on the Indian Road System. In order to solve this problem, there is need for short-term and long–term planning which would involve rationalisation and augmentation or the road network, improvement and maintenance of the existing roads, construction of new roads of all types, viz. expressways, supernational highways national highways, state highways and other roads. simultaneously there is a need for vehicle fleet modernisation, traffic management and provision of roadside amenities, including vehicle parking maintenance and communications etc. The various studies carried out in the past from time to time, have estimated road traffic and given different committees. It is interesting to know that the estimate provided by committees was far from the reality. However, it is important to know the perspective provided by these committees chronologically. Report of the working group on road transport for Ninth Five Years plan: The working group on road transport was constituted by the Planning Commission in October 1995 under the Chairmanship of Secretary, Ministry of Surface Transport (MOST) to formulate programmes for the Ninth Five Year plan. The report was submitted in August 1996. The projections have been made by taking trend growth rates over 1986 to 1996 which is 10 per cent annum for freight tannel-km and passenger-km have also been projected by linking these figures and 9.9 per cent annum for passenger traffic. The freight tonne-kms and passenger-kms have also been projected by linking these figures to the assumed GNP growth of 6 per cent, 6.5 per cent and 7 per cent. At the 7 per cent GNP growth by the year 2007 the demand for passenger transport and frieght transport would be 1236 BPKM and 3480 BTKM respectively.

The average annual rate of growth in national income and transport demand has been considered for estimation. The road freight traffic in terms of BTKM increased from 6 in 1951 to 227 in 1987. As against this, the passenger traffic recorded an increase from 23 to 722 BPKM. Accordingly by 2000 the estimated BPKM and BTKM should be 2916 and 686 Report of Steering Committee on perspective planning for transport : The Steering Committee for

Transport Planning was set-up by the Planning Commission with a view to preparing a perspective plan for transport on an integrated basis. According to the committee for arriving at the past growth of transport sector the period was divided into two parts. The periods before and after the energy crisis of 1973. As an indicator comparative annual growth rates in national income and transport sector and industrial production growth rate has been used (97and 12 per cent). At 12 per cent rate of industrial production growth projected the BPKM and BTKM. In 2000 the projected BTKM and BPKM are 916 and 3998 respectivly. Vehicle fleet modernization and Road User Change Study (1988). The base year figures (1987-88) of BTKM and BPKM have been used for projection. The report has developed three scenarios. Those are lower medium and high. According to the high scenario by the year 2003 the BPKM and BTKM would be 1030 and 3844 respectively.

Resume

The analysis of the road transport sector explains that the transport demand has been increasing by leaps and bounds. It can be resolved unanimously that the role of road sector will be more and more important to support the economic development and industrial growth. The National Highway system constituting about 1.7 per cent of the total road network is sharing 40 per cent of the total transport thereby facing severe sprain. Many deficiencies like severe capacity constraints, weak pavement requires improvement and expansion. The state and major district roads are also facing the same problems like capacity constraints, non-black topped surfaced lack of paved shoulders etc. The other district roads, village roads and roads under Panchayati Raj which provides the connectivity to the villages with population of 1,500 and above needs to be black topped. The severity of road accidents needs to be reduce through the better designing of road. The R&D facility of road construction needs to be enhanced upto the international level. Needless to mention that, within this dichotomous situation where the source of Government fund is drying up for the investment in the road sector through plan expenditure, it is essential to explore alternate sources of funding of road development.

Acknowledgement

The author is grateful to Mr. Rajeev P. Mote, investigators and Ms Sherly Mathew, Steno-typist, CIRT, Pune for their help in the preparation of tables.

References

Asian Institute of Transport : National Infrastructure Report, Development (1998), New Delhi.

Confederation of Indian Industry : Report of the National (1997) Confereance on - Transport Infrastructure (Mimeo). The Emerging Scenario, Dec., 1997, New Delhi.

Dalvi, M.Q. (1997) : *Transport Planning and Policy in India,* Himalayan Publishing House, Bombay.

Govt. of India (1998) : Basic Road Statistics of India, 1995-96, Transport Research Wing, MOST, New Delhi.

Govt. of India (1998) : Motor Transport Statistics of India,1997, TRW, MOST, New Delhi. Govt. of India (1991) : Census of India, Paper- 2, MHA, New Delhi.

Govt. of Andhra Pradesh (1994) : Statewide Strategic Options Study, A.P. State Road Improvement Project, L R & B Deptt. GOAP, 1994.

Gregory, J.W. (1931) : *The Story of the Road.*

Gregory, M. K. (1941) : *Transport Development and Coordination.*

Indian Road Congress (1996) : Seminar on Perspective Planning for Road Development in India (Vol.1&2), New Delhi.

Ministry of Finance (1996) : India Infrastructure Report, GOI, New Delhi.

Ministry of Surface Transport (1988) : Final Report on Vehicle Fleet Modernisation, Highway System Study, Dec., 1988.

Padam, S. (1999) : Road Sector in India - A Status Paper (mimeo), CIRT, Pune.

Padam, S. (1999) : Transport Sector in India : The Present Scenario, *Indian Journal of Transport Management,* Vol. 23 No.11, Nov. 1999.

Rail India Techno-Economic Survey : Comprehensive study of Road Traffic Flows in the Country (Mimeo), New Delhi.

Srivastava, S.K. (1987) : *Economics of Transport,* S.Chand & Company, New Delhi.

The Planning Commission (1999) : Various Documents (1951-1999).

The World Bank (1991) : *India : Road Planning and Budgeting - A Review,* 1991, Washington. D.C.

The World Bank (1995) : Transport Sector Long Term Issues, Report No.13192 March, 1995, Washington, D.C.

The World Bank (1989) : Policy Issues in Road Transport, Report No. 8057.

The World Bank(1996) : Five Year of Stabilization and Reform : The Challanges Ahed, Washington, D.C.

United Nations Economic and Social Commission Asia and Pacific (UNESCAP, 1993). State of Urbanization in Asia of and the Pacific, New York.

3

ROAD TRANSPORT NETWORK AND SOCIAL INTERFACE IN INDIA

Sudesh Nangia and *L.C. Mahajan*

Introduction

Transport development plays a vital role in the socio-economic development of a country. It is important that these developments are sustainable. Such developments of transport network introduced an element of dynamism and mobility in the study region for locational structures and patterns. The means of tronsport provides effective system of vital arteries which could be considered as functional organization. It serves both short terms function of satisfying demand for movement between area and long term function of helping the growth of places by inducing changes in comparative advantages as a result of changes in accessibility and relative location (Hodder, 1947) Basically, transport network is regarded as a set of a interconnected route linking of numerous destination within and outside the region and provide a pace for movement to people and commodities. It real sense, it serves manifold functions for regional development. Transport may influence on services performing by human being. In case of industry and industrial workers, the effective production and efficiency of labours are entirely impinged on availability of timely transport means, or demographic influences in form of generating demand for transport in the region. India is a developing country and requires basic need of speedy network transport in both industrially web sites and to mobilize raw materials and final products as well. Urban commuters in big cities could be best quoted in this regard as they consider transport means as a part assisting to their life. Thus, with keeping these in mind, authors in this chapter attempted to outline the significant of linkages of transport development with a number of social interfaces by quoting numerous examples and has suggested new horizon of social interface while going for transport network in future.

This study presents the variables to be considered for planning for transport network and thereby this may develop such situations to aware of things which would provide thought in proposing and implementing policy of transport development in our country.

Motor Vehicles in India

The number of vehicles have increased with increasing demand of people. People are needed vehicles as essential for performing their daily activity, either two wheelers or four wheelers. Of course the type vehicles entirely depend on the income accrued by family or venture of industry. And, therefore, more vehicles are manufactured in factories. According to the association of Indian Manufacturing, the estimates of the likely vehicles population in India by 2000 can be figured out as shown in Table 3.1. That the vehicular population is expected to grow from 27.76 million in 1994-95 to 53.1 million vehicles by 2000-2001. This implies almost two fold growth in vehicle population specially in the wake of the additional automobile production capacity to be commissioned in the coming few years. Corresponding growth in the road network remains very uncertain.

Social Interface of Road

Transport plays a very important role in the economy. Good transportation spurs economic activity by giving mobility to production factors, which permit scale economies and increased efficiency. Good transportation enlarges the area that consumer and industry can draw upon for resources (inputs) and products (outputs). Good transportation also enlarges the area to which a given plant may distribute its products economically. The resulting specialization and scale economies provide wider choice to consumers at a lower cost. New transportation investments help sustain and industrial economic growth, thereby help in tackling poverty. As to every rose there is a thorn and the same analogy holds good in the case of road transport development, vis-a-vis its social interface. Transport plays a crucial role in improving a country's economic performance and the well being of its population, the lack of which may create poverty. Transports has an impact on productive activity in different ways. For illustration, in the case of rural agricultural production, improvement in transportation facilities may lower input prices and hence production costs, improve access to credit facilities, encourage technological diffusion, extend the area of land under cultivation to marginal lands, or increase the availability of "incentive" goods. Other catalytic role of transport improvement may include increased trade and competition from

Table 3.1 : *Vehicle Population in India*

Years	Two Wheelers	Three Wheelers	Cars	Jeeps	Light M.& II C.V.	Total
1992-93	17,025,582	861,743	2,828,992	501,388	1,980,385	23,198,090
1993-94	28,711,683	941,362	3,023,397	548,013	2,118,771	25,343,426
1994-95	20,689,449	1,010,735	3,198,325	595,853	2,267,496	27,761,857
1995-96	23,305,44o	1,108,835	3,459,925	661,253	2,480,046	31,015,507
1996-97	20,183,049	1,216,745	3,759,925	733,193	2,713,851	34,606,762
1997-98	29,348,409	1,335,446	4,159,925	812,327	2,971,036	38,627,148
1998-99	32,530,305	1,446,017	4,709,925	899,374	3,253,940	43,159,562
1999-00	36,660,391	1,609,645	5,409,925	995,126	3,565,135	48,240,222
2000-01	40,260,383	1,741,122	6,166,026	1,083,947	3,848,837	53,100,314

S = Population arrived at after reducing scrapping @ 2.0 %

+ = Population arrived after reducing scrappage @ 1.5 %. Yearly sales for 1994-95 to 2000-2001 are estimated by AIAM Secretariat based on statistical trends, for arriving a population of vehicles.

* Data published by the Ministry of Surface Transport in Motor Transport Statistics of India, they include data on Trailers and other vehicles which are not include above.

	Population as on 31 March		
	1990	1991	1993
Trailers	350,529	369,007	402,939
Others	134,073	140,180	180,331

imports, in turn leading to improve production efficiency, downward pressure on social and economic linkages between the rural and urban areas and increased non-agricultural employment may also develop. Similarly, for the requisite level of growth in urban economic activity the quality of the transportation and other types of urban infrastructure is a pre-requisite. Since the time spent on travelling or alternatively the distance traversed, is a significant variable in the industrial utility functions. The cost and quality of the transport services a fect individuals leisure and access to public services, facilities for utilities. Transport systems and services also influence the housing market, which, in turn, affect the demand for transport. Transport also affects the efficiency of the labour market and labour participation rates in terms of the commutation problems of time and fare expenses. The costs of inefficient transport on the economy as well as the social life can be onerous.

There is substantial evidence to show that transport affects a country's economic performance and the success of its development ecto stand does so in many ways which are diverse as well as complex. Needless to say that an efficient and responsive transport sector is less of a burden on the economy and society and provides benefits in terms of increased productivity and reduced time as well as inventory costs. Although transportation provides the economy with numerous benefits.These positive aspects are not without associated costs, also called the social costs or externalities. Transportation affects the environment negatively in terms of pollution and as well as depletion or degradation of natural resources. Although most people feel that the benefits provided by transportation for exceed these cost the ensuing environmental cost of the future would be quite phenomenal. Good quality transportation infrastructure (rural, urban-roads and highways), transport modes (motorised and non-motorised) and their reliable services are necessary for improving economic performance and human well-being. For which better understanding of the interaction/interfaces between transport, society and the economy is required. All these inferences are born out by the data presented in Table 3.2 covering the parameters of education, urbanization, diversification of economy and health. This is also confirmed by the correlation and regression, values which are significant at 95 per cent confidence level. In

Table 3.2 : *Road length, literacy, urban population and number of hospital*

Sr. No.	States/ U.Teritories	Road length per 1000sq km (1998-89)	% of population (above7yrs) total population	% of population total population (1991)	% of population Workers secondary (1991)	No.of Hospitals (1991)
1.	Uttar Pradesh	639	41.71	19.89	8.98	735
2.	Bihar	488	35.54	13.17	4.64	298*
3.	Maharashtra	704	63.05	38.73	15.80	2104
4.	West Bangal	693	57.72	27.39	17.82	410
5.	Andhra Pradesh	505	45.11	26.84	10.48	615
6.	Madhya Pradesh	289	43.45	23.21	8.73	368
7.	Tamil Nadu	1488	63.72	34.20	16.18	408
8.	Karnataka	798	55.98	30.91	13.17	288
9.	Rajasthan	327	38.81	22.88	9.87	267
10.	Gujarat	392	60.91	34.40	17.86	1563
11.	Orissa	1115	57.14	13.43	7.51	287
12.	Kerala	5472	90.59	26.44	18.17	2924
13.	Assam	825	532.42	11.08	5.56	207
14.	Punjab	1073	57.14	29.72	14.48	230
15.	Haryana	581	55.33	24.79	13.18	78
16.	Delhi	12557	76.09	89.93	32.43	80
17.	Jammu & Kashmir	59	NA	23.83	--	67
18.	Himachal Pradesh	431	63.54	8.70	9.98	65
19.	Tripura	1341	60.39	15.26	6.41	23
20.	Manipur	295	60.96	27.69	9.66	25
21.	Meghalaya	278	48.26	18.69	3.73	15
22.	Nagaland	484	61.30	17.28	3.48	31
23.	Goa	1547	76.96	41.02	21.98	108
24.	Arunachal Pradesh	85	41.22	12.21	8.66	18
25.	Pondichary	4984	74.91	64.05	19.99	10
26.	Mizoram	162	81.23	46.10	5.07	14
27.	Chandigarh	144720	78.73	89.69	27.82	2
28.	Sikkim	217	56.53	9.12	11.12	5
29.	Andaman & Nicobar	94	73.74	26.80	25.14	8
30.	Dadra & Nagar Haveli	630	39.45	8.47	16.30	2
31.	Daman & Diu	NA	73.58	46.86	25.40	3
32.	Lakshadweep	NA	79.23	56.29	25.27	2

Source: 1. Census of India, Census Atlas National Volume 1,1991

2. Census of India

(i) Provisional Population Total April 1,1991

(ii) Rural-Urban Distribution Paper, 1991

3. Centre for Monitoring Indian Economy :

Infrastructure : Chapter 2. Pp. 41-76 *Includes Road length to Daman and Div in India.

sum, there is a evidence of strong linkages of transport to production, trade and to the welfare of the consumer and the society in general. Transport projects and the transport sector, as a whole have a sizeable effect on the public budget. There is a strong linkage and interface. The direct effects of transport on production are distinguished from the indirect ones which occur via the adjustment of prices, technology diffusion, and the impact on credit demand availability. Similarly, the impact is discernible on rural development through the strengthening of rural-urban linkages. Further transport affects personal welfare through the time spent in travelling, in carrying out the other activities and also the flexibility of an individual in choosing between the location of residence and a job. Transport also affected an individual's community's access to public facilities as well as the operation of the labour and housing markets. It also covers the public finance effects of the transport sector through the input linkages. As social interface it affects all and one through its environmental fall out, involuntary rural mass transfers to urban-areas with cognizant deprivations.

National, regional and local transportation requirements are influenced by demographic trends. These include population, its size, its distribution among different regions, age, sex, workforce in rural and urban areas etc. Demographic characteristic infuences the demand for particular modes, and the distribution of transport demand by place as well as by time. Besides, the propensity to travel and pattern of mobility is affected by the personal disposable income patterns, ways of living and socio-economic pursuits including the employment patterns. Above all, migration of population emanating from economic development differentials between regions accentuate urbanization which is turn influence the travel and movement flow patterns. Further, the structural shift in the industrial development of a country from a predominantly a change in the supply and distribution of capital and labour. Consequently, as news growth centres emerge, the workforce shift from agrarian bases to industry and manufacturing centres thereby causing a consequential shift in the supply and distribution, of labour and capital. All these structural shift lead to massive rural population migrations, alternatively called the phenomenon of urbanization, necessitating changes in the transport demand patterns for the rehabilitation of displaced persons and their socio-economic well-being. Absence of affordable transport supply in adequate volume leads to the development of slums and encroachment on public land in or easer the centre business districts. Alternatively the poor are shifted to the peripheral areas which involve high money and time costs for daily commuting.

Road Development

It is extremely difficult to provide any new infrastructure in a densely populated country like India without displacing individual business or residences. The resulting revoluntary resettlements is an external cost associated with the provision of transport services to the extent that the individuals displaced suffer dislocation with a net loss in income and assets. Beyond the quantifiable costs of involuntary resettlement there may be a host of other associated costs. By far the most difficult area is involuntary resettlement is in the urban transport-sector. Any attempt at improving the flow of road vehicles in, through or around urban areas through bye-passes quickly encounters the constraint of land acquisition from individuals who have legal rights as well as those who are squatters who seems to have gained political god fathers. Presently, the Ministry of Surface Transport, through the National Highway Authority of India has plans to build 10,000 km. long network of expressways by the year 2015. This is besides the programmes of the state Government for the development of other types of roads in the country. This will involve massive land acquisition leading to involuntary resettlement of the displaced population and oustees, a case of an adverse social interface. In view of the massive if restructure development programme and prevailing scarcity of Government resources, large scale participation of private sector in road building is envisaged. The private sector is wary about the viability of express way investment recovery through the toll tax collection due to the tendency of the road users to use the parallel existing (toll tax free) highway as well as the likely shortfall of the projected volume of traffic materializing on the toll road. Accordingly, the present consensus is in favour of the concession to use land development for township/Industrial estate, tourist resort development along the new expressway corridor. For illustration, under this scheme, Bharat

Forge Ltd. has proposed to develop five townships of the Bangalore-Mysore Expressway in lieu of the toll collection from traffic plying on the proposed Bangalore-Mysore expressway. The construction of townships/Industrial estate/tourist resorts would involve dislocation of the local population both from the areas falling directly under the expressway as well as the land requirement of the proposed townships.

In this case, for an expressway length of 161 km (comprising 12 km expressway length and 41 km peripheral roads about 15000 acres of land is to be acquired for township development by the private investors for construction of the proposed expressway. In other words, about 94 acres of land would be acquired per km length of expressway construction. The various proposals to involve oustee-friendly approach can be as below :

(i) Fair compensation for the land acquired,

(ii) At least to provide one job in the case of house holds without land, and

(iii) Leaving the village intact as well as insulated from the influence of the proposed new townships

Population Migration

The population migration may be generally classified into two, namely, refugee migrations in the wake of communal political, atrocities and upheaval, and population migrations conditioned by socio-economic impulses. Here only the latter category and its resultant influences in India, is discussed. Migration may be rural-urban, urban-rural and rural-rural. The rural-urban migration is quite significant. This takes place from the rural areas toward mining, industrial and manufacturing centres promising employment and job prospects to migrants released from the agriculture sector and looking for non-agricultural occupations. In the rural areas growing population or dwindling agricultural jobs due to mechanization may generate large masses of surplus landless or marginal agriculture labour force heading towards urban, mining. Industrial and metropolation areas for a living. These displaced persons start living in peripheral and satellite towns and villages or in shanties in the central districts. In brief unemployment, underemployment, the desire for freedom from feal oppression and an urge for better standards of living constitute the driving force for people to migrate from one area to another. The pattern and quantum of transport demand due to human migration continues to be on the rise.

Growth and Trends in Urbanisation

It is apparent from the demographic trends derived from the analysis of the previous Census data that the pace of urbanization has been significantly increasing over the past decades as a consequence of the growth in population, lack of employment opportunities and under-employment in the rural areas in agricultural sector in particular. Growth trends in urbanisation in India since the dawn of the Twentieth Century, is presented in Table 3.3. It is apparent that the trend of urbanization is upward. Percentage share of the country's population in the towns of over one lakh persons is fast rising. While the share of the towns (50,000 to 1,00,000) is nearly stagnent, the share of the rest is fast declining.

Table 3.3 : *Trends in Urbanisation (1901-1991)*

Years	% of urban population in total population	percentage of urban population by each size class						
		I	II	III	IV	V	VI	All
		(1,00,000- and above	(50,000- 99,999)	(20,000- 49,999)	(10,000- 19,999)	(5,000- 9,999	Less Than 5000)	Classes
		Rising	Stagnant	Stagnant	Falling	Falling	Falling	Stagnant
1901	10.8	22.9	11.8	16.5	22.1	20.4	6.3	100.00
1911	11.2	25.3	12.4	16.9	18.9	19.0	7.4	100.00
1921	13.9	35.4	11.8	17.7	16.3	15.4	3.5	100.00
1941	17.3	41.8	11.1	16.7	14.0	13.2	3.2	100.00
1961	18.0	48.4	11.9	18.5	13.0	7.2	1.0	100.00
1971	19.9	55.8	11.4	16.3	11.3	4.7	0.5	100.00
1981	23.7	60.4	11.5	14.3	9.5	3.7	0.6	100.00
1991	25.7	65.2	10.9	13.2	7.8	2.3	0.3	100.00

•Exclusive Assam.

Source :
1. Cencus Report of India.
2. Transport Planning Policy Framework (1988), Planning Commission.
3. Conde, Julien : Rapid Population Growth in Developing Countries (1978) CECD (Paris)

For regionally balance progress of urbanization and for counteracting the diseconomics of the growing mega cities, smaller cities need be revitalised. In Table 3.4 the number of towns/town groups, population size and decadal growth rate of population according to the size class of town is given for the period of 1951 and 1991.

Table 3.4 : *Variations between 1951 and 1991 in urbanisation*

Years	percentage of urban population by each size-class of towns						
	I (1,00,000- and above	II (50,000- 99,999)	III (20,000- 49,999)	IV (10,000 19,999)	V (5,000- 9,999	VI Less Than 5000)	All Classes
A) Number of Towns and Towns groups							
1951	38	93	352	652	1,175	608	3,018
1961	113	138	484	748	761	218	2,462
1971	148	183	582	874	678	178	2,643
1981	218	270	744	1,053	761	255	3,301
1991	NA	NA	NA	NA	NA	NA	NA
B) Population (million)							
1951	25.8	6.3	10.1	9.0	8.4	2.0	61.9
1961	38.2	9.4	14.6	10.3	5.7	0.7	78.9
1971	60.9	12.4	17.8	12.4	5.1	0.5	109.1
1981	95.1	18.2	22.6	14.9	5.8	0.9	157.5
1991	138.8	23.3	28.1	16.5	5.5	0.6	212.9
C) Growth (%) of urban population							
1951	48.1	49.2	20.4	14.4	-32.1	-65.0	27.5
1961	52.2	45.3	29.2	20.2	-8.8	-22.9	38.2
1971	56.2	46.8	27.0	20.2	13.7	80.0	44.4
1981	46.9	28.1	25.3	*	*	*	*

* Excluding Assam

Source : Census Report of India - Part IV to V

According to the demographic researchers and analysis of the Census data, the following observations are discernible. Migration of rural labour has been mainly caused by a flight of the landless, including unemployed labour and marginal farmers, which got fill up in the wake of frequent floods, drought leading to crop destruction and failure as well as epidemics. Volume and pattern of migration is certainly large and the migrants include all categories of the rural society viz the poorest, the middle and even the richest ones for reason of livelihood and better civic amenities.

Landuse and Management

Land abuse/mismanagement is also a concern emanating from the transportation activities. The unnecessary destruction of land, vegetative cover and wild life threaten the environment in the wake of large scale expansion of transportation system through out the length and breadth of the country. India has a mass area of 3288 thousand square km. The country is home to 883.6 million population (mid, 1992) implying thereby 273 persons per sq. km in 1992. The land mass area is static whereas the population growth is quite dynamic, with obvious influences on the landuse. According to the Report of the Steering Committee on Transport of the Planning Commission (August, 1988) popularly called the Luthra Committee Report, the increasing mismanagement of land or land abuse is due to the railways and roadways mainly because each of these modes of transport requires a dedicate gateway mainly because each of these modes of transport requires a dedicated guideway or corridor of its own. Moreover, increasing sophistication and demand for both rail and road transport is likely to lead to the proliferation of such guide way roads qualitifying to be highways, expressways and the mega ways and railways into dedicated lines for the movement of high capacity freight trains and high speed passenger trains, into double, triple and four land corridors. In the context of India, over the sprawling national geography, road construction is an emergent national nescessity for economic development for which several programmes would need to be taken up by the Government in the private or joint sector under the liberation regime. Similarly in the rural areas the road building programmes are being intensified under the Minimum Needs Programme for bringing villages of different population sizes within easy reach of all weather road network in order to bring them into the mainstream in a phased manner.

Land Requirements – Modal Variations

The requirements of land for the construction of new roads or railway lines according to the per unit right of way land requirements and the length of the network planned for investment are significantly high. Both these modes of surface transport occupy 3.1 per cent of the country's total area and about 4.6 per cent of arable land. India has a total rail length of 62,456 km route in plains, hills deserts, coastal areas etc. Railway line construction disturbs environment including landuse in many ways. It is not only the lay of the land that gets disturbed when a new railway line is built. What is even worse is that the natural drainage across the land, is also interfered with. embankments are raised and frequently the free flow of water across the embankment is obstructed. As a result, there may develop dry pastures starved of water on one side of the embankment, with flooding of the land table and water stagnation on the other. Both bringing about adverse consequences to the agronomy as well as the envionment including mosquito menace. Further rail construction leads to deforestation in three ways viz, use of wooden sieepers on track, cutting of trees for new lines and increased illegal wood felling and the problem is quite serious in the hill areas wheel the hill formation gets adversely affected through blasting, tunnelling etc. Better land management under the railway construction calls for different corrective steps, as below :

(i) By avoiding borrow pits for constructing newlines.
(ii) Use of concreate sleepers in place of wooden sleepers and compensatory afforestation by planting tree along the new lines.
(iii) Management of borrow pits for pisciculture or tree plantations.
(iv) Providing additional flood openings and extension of existing ones by constructing management of the above constructive steps could be provided in the investment budget.

The country has the advantage in that the overwhelming proportion of passenger and freight by railways moves along the arterial and therefore their evisaged doubling and trebling along these routes should be preferred for proper a land management over the opening up of the new areas and interlands from the view point of the extent of environment pollution.

Landuse in the Road Sector

It is not commonly known that roads occupy much more land area than the railways. Road occupy about nine times the area occupied by the railway lines singley because road network (1.8 million km) is 18 times longer than the rail network (only 0.1 million km). Consequently, what has been said earlier about railways, applies even more forcefully to the roadways. In the Steering Committee Report "Perspective Planning for Transport Development" of the Planning Commision, the following have been suggested to avoid or minimise the land abuse while planning for new roads and upgradation of the existing ones :

(i) Dislocation of natural drainage of the land and flows of river and streams in the vicinity.

(ii) Use of flash abundantly available from the nearby thermal plants instead of digging huge narrow pits.

(iii) Soil or sediment erosion likely to lead to chemical and biological pollution such as turbidity of stream, temperature rise etc.

(iv) Adverse effects on physical or natural features of the area like ecological formation, scenic beauty, spots (if any) archaeological and historical landmarks, sancturies etc.

(v) Effect of unplanned downhill dumping of waste material of hill road construction activity resulting in added ecological damage to the downhill slope and to mitigate the same through appropriate revision of existing technologies of landside repairs.

(vi) Construction of new roads in the hilly areas resulting in illegal besides encouraging illegal wood extraction.

(vii) In the rural areas, where lot of emphasis is being laid under the new plan schemes of rural development. The road building programme should be carefully executed to preserve natural drainage, flow of stream and rivulets, soil and forest cover, natural geology etc. There is a lurking fear of rural road development taking place in a haphazard manner.

Landuse management gets adversely affected by road/rail construction. It manifests itself in the loss of forest and vegetation cover, erosion of soil, destabilisation of hillside terrain or disorganisation of drainage system including the flow of streams/ rivulets. Possibility of flash floosh and leaching of soil nutrients, a point of overwhelming importance in the adoption of proper scientific landuse management of avoidance of a esthetic and ecological degradation. Since land is our most scarce resource, its use cannot be allowed indiscriminately. This makes it imperative that every new road or railway line construction is evaluated carefully taking into account the use of land which is a starve resource and viable alternatives considered wherever feasible. This calls for new regulation, fines and the requirement to pay damages for land abuse vegetative cover destruction and water flow obstructions. It is hoped, that. It will force greater concern for proper landuse and management. The urban transport system are beset with a number of social problems due to (i) Traffic congestion and parking problems, (ii) Encroachment of roads by commercial residential landuses, (iii) high day time density of population and traffic of time-loss in movement (iv) Air pollution, noise pollution etc. leading to several lung an dear diseases, and (v) Accidents and serious problems to pedestrians, slow traffic leading to fuel and time loss.

Hazards of Road Transport

The hazards of road transport can be listed which are generally prevailingly in the way of road transport and thereby somewhat disturbance is created in road transport. The major hazards can be summarised as :

(i) The rate of accidents and loss of human lives continue to be very high on our roads. Steps are required to secure the safety of reducers.

(ii) People affected by the hazards of transport like accidents (resulting in loss of limbs/life) should be suitably compensated by the transport authorities.

(iii) Transport insurance scheme should be automatically applicable to all those who use every mode of transport.

(iv) Special funds should be earmarked to help the victims of transport disasters.

(v) Medical expenses in such cases should be fully victims of transport disasters.

(vi) Some ratio between transport network and vehicular traffic to be maintained to minimize social damages.

(vii) Careful planning of flyovers, subways to reduce the impact of traffic noise, vibrations, air pollution and congestion on the residential areas, in the proximity of the road network.

(viii) Innovation of new technologies related to noise-free and pollution free vehicular traffic.

(ix) Special lanes for slow moving traffic/pedestrian traffic to minimise death/injury caused due to accidents.

(x) Control use of land near the transport lines (low landuse).

(xi) Greening of the land (with plantation, recreational parks etc.) near the national/state highways to minimise noise/ air pollution for the neighbouring settlements.

(xii) Protection of right of the local population or use of natural resources while introducing road network into the forest area as an other resource rich regions.

Encroachments

The construction of road capacity is a costly preposition and therefore its use should be pitomised. Indian road network suffers from encroachment in many sections. These encroachments are of permanent, semi-permanent or temporary nature, Road encroachments adversely affect the free flow of traffic, an vehicle operating costs besides being a major source of congestion inducting road accidents. According to the Motor Vehicles Act 1985 temporary encroachment of carriage way which affects the free flow of traffic is punishable. Encroachment should be made punishable by the State Government also in order to dissuade masses from this malpractice.

The road network and maintenance in India is more labour oriented compared to other modes of transport. Main power employment is involved in construction as well as its maintenance. Employment generation is universal in all parts of the country including the rural area where nearly half (about 47%) of the country's total road network lines. This creates rural employment for the construction as well as the maintenance phase. Under 2001 A. D. Perspective Plan the road length is planned to grow by 33 per cent in the aggregate and rural roads by 131 per cent for 955

(000) km in 1992 to 2212 (000) km by the desire rural employment in the process. At least one km out of every 3 km of our National Highways is in need of maintenance due to cumulative neglect of the past decades. The corresponding proportion of the State Highways in need of attention varies from state to state but on the whole it is estimated to be 3 km out every 5 km. The situation in the case of major district road and road in more pitiable. Barring the expressways and National Highways construction of new roads will be create lot of employment of the skilled, semi-skilled and unskilled labour force. Proper maintenance of these facilities would created recurring employment all over the country, all the year round.

Road User Education

Traffic condition network are chaotic, claiming precious lives valuable property. Accident scenario may worsen due to the growing mismatch between the increasing vehicle population and limited road capacity. But the chaos on the road are more due to the lack of adherence to traffic discipline. This is more due to their ignorance about the rules of Road Regulation as well as to poor enforcement of traffic discipline. To achieve better road traffic performance and safe environment there is an imperative need for road users education. This awareness need be included at the school level following the principle of "catch them young". Subject of civic sense need better be introduced as a compulsory paper as part of school curricula wherein they should be taught their duties, responsibilities and rights in the areas of road environment. Curriculum may include rule of road regulations, environment protection energy efficiency offences and penalties for traffic discipline violation, regard for public property and due consideration for other road users. Our emphasis on road education among the roadusers, other than the vehicle crew (driver, conductor etc.) also need be stressed.

Traffic enforcement is key to safe driving. Besides this traffic police and transport authorities role in traffic disciplining, the roadusers participation in accident abatement need be encouraged. Because it is wearer alone who better known where the shoe pinches. This will also create the much desired sense of involvement.

It is a general complaint that the traffic police and transport authorities are under staffed and are proving inadequate to handle the automobile boom. It is therefore desirable that some of the functions and duties of these authorities be privatised for lightening load of these authority and thereby enable team to concentrate on more vital issues. Before privatization, the modes operation need be debated in public seminars, synopsis etc.

Concluding Remarks

It is safe to say that pace of transport development is entirely based on existing social interfaces for fully utilization of natural resources in study region. The vast country like India needs to be well-developed transport network not only to urban centres or big cities but it should to down to rural India for its fast development in all sphere. And therefore, before taking any step toward transport development, it is necessary to consider the environment and social interface of the region. The given situation and human response can obtain best or sustain development in transport network. In brief, social interface should be given thought as one major integral part while proposing and implementing transport policy for its development in our country.

References

Govt. of India (1988) : "Perspective Planning for Transport Development" Report of the Steering Committee, Planning Commission.

Govt. of India (1981-85) : "Report on Transportation (1982)", Railway Reforms Committee, Ministry of Railways, New Delhi.

Govt. of India (1992-97): "Eighth Five-year Plan", Planning Commission.

Govt. of India (1991) :"Census Report of India," 1991.

Govt. of India (1995) : "Economic Survey."

Govt. of India (1995) : " Building Super National Highways to Progress", National Highways Authority of India, Ministry of Surface Transport, New Delhi.

Mishra, B. N. and Tripathi, S. N. (1991) " The Level of Transport Development in Basti District, *Geographical Review of India*, Vol. 53, No. I, Calcutta.

Pandey, N. P. (1990) *Geography of Transportation*, Inter-India Publications, New Delhi.

Ray, S. K. (1995): *Transport Planning for Developing Countries*, Prentice Hall, New Delhi.

Singh, D. N. (1975-76) : *Transportation and Regional Development with particular reference to India - A Geographical Perspective*, Vol. 2, pp 47-50.

Vaidya, B.C. (1998): *Readings in Transport Geography - A Regional Perspective*, Devika Publicationas, New Delhi.

4

RAILWAY FREIGHT RATE AND REGIONAL DEVELOPMENT IN INDIA

Mondira Dutta and *Bupinder Zutshi*

Introduction

The development and expansion of the railway network in India, during the British period was primarily governed by exogenous forces and was not initiated in response to the requirements of the internal economy. The requirement of the colonial suction mechanism warranted the British to formulate a tariff policy which was purely a port oriented freight rate policy.

After independence, the planned development programme launched entrusted the task of carrying a major proportion of goods to Indian railways. This was done as the movement of goods, by rail mode, provides economic advantages. The gestation period, as well as the fixed costs for the railway system is extremely high, therefore, this facility had to be utilised effectively and efficiently to cover the costs. The cost disadvantages and unfavourable passenger freight earnings due to Government policies of subsidization has made railways to survive, mostly on the movement of bulk commodities over long hauls. This is essential to achieve the competitiveness in the operation and to generate more funds for the modernising the system. Moreover, maintaining of high share of bulk commodity movement, by railways is also desirable in a country like ours from the point of view of maintaining ecological balance and preserving and reducing dependence on fossil fuels. It is believed that railways are most suited to carry heavy and bulky goods across the vast territories, because of their high speed and large carrying capacty. "Indian railways has been and will continue to occupy a key position in the strategy of building up the basic infrastructure required for a growing and diversified economy". (Ghose, 1918).

The magnitude and the nature of commodity freight movement is influenced by several factors like : inter-linkages, public sector investment the industrial licensing policy and the freight rate structure of the movement of goods. The freight rate policy reflected by transport cost plays an important role in motivating large movement of commodities especially for long distances. Freight rate structure also plays an important role in motivating location decisions for opening and diversifying the economic activities. The freight rate policy, pursued, by the British was largely responsible for the port oriented movement of commodities, especially the raw materials form the hinterlands. Thus, the policy was instrumental in the suction mechanism leading to the depletion of raw materials.

Among the various studies related to the development of railway freight rates and the government regulations over them, The most pioneering one has been by Ghose (1918), Some of the other studies were done by Mehta (1932), Antia (1937), Tiwari (1971), Some of the recent studies are that of Lefeber and Choudhary (1986) and Raza and Aggarwal (1986).

Objectives

The present treatise attempts to :

(i) To analyse the major features of the railway freight rate structure, pursued by Indian Railways.

(ii) To examine the freight rate changes for bulk commodities between 1948-1992.

(iii) To study the temporal changes in the proportion share of freight and revenue earned by major bulk commodities during 1969-92, and

(iv) To propose changes in the freight rate for the bulk commodities.

The freight rate by the railways should also include the other costs like the charges of handling the goods to and from market to station, as these charges does not apply to road transport. This fact along with the delays in the movement of goods by the railways, strengthens the users option for the use of road mode of transport.

The present analysis, however does not take the additional cost as indicated above into consideration.

Railway Freight Rates–A Historical Overview

The development and the expansion of the railway network in India was unfortunately not initiated in response to the requirements of the internal economy, but was primarily governed by exogenous forces. The requirements of colonial suction mechanism, warranted the British to formulate a tariff policy which was purely a post oriented freight rate policy. The railway freight rates were not based on any scientific considerations. The freight rate policy remained strictly under the whims of the private sector. The structure of freight rate policy has, however, undergone several phases of changes after the Independence, which needs a careful study.

The essential feature of the railway freight rate policy since Independence was to promote the mobility of basic raw material for the industrial and agricultural development of the country Keeping this principle in view, varying freight rates were charged for different commodities. The rates were highly subsidised for promoting agricultural and basic heavy industrial development. Simultaneously, commodities which could withstand a higher transportation cost were categorised as a High Class. At present the railway tariff structure provides Wagon load rates under different classes according to the cost and value of the service concept. As the train load have cost advantage to the railways, the part and full train load carry concession by lowering the class. For instance Wagon load of cement which is carried at Class 135 will be carried in Train load at Class 130. Another feature of the railway freight rate is the taper in the freight rates, with distance in such a manner that the total transport charges increase with distance but at a declining rate reflecting telescopic freight rate principle, taking a shape of curvi-linear relationship. A historical overview of the freight rate policy indicated division of commodities into five classes prior to 1920's and the important features were:

(a) Freight rate selected as per the whims of the company.
(b) Less freight rate for port oriented raw materals.
(c) Opening of new railway lines and bringing about competition.

(d) Tapering system of rates were introduced.

After April, 1922, new classes of commodities were identified however, classes were not evenly spread and the proper relativity did not exist between the rates at railway risk and owners risk. Other features of the freight rate policy till Independence of the country were :

(1) Classification was arbitrary and their increase in numbers were insignificant.
(2) The continued existence of "Exceptional" and "Adjusted", class rates was the major instrument in the hands of railway companies by which the rates were manipulated as per the requirements of the Company.

After 1948, keeping in view, the national objectives tariff structure stipulated lower rates of transport for social and economic development utility commodities. Commodities of bulk movement were given sharper taper than other classes of commodities. The number of class of commodities were sharply increased to 7 by August, 1967.

The major policy shift in the railway tariff rates were introduced in 1983 whereby both number of class of commodities and their freight rate underwent significant changes. Table 4.1 depicts the disance slab and number of commodity classes identified for the three selected years. With a marginal change of initial slab from 1-50 km in 1982 to 1-75 km in 1983 and 1-100 km in 1991. The other distance slabs remained unchanged during the three selected periods. In case of number of commodity classes, significant changes were incorporated both by enhancing the number of commodity classes, significant changes were incorporated both by enhancing the number of classes as well as by increasing the rates class. The number of classes increased steadily from 29 in 1982 to 34 in 1983 and to 55 in 1991.

Table 4.2 depicts the tariff rates per quintal for the selected commodity class for specified distance slabs in 1981 and 1991 (same commodity type was selected for comparison of freight rates). The percentage increase of tariff between 1981 and 1991 for the lowest class of commodity in the distance slab of less than 100 km

Table 4.1 : *Commodity Tariff (distance slab and classified Classes)*

Year		Distance		Classified No. of Classes
1982	I	1.50 kms		32.5,35,37.5,40, 42.5, 4.5
	II	50	Onwards with interval of 5 kms upto 260 kms	47.5, 50, 52.5 55, 60, 62.5
	III	260	Onwards with interval of 10 kms upto 800 kms	65, 67.5, 70, 80, 82.5 85,
	IV	800	Onwards with interval of 25 kms upto 2400 kms	87.5, 90,92.5, 100, 105, 110
	V	2400	Onwards with interval of 50 kms upto 5000 kms	115, 120, 130, 150
			Total = 29	
1983	I	1.75	kms	I 65 Onwards with interval of 5 kms upto 190 km
	II	50	Onwards with interval of 5 kms upto 250 kms	II 190 Onwards with interval of 10 kms upto 240 km
	III	260	Onwards with interval of 10 kms upto 800 kms	III 240 Onwards with interval of 20 kms upto 300 km
	IV	800	Onwards with interval of 25 kms upto 2400 kms	
	V	2400	Onwards with interval of 50 kms upto 6000 kms	
			Total = 34	
1983	I	1.100 kms		80A, 80, 85A, 86, 90B 90X 90, 95B, 95, 100B,100c 105k,
	II	50	Onwards with interval of 5 kms upto 250 kms	105,110,115k, 115, 120, 125B, 125X, 125,30B, 130, 135k,
	III	260	Onwards with interval of 10 kms upto 800 kms	220B, 220, 230, 240D, 240, 250, 300,190A, 190D,190,185,
	IV	800	Onwards with interval of 25 kms upto 2400 kms	135X, 135, 140k, 260D, 260, 280, 300D,200D, 200, 210,
	V	2400	Onwards with interval of 50 kms upto 6000 kms	140X, 140, 145, 150, 160C, 160,165, 170, 175,180, 185A
			Total= 55	

Source : Goods tariff, Goods Rate Table (1982-1991) Indian Railway Conference Association.

Table 4.2 : *Commodity Freight Tariff*

Distance Slab	1981 (Rs.p.) per quintal							1981 (Rs.p.) per quintal						
	Class types							Class types						
(in kms.)	32.5	50	60	80	90	100	150	80A	90B	100	125B	185A	200	300
100	1.86	2.64	3.10	3.98	4.46	5.35	7.14	2.84	3.98	4.76	5.12	5.51	8.38	12.00
500	4.88	7.27	7.86	11.31	12.64	14.35	20.76	9.38	13.41	16.37	18.14	20.49	31.44	46.50
1000	9.72	14.90	17.42	22.99	25.79	31.35	42.55	17.79	25.51	31.31	34.94	39.83	61.18	91.07
1500	13.36	19.91	23.77	31.46	35.32	43.04	58.44	25.15	36.14	44.41	49.65	56.79	87.31	130.31
2000	16.15	24.47	29.23	38.79	43.54	53.05	72.12	30.73	44.17	54.32	60.83	69.67	107.11	159.31
3000	20.09	30.56	36.52	48.50	54.99	66.42	90.37	38.99	60.36	69.01	77.40	88.88	136.57	204.95
4000	22.13	33.68	40.31	63.52	58.49	73.37	99.80	46.93	67.55	83.14	93.29	107.14	164.79	246.44
5000	24.18	36.83	44.09	58.56	65.80	82.28	109.23	54.87	83.31	106.84	109.19	125.50	193.04	288.00
6000								62.81	90.46	111.40	114.75	143.85	221.29	331.17

Source : Goods Tariff, Goods Rate Table (1981-1991) Indian Railway Conference Association, New Delhi.

was 53 per cent however 88 per cent tariff increase was recorded for the same class of commodity in the distance slab of 1500 km. In case of highest class of commodity the corresponding percentage increase in tariff was 68 per cent and 123 per cent for the distance slab the less than 100 km and 1500 km.In case of 5000 km distance slab, the tariff increase was 126 per cent for the lowest class and 167 per cent for the highest class of commodity.

The highest class commodity (class 300) recorded 4.95 times more tariff rates compared to the lowest commodity class (80A) for the distance slab of 500 km. The difference of the tariif rates for the two class of commodity widened up to 5.24 times for the distance slab of 5000 km. The incidence of taper in the freight rates with increase in distance was another feature of the freight rates. An increase of 50 times distance recorded only 13 times increase in the freight rate for class 80A commodity and 15 times increase in the freight rate for class 300 commodity in 1981. The taper in the freight rate was changed in 1991, as the same distance variation registered 19 times increase in the tariff of *class 80A* and 27 times increase in the tariff for the *class 300* commodity in the year 1991. This indicates the reduction in the degree of taper between 1981 and 1991, though the taper has still been maintained. The taper benefits are more for the commodities with larger distance leads, like, food products, manure and finished products .

Railway Commodity Freight Structure

The Indian railway have classified the commodity freight for total commodity and revenue earning commodity. The present analysis is done for the revenue earning commodity only. Table 4.3 presents commodity tonnage. Net Tonne Kilometre (NTKM). Average Lead and Total Revenue earned by the railways through transport of commodities between 1950-51 to 1990-91. Railways recorded only 72.3 million tonnes of goods for transport in 1950-51 which showed 4.35 times increase between 1950-51 to 1990-91. The increase in the tonnage was not uniform throughout the last four decades of planned development. The total tonnage recorded highest growth between 1950-60's and 1980-90's. This is purely a reflection of prevailing economic scene like; development of new industries especially iron and steel, machinery petroleum, fertilisers and chemical refinery during 1950-80's The tempo initiated earlier, recorded a decline between 1970-80's. However, it again gathered momentum from 1980-81. Net tonne kilometre freight recorded similar picture, as was the case with total tonnage, with its ups and downs. The declining trend during 1970-80's for both total tonnage and NTKM freight could be attributed to the slackness in the economic scene, owing to unfavourable natural conditions, political disturbances and difficult fiscal climate.

Average distance lead of the commodities by railways recorded, a slab increase till 1980, however it recorded decline after 1980. This could partly be explained by the growth of industrial complexes throughout the space and partly by the presence of large share of mine ores, coal and cement in the total freight. These commodities from road users especially for finished products which move for long distances.

Railways earned Rs. 11,659 millions (Refer Appendix-I) from the freight in 1950-51. The increase in the earnings during 1950-60 was 68 per cent, however, subsequent two decades of 1970-70 and 1970-80 did not record any significent increase in the revenue earnings. In fact 1970-80 recorded negative growth of 2.25 per cent in the earnings. The revenue earned recorded phenomenal increase of 274 per cent during the decade 1980-90's. This is largely attributed to the changes in the freight rates from 1983 onwards following the recommendations of the Railway Tariff Committee. Moreover, total tonnage and NTKM also increased substantially during this period. This indicates that freight rate increase did not effect the Total tonnage and NTKM. (Table 4.3) and (Fig. 4.1) Revenue earned is worked on 1987 prices till 1980-81, however for 1990-91, actual prevailing rates have been stated.

Commodity Composition of Railway Freight 1991

The commodities of coal, iron ore, cement, mineral oil, chemical manuer, iron and steel, mine products and food products constitute 93 per cent of the Total tonnage, 88 per cent of NTKM and 89 per cent of Revenue earned by the railways in 1991. Coal, iron ore, cement, mineral oil and chemical manure alone represented 76 per cent of the total tonnage, Net tonne kilometre recorded slightly less proportion for the above stated commodities, as they move for short distances. On the other hand food products and chemical manure registered higher net tonne kilometre proportion than the total tonnage of the freight (Table 4.4) and (Fig. 4.1)

Table 4.3 : *Railway Commodity Freight Parameters (1950-91)*

Year	Total tonnage		N.T.K.M. (Net tonne kilometre)		Lead		Revenue Earned	
	Million	% Growth	Million	%Growth	KMS.	%	Rs. Million	*% Growth
1950-51	73.2	–	37565	–	513		11659	–
1960-61	119.8	63.9	72333	92.55	603	17.54	19549	67.68
1970-71	167.9	40.15	110696	53.03	659	9.28	22583	15.52
1980-81	195.9	16.67	147652	33.38	754	1441	22074	-2.25
1990-91	318.4	62.53	235785	59.68	741	-1.72	82470	273.50

Source: 1. Indian Railway : Annual Report and Accounts 1990-91
2. Indian Railway : Facts and Figures 1990-91
3. Indian Railway : Years Book 1990-91
4. Government of India : Ministry of Railway, Railway Board, Goods revenue Statistics, 1965-90
5. Statistical Abstract of India, 1957-58, 1970, 1972, 1985, 1987, Government of India, Department of Statistics, Central Statistical Organization, New Delhi.

The revenue earned from the freight by the railway was highest for coal (37.76 per cent). Mineral oil and iron and steel finished products recorded higher proportion for the revenue earned, as compared to NTKM and total tonnage. These commodities are charged high freight rates. The proportion of revenue earned for food products and manure was very low compared to NTKM and total tonnage. This could be attributed to the subsidised freight rates for these essential commodities in order to encourage agricultural development.

Bulk Commodity Freight

The freight rate charged by the railways per ton, have been analysed

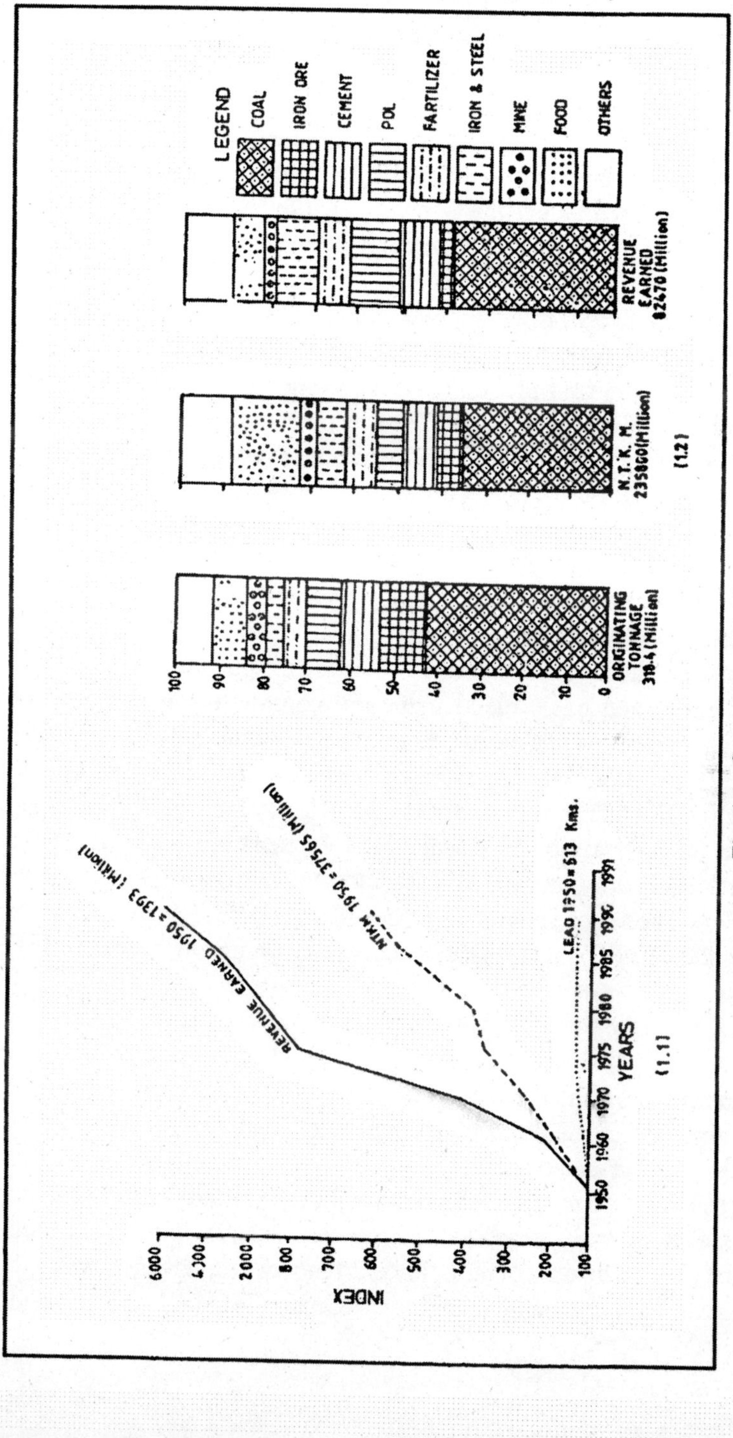

Fig. 4.1: *Commodity Freight 1991*

for the three selected legs of upto 500 km. 800 km. and 1000 km. from 1948 to 1992. The analysis has been attempted for the bulk commodities namely, coal, cement, chemical manure, mineral oil, iron ore, iron and steel products, mine products and foodgrains. These commodities constitute nearly 90 per cent of total tonnage of freight transported by the railways. The prices charged per ton have computed as per the General Price Index of 1987, however, in case of 1990 and 1992, actual prevailing prices have been considered.

The freight rate structure indicates, two clear-cut phases of freight rates charges by the railways (i) 1948-1982 (ii) 1983 and onwards. The first phase recorded insignificant changes both for the class of commodity and for the tariff rates. The rates were fixed mainly to strengthen the movement of raw materials for rapid industrialisation and agricultural development. In real terms freight rates were charged less for most of the commodities till 1980 as compared to 1948 (Based on General Price Index of 1987). Three commodities include coal, cement, iron and steel, chemical, manure mine products and food products. (Table 4.5 and 4.6). Hence, major emphasis during the first phase was national rebuilding, development of heavy and basic metal industrial complexes, infrastructural development. This period subsidised freight rate on essential industrial and agricultural raw material. The freight rate policy adopted till this phase developed several heavy industrial complexes. But modernisation and technical development of railway received back seat owing to over subsidisation.

The major policy shift in the railway rates were recommended by the (RFFC), and accepted by the Government in 1983. The class of commodities, distance legs and freight rates were substantially changes. This resulted in the significant increase of revenue earned by the railway, ensuring modernisation phase for the railways. The increase in the freight rates were between 1.5 to 2 times for the selected bulk commodites and 3 to 4 times for the other higher class commodities. The railways increased 74 per cent revenue due to commodity freight hike in the year 1983. The concept of *train load and wagon load* was introduced for the first time to encourage bulk transport movement, however, taper in the rates did not record significant changes (Table 4.5 and 4.6) and

Table 4.4 : *Freight Parameters Showing proportion of commdity types (1990-91)*

Commodity	Total Tonages (000) S	Percent to Total	Net tonne kms. Million	%to Total	Earnings (Rs.Crores Rs. (Crors)	%to Tatol
Coal	135162	42.45	85902	36.43	3114.11	37.76
Iron Ore	34481	10.83	12585	5.34	346.20	4.19
Cement	28876	9.07	18935	8.03	710.53	8.62
Mineral oils	24988	7.85	15123	6.41	961.74	11.66
Chemical Manure	18358	5.76	17265	7.32	511.94	6.21
Iron '& Steel mine Products	12480	3.92	13981	5.93	744.60	9.39
Limestone & Dolomite	8672	2.72	4615	1.96	159.99	1.94
Stones other than Margble & Gypsm	3098	0.97	1122	0.48	40.10	0.49
Other ores other Than iron ore	3151	0.98	1505	0.63	62.69	0.76
Non Rerous M Metals	798	0.25	617	0.26	22.95	0.28
Food products	25348	7.96	35594	15.10	664.96	8.06
Others	22700	7.12	28000	11.87	864.60	10.48
All Commodities	318400	100%	235800	100%	8247.00	100%

Source : 1. Indian Railways : Year Book, 1990-91.
2. Indian Railways : Facts and Figures, 1990-91
3. Indian Railways : Annual Report and Accounts, 1990-91.
4. Govt. of India, Ministry of Railway, Railway Board Statistics 1985-90

(Fig. 4.1 and 4.2).

The freight rates were continuously changed after 1983, as the rates were enhanced in 1985, 1989 and 1992. However, food products and chemical manure were the exceptions as their rates have not been increased since 1985. This has been done to encourage the farming community to produce specialised agricultural products and to encourage green revolution. The focus shifted to green revolution to meet the food requirements of the country. This necessitated large scale movement of chemical fertilizers. The average lead of chemical fertilizers increased sharply.

Table 4.5 : *Bulk Commodity Freight Rate per Ton (according to 1987 price-index)* (1948-92)

Commodities	Lead	1948	1960	1970	1980	1985	1990	1992**
Coal	500 kms	76.17	100.85	82.04	62.39	115.08	178.00	217.40
	800 kms	90.15	121.05	117.21	101.05	175.84	281.30	343.20
	1000kms	99.36	134.09	134.79	119.81	212.35	342.90	395.70
Cement	500 kms	148.16	152.70	113.69	85.42	129.47	191.50	240.80
	800 kms	184.15	213.78	159.79	138.71	207.64	302.70	381.10
	1000 kms	210.94	247.09	187.92	164.58	262.08	369.40	465.00
Chemical	500 kms	213.45	163.11	126.97	74.33	119.84	164.80	186.80
Manure	800 kms	293.81	227.66	178.55	120.66	191.52	259.70	293.70
	1000 kms	347.38	263.06	207.07	143.12	230.72	316.50	357.90
Mineral Oil	500 kms	415.19	441.44	285.99	179.93	299.04	418.20	532.00
	800 kms	641.20	574.71	406.72	296.05	485.29	657.30	851.30
	1000 kms	793.55	757.95	480.56	352.20	589.12	805.30	1043.00
Iron Ore	500 kms	125.56	131.87	100.41	69.92	105.28	162.80	208.40

(Contd.)

(Table 4.5 *(contd.)*

800 kms	150.67	148.81	140.65	113.70	176.17	255.90	329.00	
	1000 kms	192.52	208.22	165.26	134.74	202.72	311.60	400.70
Iron & Steel	500 kms	224.33	258.20	186.75	127.06	211.90	291.50	402.90
	800 kms	343.20	381.05	264.11	135.73	374.64	464.00	642.30
	1000 kms	419.37	447.69	311.78	149.52	415.40	567.30	786.00
Mines	500 kms	372.50	367.17	259.42	149.94	192.52	264.70	338.00
	800 kms	554.14	542.08	368.43	245.60	310.68	420.80	537.90
	1000 kms	673.00	637.18	435.24	292.22	376.76	514.60	657.70
Foodgrain	500 kms	152.34	122.16	93.17	54.29	90.72	110.40	110.40
	800 kms	213.45	170.74	131.27	87.83	144.14	172.40	172.40
	1000 kms	248.61	197.81	153.93	104.18	173.71	209.30	209.30

* Actual prevailing prices were considered.

Source : Goods Table and Goods Rate Table, Indian Railway Conference Association -1948-92

Table 4.6 : *Bulk Commodity Freight Rate per Ton Growth Index (1948-92)* *(Based on general price Index of 1987)*

Commodities	Lead	1948	1960	1970	1980	1985	1990	1992**
Coal	500 kms	100	132	108	82	151	234	285
	800 kms	100	134	130	112	214	345	398
	1000kms	100	135	136	121	214	345	398
Cement	500 kms	100	103	77	58	87	129	163
	800 kms	100	116	87	75	113	164	207
	1000kms	100	117	89	78	124	175	220
Chemical	500 kms	100	76	59	35	56	77	87
Manure	800 kms	100	77	61	41	65	88	100
	1000kms	100	75	60	41	66	91	103
Mineral Oil	500 kms	100	105	80	56	84	130	166
	800 kms	100	99	93	75	117	170	218
	1000kms	100	108	86	70	105	162	208
Iron Ore	500 kms	100	115	83	57	94	130	180
	800 kms	100	111	77	40	109	135	187
	1000 kms	100	95	65	43	56	76	98
Iron & Steel	500 kms	100	80	61	63	60	72	72
	800 kms	100	80	61	41	68	81	81
	1000kms	100	80	62	42	70	84	84
Mines	500 kms	100	367.17	259.42	149.94	192.52	264.70	338.00
	800 kms	100	80	368.43	245.60	310.68	420.80	537.90

* Actual prevailing prices were considered

The freight rates for the other bulk commodities recorded significant increase between 1983 to 1992. The highest increase in the freight rate was in case of coal, iron steel, cement, iron ore, mineral oil and mine products respectively. The average distance lead was recorded highest for food products and chemical manure i.e. 1407 km. and 335 km. respectively in the year 1991. Thus, these two commodities were subsidised on both counts i.e.; tariff rates and taper in tariff. On the other hand, tariff rates for mineral oil and iron steel products were increased sharply after 1983 and currently these two commoditiese are classified as high class commodities for tariff purposes. (Table 4.5 and 4.6) and (Figs. 4.2a and 4.2b).

The freight rate policy followed in the second phase affected the development of industrial complexes. Most of the new industrial complexes were located near to the raw material source in order to curtail transportation costs. The average lead for bulk commodities like coal, iron ore, iron and steel and mine products is much less than the average lead for other bulk commodities This shows that the bulk commodities move over shorter distance. The analysis thus shows that there is a relative concentration of basic manufacturing activities near the source of raw material. The finished products move over longer distances to meet the final demand in the economy. Thus the freight rate policy affected the location of new industrial complexes Dutta, M.(1979).

The proportion of the selected bulk commodities had increased significantly to the total freight transport by the railways, between 1969-70 to 1990-91. This reflects the shift of non-bulk commodities like: jute, textiles, sugarcance, oilseeds tea etc. from the railway freight to other mode of transport. This signifies the competition for the movement of finished products and high class freight commodity from roadways and pipelines. Thus a share of bulk commodities i. e. coal. cement, chamical manure, mineral oil. etc. in the total freight have gone up from 65 per cent in 1970 to 84 per cent in 1991. Coal continues to be the major commodity to be transported by the railways. Food products and mineral oil has also recorded increase in the proportion of NTKM, but the increase was insignificant, inspite of the fact that freight rate charged did not record increase in case of food products. This indicates the preference of suppliers to move food products through roadway,

Table 4.7: *Proportion of total Tonnege and Revenue Earned for Bulk Commodity Type (1969- 1991)*

Commodities		1969-70	1975-76	1980-81	1985-86	1989-90	1990-91
Coal	NT KM	28.02	24.39	23.93	28.79	37.32	36.43
	Revenue	18.83	22.44	20.43	15.21	37.72	37.76
Cement	NT KM	5.65	6.39	4.86	5.96	7.42	9.37
	Revenue	5.08	6.63	5.25	6.52	8.11	8.6
Chemical	NT KM	3.39	3.54	6.04	7.57	7.56	7.34
Manure	Revenue	3.15	3.42	5.32	6.50	6.26	6.20
Mineral oil	NT KM	4.73	5.22	7.89	5.57	6.83	6.40
	Revenue	7.66	6.59	11.83	11.11	12.24	11.66
Iron ore	NTKM	4.96	5.31	4.93	4.11	3.70	3.19
for Export	Revenue	6.85	7.67	6.98	5.32	4.58	4.19
Finished	NTKM	5.60	5.75	5.79	5.29	5.79	5.93
Product from Steel	Revenue	—	—	—	—	9.26	9.39
Food Products	NTKM	13.10	11.47	16.46	16.63	13.80	15.09
	Revenue	8.78	7.77	9.92	9.34	7.96	8.06
Total	NTKM	110696	134874	147652	196600	229600	235800
Commodities (Million	Revenue)	5489	10957	15509	42321	74604	82470

as the movement is faster and more efficient. Thus, railways must improve the efficiency of mineral oils have increased the competition for the movement of petroleum products (Table 4.7) and (Fig. 4.3).

The proportion of revenue increased showed an abrupt increase for coal, especially after 1983. NTKM freight for food products and chemical manure although recorded significant increase, but the proportion of revenue earned from these commodities decreased. due to lower tariff rates. Revenue earned for petroleum porducts did not show significant increase, inspite of the fact that tariff rates were increased sharply. This could be attributed to the competition for movement from pipelines and roadways, as modes of transport are more efficient and faster (Table 4.8) and (Fig. 4.4)

Inspite of steep hike in the freight rates for bulk commodities like: cement (207%) and coal (142%) in 1983, their NTKM increased by 59 per cent between 1980-81 to 1990-91. This signifies persistent choice to move these bulk commodities by railways, inspite of steep freight rate increase. However NTKM for the finished products like : iron ore for export, finished product from steel, petroleum products, and food products recorded very less percentage increase in the freight during 1980-81 to 1990-91 (Table 4.8) and (Fig. 4.4). The revenue earned showed 7.70 and 7.57 times growth respectively for coal and chemical manure during 1969-70 to 1990-91. This was closely followed by cement and mineral oils. The least revenue earned increase was for iron ore for export and for food products.The average distance lead for the selected bulk commodities was 660 kms. in 1990-91 coal, iron ore and mineral oil recorded very low increase in the distance lead between 1969-91. Thus *concentration of manufacturing activities near source of raw material was reported during the second phase of tariff policy.* However, cement, chemical manure and food products registered high growth in the distance lead during this period. Less fright rate and taper in the freight have encouraged long distance movement for food products and chemical manure. (Table 4.8) and (Fig. 4.5).

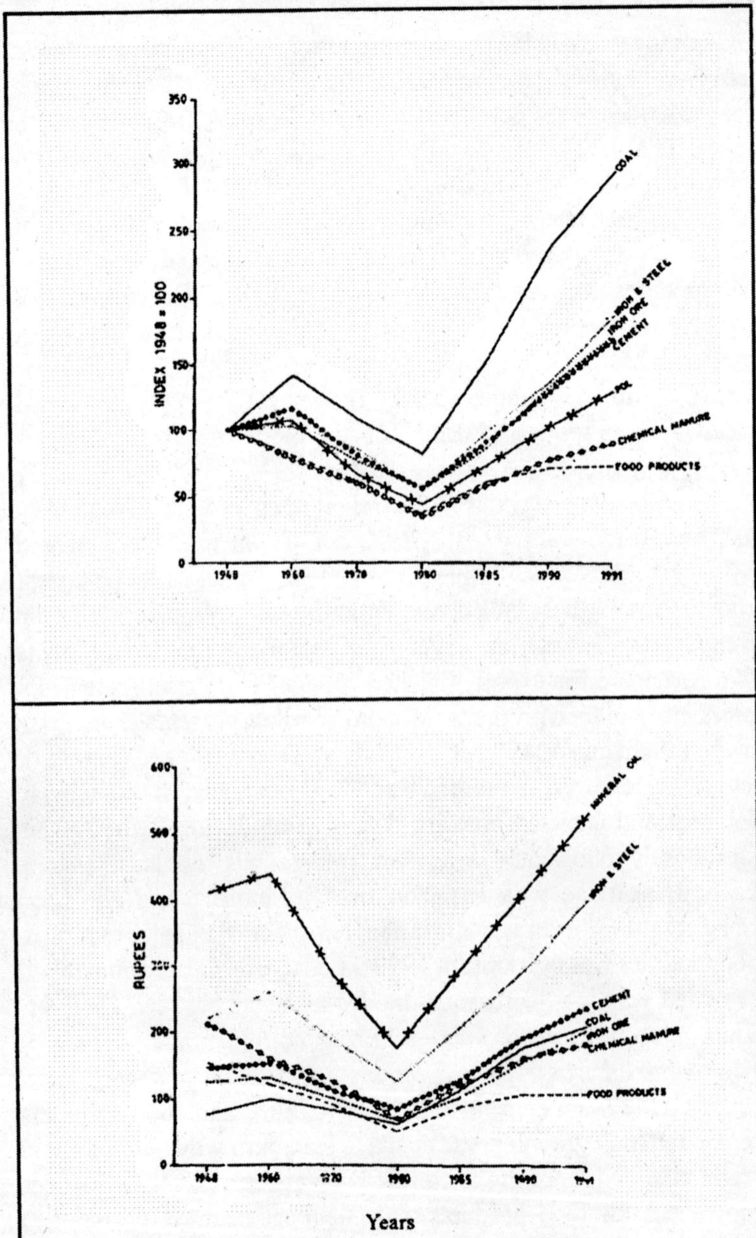

Fig. 4.2 : *Index of Freight Rate per ton for 500 km. Load Major Bulk Commodities 1948-91 (Base on General Price Index of 1987*

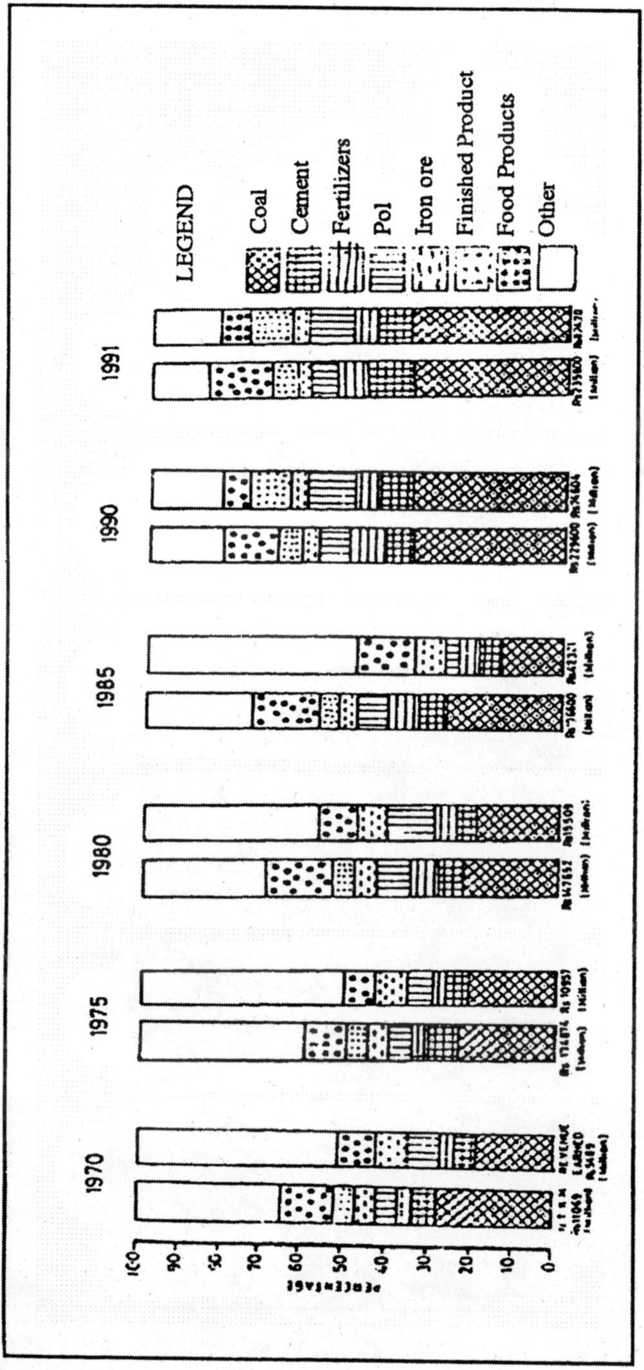

Fig 4.3 : *Commoditywise (proportion) Originating NTKM and Revenue*

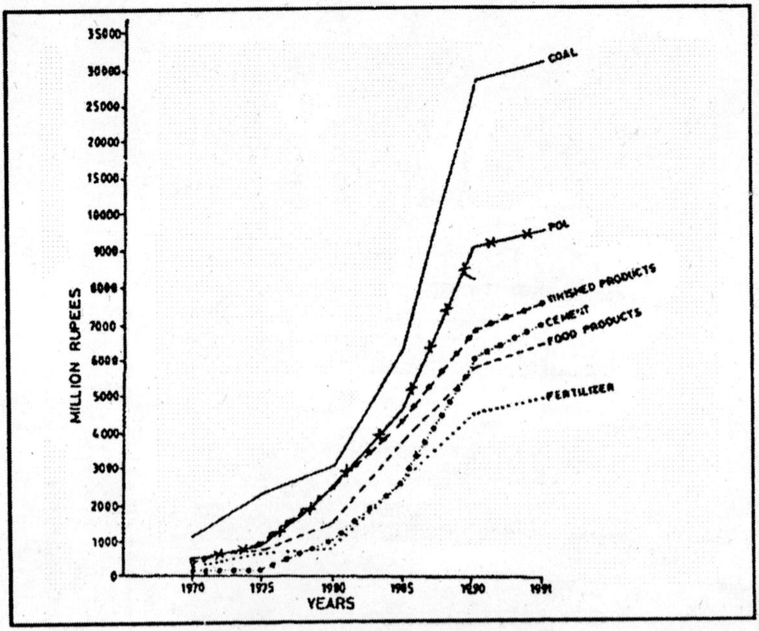

Fig. 4.4 : *Bulk Commodity Freight Net Ton Kilometers K70-71*

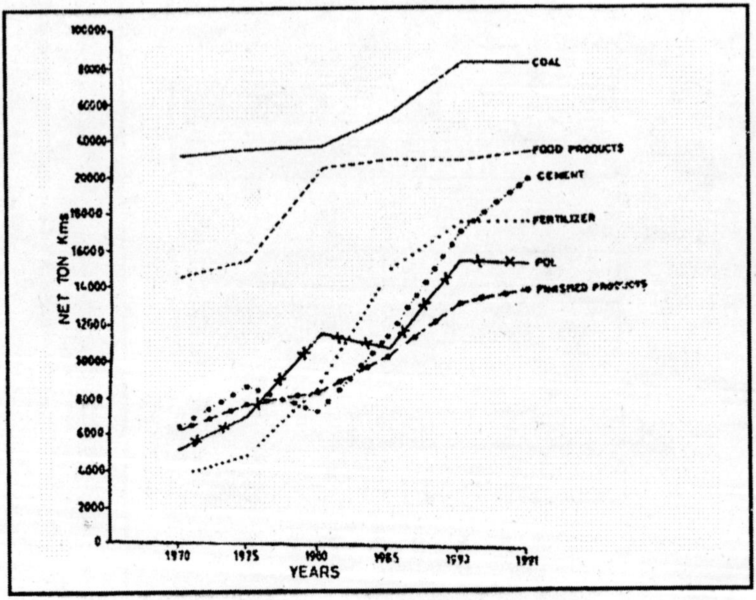

Fig. 4.4 : *Bulk Commodity Freight Revenue Earned (Rs. Million) 1970-91*

Table 4.8 : *Growth pattern of Parameters for Bulk Commodity types (1969-91)*

Commodity	Originating Freight	1969-70 % Growth	1969-70 Growth Index	1980-81 % Growth	1980-81 Growth Index	1985-90 % Growth	1985-90 Growth Index	1990-91 % Growth	1990-91 Growth Index
Coal	NTKM	31025 (Million Tonnes)	100	13.90	113	60.17	182	51.75	276
	Lead	581 kms	100	-2.23	97	11.61	109	-0.03	109
	Revenue Earned	4043 (Rs. Million)	100	11.32	111	60.25	178	331.7	770
Cement	NTKM	6256 (Million Tonnes)	100	14.91	114	63.02	1787	88.61	353
	Lead	585 (kms)	100	17.50	127	-12.46	111	11.02	123
	Revenue Earned	1091 (Rs. Million)	100	6.04	106	167.3	283	129.71	651
Manure	NTKM	3760 (Million Tonnes)	100	137.28	237	67.01	396	16.18	460
	Lead	800 (kms)	100	35.97	135	-1.0	134	-14.14	115
	Revenue Earned	676 (Rs. Million)	100	73.37	173	112.88	456	66.14	757

(Contd.)

Table 4.8 (Contd.)

Commodity	Originating Freight	1969-70 % Growth	1969-70 Growth Index	1980-81 % Growth	1980-81 Growth Index	1985-90 % Growth	1985-90 Growth Index	1990-91 % Growth	1990-91 Growth Index
Pol	NTKM	5492 (Million Tonnes)	100	32.79	132	10.87	147	-6.78	137
	Lead	594(Kms)	100	31.31	131	-24.61	98	2.89	101
	Revenue	1646	100	58.26	158	102.26	320	82.52	584
Iron ore	NTKM	5492 (Milliontonnes)	100	32.79	132	10.87	147	-6.78	137
	Lead	560(Kms)	100	17.67	117	-13.35	101	-36.07	65
	*Revenue Earned	1470 (Rs. Million)	100	4.62	105	64.23	172	37.05	190
Food Products	NTKM	14505 (Million Tonnes)	100	67.58	167	34.58	225	8.80	245
	Lead	961(Kms)	100	37.98	137	2.33	141	3.68	146
	*Revnue Earned	1885 (Rs. Million)	100	16.02	116	102.46	235	50.18	353
All Commodities	NTKM	110696 (Million Tonnes)		33.38	133	+33.15	178	19.93	213
	Lead	660(Ksm)		14.24	114	+0.79	115	-.025	112
Earned	*Revenue (Rs. Million)	21464		2.61	103	115.23	221	73.98	384

* According to 1987 prices excepting for the years 1989-90 and 1990-91 whose current prices have been considered.

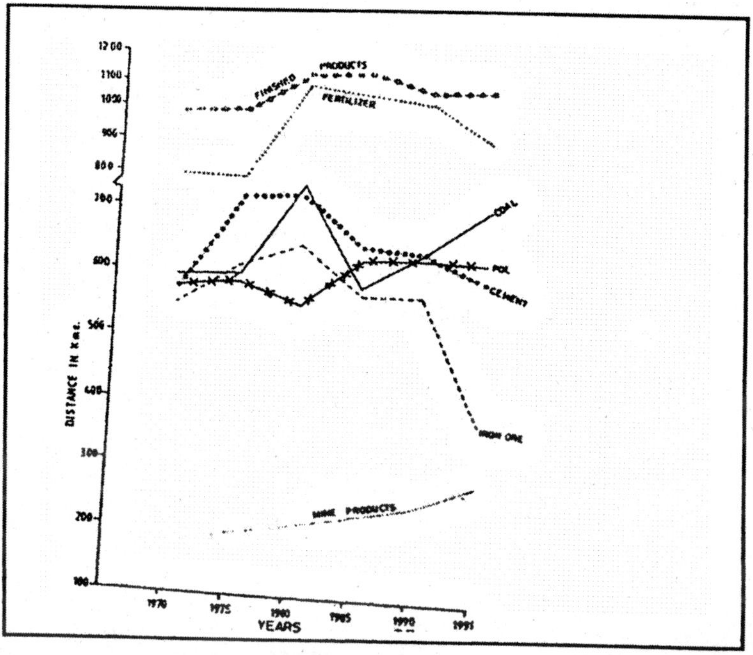

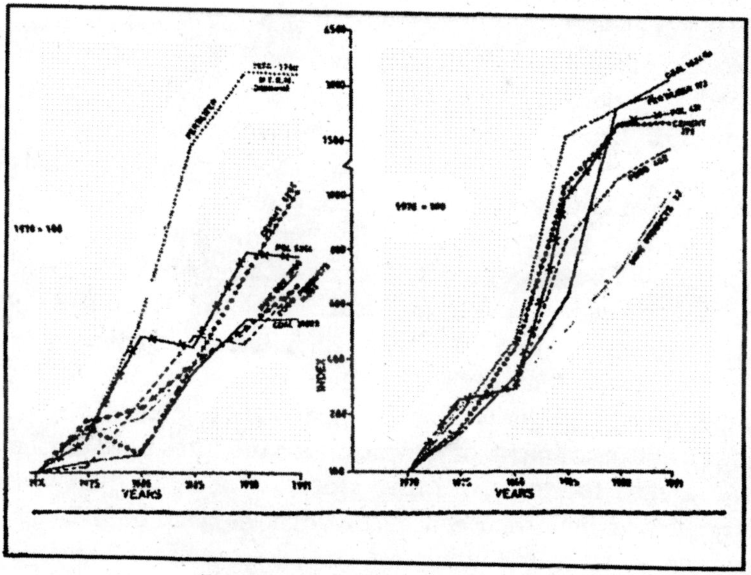

Fig. 4.5 : *Bulk Commidity Lead (KMS)*

Concluding Remarks

The initial phase of freight rate after independence was dominated by differential rate structure, to promote the mobility of basic raw materials of basic raw materials for the rapid industrial and agricultural development of the country. Such a policy resulted in the subsideration of railway tariff rate. This was done through low freight rate, taper in the freight rate and indirectly through freight equalisation.Continuance of such a policy was thought as a catalytic agent, resulting in high degree of regional disparities by offering location advantages only at few places. This also resulted in poor services and restricted modernisation and technological development of railways.

Freight rates were increased in 1983. Although sharp hike in freight rates did not effect the commodity movement of the bulk commodities, yet it was felt that freight rate enhanced for cement, finished goods, mineral oils and other high class identified commodities have already reached a saturated stage, which has resulted in the competition from other modes of transport. The railways are no longer in a position to maintain the monopoly in the matter of inter-regional freight point of view has shown a gradual decrease in the total volume fright movement. The railways must encourage the movement of finished goods which are more profitabie. Excessive delays in the transhipment must be reduced and attractive packages for the movement of these finished goods to and from station to markets must be ensured.

Proper evaluation of costing and freight charging of commodities by various modes of transport should be worked out, so that proper freight rates are fixed. Haphazard increase in the freight rate high class commodities will further result in the loss of their originating tonnage. The burden of subsidisation and freight equalisation for majority of commodities must reduced.

The movement of essential commodities should be encouraged for some, till diversification of the economy is strengthened. Improvement in the fleet of bogies must be given top priority so as to encourage the freight movement. Although transportation rates are less for the railways, yet users prefer commodity movement through road transport especially for finished goods, agricultural products and perishable items. In order to recover the initial exorbitant investments, the railways will have to depend on commodity freight therefore conducive climate must be created for encouraging freight by offering better incentives and services, even at high freight rate. The economy has the resilience to withstand the higher charges.

The location decisions for the expansion of new industrial complexes are directly affected by the freight rate policies. Thus effective freight rate policy could reduce the disparities in the location of industrial complexes.

Notes

Anita, P.P. (1932) : *Inland water Cost.*

Dutta, Mondira(1979) : Commodity Flow and Regional Economy, M. Phil. Dissertation, CSRS/SSS, JNU, University, New Delhi.

Ghose, S.C. (1918) : A Monograph of Indian Railway Rates.

Govt of India (1061-91) : *Goods Revenue Statistics*, Ministry of Railways, Railway Board, New Delhi.

Indian Railway (1990-91) : Facts and Figures.

Indian Railway (1990-91) : Annual Report and Account.

Indian Railway Year Book for 1990-91.

Lefeber & Chaudhary, Datta (1971) : *Regional Development Experiences in South and Southeast Asia*, Moutoh, Paris.

Mehta, N. B. (1927) : *Indian Railway Rates and Regulation.*

Raza, M. and Aggarwal (1986) : *Transport Geography of India*, Concept Publishing Co., New Delhi.

Statistical Abstract of India, 1957-58,1970,1972,1981,1987, Govt of India. Ministry of Planning, Dept of Statistics, Central Statistics Organisation, New Delhi.

APPEANDIX-I

INDEX UNDER OF WHOLESALE PRICES IN INDIA BASE 1949-100

Years	Index
1949	100
1950	105
1951	119.5
1955	87.5
1961	120.6
1970	214.2
1976	364.7
1981	568.9
1985	747.3
1987	837.0

Source : Statistical Abstract of India 1957-58, 1979, 1972, 1987 Govt. of India, Ministry of Planning Dept. of Statistics, Central Statistical Organization, New Delhi.

5

ROAD NETWORK WITH SPECIAL REFERENCES TO EXPRESSWAYS IN INDIA
Some Possibilities and Alternatives

Amresh M. Deshpande and *B.C. Vaidya*

Introduction

Roads play a major role in the economic development of our country. They are considered as an integral part of the transport network of the nation. The road network in India consists of nearly 21 lakh kilometres (Ramakrishnan, R., 1995). But as far as the quality of roads is concerned, our country lags behind many other countries of the world. The history of road development in India can be traced to as early as 3500 B.C. the noted metalled road belonging to the period of Harappa-Mohenjodaro civilization was excavated at Rajgir near Patna. During the regime of the Guptas and the Mauryas, a good network of roads existed in our country in the medieval ages. Shershah Suri and later the Mughals contributed to the extension of road mileage. The British were instrumental in reviving the idea of road construction and formed the Central Public Works Department (CPWD) during the time of Lord Dalhousie to lookafter this work. The Indian Roads Congress (IRC) was formed in 1930 to give recommendation on the quality of roads. In December 1943, a conference of the Chief Engineers was held in Nagpur where a ten-year plan was formulated for the construction and development of roadways in India. This plan is known as the Nagpur Plan (Ahuja and Birdi,1967). Depending upon the category of roads, the Central and State Governments or even local bodies have financed to roadways in our country so far. The taxed levied by the Government on vehicles and excise duties on petrol and diesel are the major source of revenue for this purpose. This paper attempts to examine the present situation of the road network in India and to highlight on expressways.The objective of this paper is to examine the present situation of the road network and to highlight the policy formulated by the Government of India regarding the construction of expressways in our country. An attempt has also been made to suggest suitable alternatives to proposed expressways.

Existing Road Network in India

Roads in India can be classified into four categories as —(i) National Highway, (ii) State Highway, (iii) District Road, and (iv) Village Road. The national highways connect the major cities, capitals of the States, ports and industrial centers in India. The departments of the Central Government maintains the national highways. The state highways link important cities and towns within a particular state and act as feeders to the national highways. The departments of the State Governments are responsible for their maintenence. The district roads are lookafter by the district authorities link the areas of production and the markets within the district. The village roads are under the control of the Zilla Parishad and connect one village to another. These roads are used for the daily movement of the local productions as well as the people. Table 5.1 illustrates the growth of roadways in India since 1951.

Table 5.1 : *Vehicle Population and Road Length in India*

Sr. No	Years	No. of Vehicles (Lakhs)	Road Length (Lakh kms)
1.	1951	3.06	3.99
2.	1961	6.65	5.25
3.	1971	18.65	9.18
4.	1981	53.91	14.81
5.	1991	21.40	21.40
6.	1992	235.07	21.65
7.	1993	252.99	22.00

Source : R. Ramakrishnan, 1995.

The condition of the national and the state highways in our country is poor. A study by Lea Associates South Asia Private Limited has examined the condition of the highways in India has found that 58 per cent of the highways are in a 'poor' or 'very poor' condition, while 21 per cent are 'good' or 'very good' (Aggarwal, 1999). The existing national highways form less than 2 per cent of the total road traffic (Ramakrishnan, 1995). Vehicles travel at an average speed of 30-35 kilometres per hour on these roads. This leads to loss in term of time and fuel vehicle owners and ultimately to the nation lack of funds for the maintenance and upgradation of the roads has been cited as the main cause for this situation. Table 5.2 shows the yearwise allocation of funds for the development of highways in India.

Table 5.2 : *Fund Allocation for Development of National Highways in India*

Sr. No	Years	Fund Allocation (Crores of Rupees)
1.	1950-51	02
2.	1960-61	10
3.	1970-71	22
4.	1980-81	100
5.	1990-91	390
7.	1994-95	633

Source : R. Ramakrishnan, 1995.

Expressways

Expressways can be best described as high-speed, high-quality toll roads. They have often been called as Super National Highways. The expressways are access controlled i.e. they have fixed entry and exit points. A majority of the expressways all over the world are constructed on the BOT basis (Build-Operator and Transfer). The operator of these roads has right to build and maintain the expressways for a specified time period (usually 30-40 years). The operator has the rights for toll-collection. After the fixed period, the project has to be handed over to the Government agencies. Then, it is the responsibility of the Government to maintain them later on. Privatization of the construction and operation of expressways has certain advantages as well as disadvantages. The advantages can be noted as : (i) This provides alternative sources for the development of infrastructure. This reduces the burden on the state, (ii) The privatization of expressway cuts down time and cost overruns, (iii) Innovative technologies are used for their construction, and (iv) The toll can be collected efficiently. However, there are some disadvantages of this system too, namely, (i) Large capital investments and long gestation periods for recovering the investments, (ii) Expressways are feasible only along routes where the traffic density is high, (iii) The toll rates are high, and (iv) The returns on the investment are not guaranteed considering the high risk factor. The Government plays an important role in these projects and acts as a guarantor. It also helps the developer in land-acquisition and may also participate in equity. Thus, the Government has the overall responsibility of ensuring that the venture is a success.

The Expressways Policy of the Government of India

After realizing the need of private sector participation in road projects and after observing the performance of private sector activities in other countries in this field, the Government of India initiated her action for the involvement of the private sector in road projects (Kapila and Mamtani, 1995). In 1995-96, the National Highways Acts was amended to permit private investments on road project. This amendment also allowed the recovery of investment by means of charging a toll for a specified time period. In this regards the Government has set up the National Highways Authority of India as the regulatory body for this purpose. The original plan was formulated in 1995 proposing the construction of about 11,000 km of expressways that would link all the major cities and ports of our country. It is proposed to allow foreign investment in big way.

The policy on expressways announced in 1998 proposed the construction of 7,000 km long expressways from Kashmir in the north to Kanyakumari in the south and Silchar in the east to Saurashtra in the west. These are expected to be operational by the year 2005. The important feature of this policy is to encourage the private sector and foreign investors to participate in such projects. The operators will be allowed to have as much as 100 per cent equity in these ventures and will be provided with several incentives

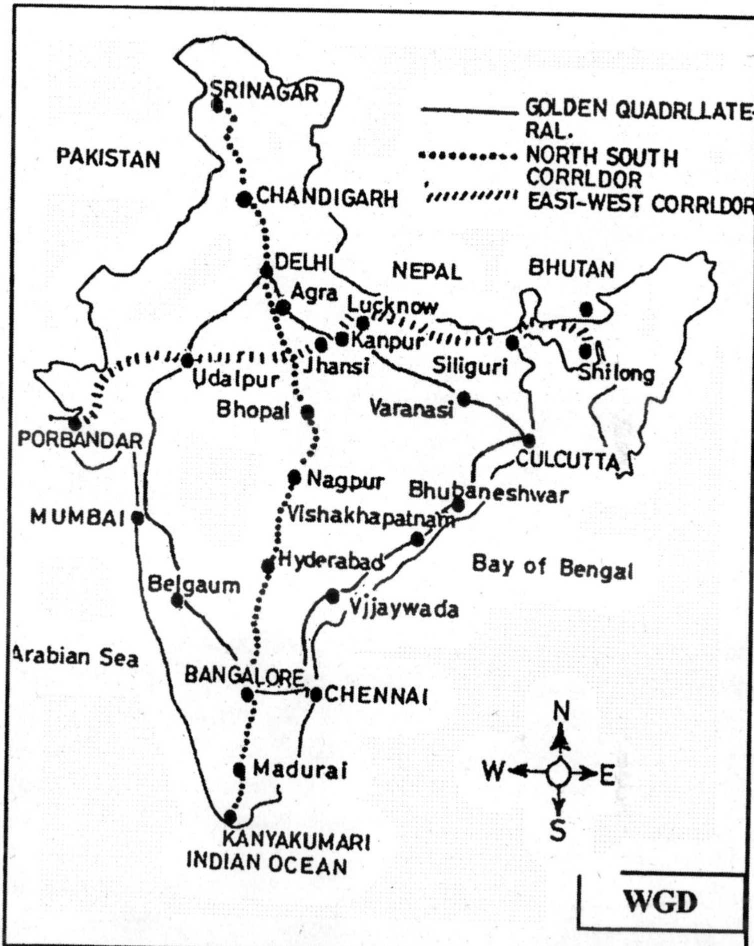

Fig. 5.1 : *National Interated Highways Development Projects*

to investors such as a free hand in toll collection, development of real estate and permission to set up petrol pumps along the expressways. Moreover, these projects will be on a build-operate-transfer basis.

The main objectives of this policy are to revolutionize road transport in our country, creating new jobs and providing new avenues to the construction industry. Table 5.3 shows the pattern of the toll collection for the proposed Bangalore-Mysore infrastructure corridor.

Table 5.3 : *Toll for Selected Trips Along the Proposed Bangalore-Mysore Expressway*

Sr. No.	Mode of Travel	Toll for Selected Trips (Rupees) A	B	C	D	E
1.	Buses	250	25	55	100	70
2.	Trucks	100	10	22	40	28
3.	L.C.Vs	75	8	17	30	20
4.	Automobiles	60	8	13	24	17

Source : Project Summary Report of the Bangalore-Mysore infrastructure corridor in Karanata State, India

Note : Two wheelers that use partial sections of the expressway will be charged a standard toll Rs. 20.

A= Bangalore outer peripheral road to Mysore Link Road.
B= Bangalore outer peripheral road to Township site # 2 and Biddadi
C= Township site # 2 and Biddadi to Township site # 2
D= Township site # 2 and Channapatna to Mandya
E= Mandya to Mysore Link Road.

Problems Faced by the Investors

The participation by the private sector in infrastructure projects in India has not been vary successful so far. The implementation of the policy announced by the Government is seldom done in right manner. Not so long ago, the centre allowed private companies to set up cellular phone services in certain parts of the country. A policy of two operators per telecom 'circle' was formulated. Cellular phone operators set up their establishments hopping to earn profits within no time. But they landed with losses amounting to crores of rupees, indicating wrong estimates of demand. Numerous twist in the Government's policy added to their light. The policy reversals in the civil aviation and the power sectors are well known. As mentioned earlier, expressways are a long-term investment. Given the high interest rates in our country, very few Indian promoters will be willing to borrow from the local markets. High investments mean that the toll rates will be high. The toll revenue from trucks and privately owned vehicles is the main source of income from these expressways. In a few years' time, the Administrative Price Mechanism in the petroleum sector will be phased out. This will mean that diesel prices will rise and this will have an adverse effect on the number of vehicles that will use the toll roads.

It was initially estimated that 28,000 crore rupees would be required to construct 7,000 km of expressways. But the planners went wrong in estimating the costs. They assumed that the cost of construction of an expressway would be four crore rupees per kilometer. But a six-lane expressway costs ten crore rupees per kilometer, thus pushing up the total cost of the project to 70,000 crore rupees. Another 42,000 crores will be required for four-lanning the highways that link the four metros, namely, Mumbai, Delhi, Kolkata and Chennai. An additional 12,000 crore rupees would be needed for constructing bridges, bypasses and for two lanning of the highways along some routes. Thus, the total investment required for roads is 1,24,000 crore rupees! (*Business World*, January 1999).

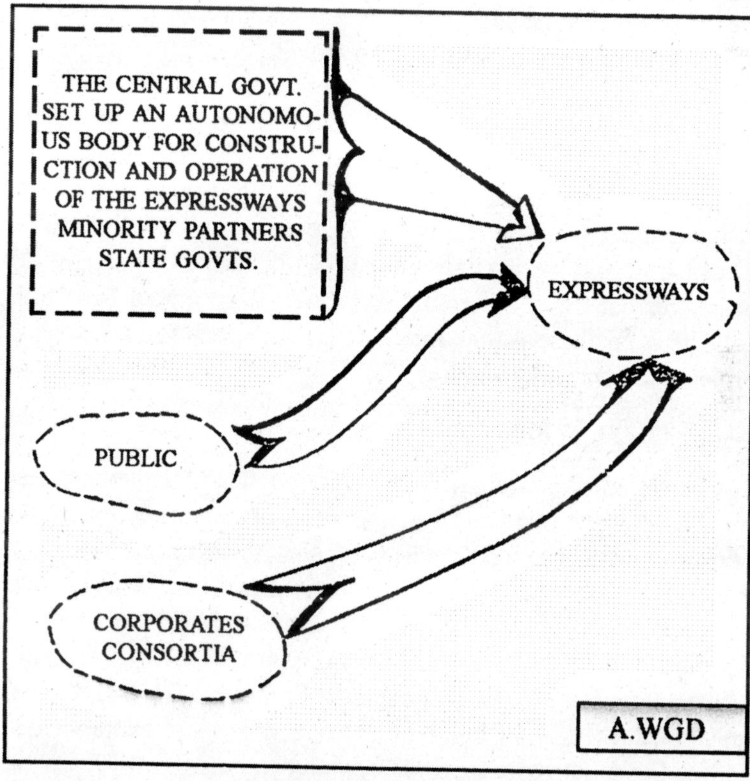

Fig. 5.2 : *Financing for Expressways*

The estimated date for commissioning of the expressways in our country is the year 2005. But before their work can begin, detailed survey report, estimates of traffic growth and land to be acquired will have to be prepared. Besides this, getting clearances from the ministries of environment and defence will also take a long time. The 'model agreement' for the first major four-laning of a highway on Build-Operate-Transfer basis is very onesided with all sorts of conditions and clauses that seek to penalize the private operator. The penalties include, for instance, ten lakh rupees per day if the operator does not reach certain project 'milestones' in time and three lakh rupees per week if the project is not completed within the stipulated period. On the other hand, the National Highways Authority of India (which is responsible for the acquiring the land on behalf of the private operators) is liable to pay a fine of 1,000 rupees per month for every 1,000 sq. meters of land that it is unable to acquire by the time it is needed by the operators. There are only two contenders for the contract at present (Aggarwal,1999). This could very well mean that no private contractor would be willing to participate in expressway project.

Morever, the proposed expressway network does not touch the high-density traffic corridors such as the Delhi-Lucknow-Varanasi route or even Mumbai, India's financial capital. Analysis estimate that expressways are feasible only along routes that have a traffic density of a minimum of 30,000 vehicles per day or passenger car units. Therefore, the revenue generated from toll collection on these routes may not be sufficient to make these projects economically viable.

Expressways in Maharashtra

The Government of Maharashtra has formulated an ambitious plan for connecting Mumbai and the major cities of the state by a network of expressways. The work on the first one of these, i.e. the Mumbai-Pune expressway has already commenced and is expected to be completed by January 26th, 2000. The other projects include the Mumbai-Nasik and Mumbai-Sawantwadi expressways. The Maharashtra State Roads Development Corporation (MSRDC) has undertaken the task of building and operating these expressways MSRDC was set up by the State Government in August 1996 with an equity base of five crore rupees. It is fully owned by the State Government, and has raised over 1,170 crore rupees through non-convertible redeemable bonds, which are India's biggest bonds for infrastructure (*Business India*, February, 1999). This corporation has also undertaken the task of constructing 55 flyovers in Mumbai. The Table 5.4 reveals the proposed network of expressways in Maharashtra.

Table 5.4 : *Proposed Network of Expressway in Maharashtra*

Sr. No	Years	Routes	Length (km)
1.	1995-2000	Mumbai-Thane-Nasik	145
2.	2001-2015	Thane-Pune	140
3.	2001-2005	Pune-Belgaum	333
4.	2006-2010	Mumbai-Balsad	191
5.	2006-2010	Nasik-Balsad	408
6.	2010-2015	Agra-Nagpur	1270
7.	2010-2015	Pune-Solapur	242
8.	2010-2015	Dhule-Nagpur	517

Source : *Lokmanya*, 16th November 1994. Govt. of Maharashtra, Mumbai

Concluding Remarks

The present chapter has examined the transport network in India and provides new horizon to the expressways. The expressways are needed for all-round development of the country. But this calls for spending huge amounts of money for the construction of the expressways. Taking the present economic situation of our coutry, the decision to construct the expressways seems to be hurriedly taken. Instead of this old highways could be widened through private sector participation so as to ease the traffic and movement of goods. Actually there is a great need to maintain and upgrade the existing highways. Considering, this view it is suggested that the construction of the proposed expressways be postpones till 2025. After reviewing the traffic situation within and between India's larger cities, the new construction work should be initiated. It is also suggested that the expressway from Mumbai to Pune be treated as a pilot project in this area. This performance of this expressway on basis of the toll revenue it is able to generate and the traffic it necessities be assessed for a period of five years. And, only then the work on linking the big cities by expressways in entire India should be started.

References

Aggarwal, Y. (1999) : "Glaring Potholes Maharashtra Expressway Project", *Times of India*, March 8th, 1999.

Ahuja, T.S. and G. S Birdi, (1967) : *Roadways, Railways, Bridges and Tunnels.*

Bangia, Ashok and W. C. Taylor, (1995) : "Highways Privatization : The answer to India's Transportation Needs" *New Horizons in Roads And Road Transport-Vol II ICORT-95,* New Age International Limited, New Delhi.

Business India (Magazine) : "Just flyover Mumbai" February 8-21, 1999.

Business World (Magazine) : "The Roads to Nowhere." January 7-21, 1999.

Kapila, K.K. and I. J. Mamtani, (1995) : "Priority of Four-Lanning of existing Two Lane High-Density Highways under the Private Sector," *New Horizons in Roads and Road Transport – Vol.* ICORT-95, New Age International Limited, New Delhi.

*Outlook (*Weekly*)* : "The Road to Utopia", December 7th 1998.

Project Summary of the Bangalore-Mysore Infrastructure Corridor in Karnataka State, India the Kalyani Group.

Ramakrishnan, R. (1995) :"Relevence/Risks on Privatization of Roads", *New Horizons in Roads and Road Transport,* Vol.II, ICORT-95, New Age International Limited, New Delhi.

Vaidya, B.C. (1998) : *Readings in Transport Geography – A Regional Perspective.* Devika Publication Private Limited, New Delhi.

SECTION-II

METROPOLITAN TRANSPORT SYSTEM

6

DEVELOPMENT OF TRANSPORT IN CHENNAI METROPOLITAN AREA

G. Marthandan and *S. Subhiah*

Introduction

Transport system in a region may be considered as a mirror of economic development and material prosperity of that region. It paves way for the basic infrastructure for the growth and total developmeent of that region. The development of the transport system has been influenced mainly by the accelerated urbanization process, characterised by the dispersion of activities and uncontrolled occupation of urban space. Transportation network of a region plays a pivotal role in reducing the spatial disparities and bringing about a balanced and intergrated development. It helps in the proper exploitation of the available resources of a region and to move towards an economic equilibrium. Transportation is the movement of people or goods between places of origins and places of destinations within an area. The character of transportation as a whole at any particular time is altogether determined by its interrelations with physical and social forces and conditions. The transportation involves different modes at different times for movement of passengers or goods to satisfy different purposes. This paper highlights the development of transport within Chennai Metropolitan Area (CMA).

Study Area

Chennai (formerly known as Madras), the gateway of south India, was established in 1639 by East India Company as the Chief Trading Centre. The origin of Chennai was from a small fishing village called Chennapatra located on the Cholamandalam Coast of Tamil Nadu. It comprised of scattered settlements separated by long distances. Each settlement grew around a temple and has its own history. Some of the well-known small settlements were in Purasawalk, Mylapore, Thiruvallikeni, Thiruvatteeswaranpetta, Egmore, Nungambakkam and Saidapet. The suburban areas which include Tiruvottiyur, Velachery, Saint Thomas Mount, Pallavaram, Tirusulam, Tiruneermalai, Mangadu, Padi, Poonamallee, Kunnathur, Ayanavaram, Vyasarpadi, Villivakkam, Ambatturm Koyambedu etc. were self-contained and had their own agricultural production and household industries.

The nucleus of Chennai rose from 67sq. km to over 172 sq. km. to become the Chennai City. The urban agglomeration which covers 531 sq.km. including city and a number of major towns and Panchayats. The Chennai Metropolitan Area (CMA) which includes the city, 5 municipal councils, 5 townships, 26 town Panchayats and a large number of rural settlements, spreads over 1167. sq.km. The extent of Chennai city and Metropolitan Area are shown in Fig. 6.1.

Population Growth

In spite of the fact that man is much more flexible than the lower animals in adapting to a large number of different environment. human beings have found certain environments much more congenial than others. The environment factors affecting population distribution at a given point in time are; (a) climate, (b) location of water, soil, energy and mineral resources, and (c) transport relationships. With the then technology available, cost of water transport was very much cheaper than transportation by land. This induced better growth of population along the coastal lines than the interiors. Thus the population of Chennai city which was established along the coastal lines, has grown from meagre 19,000 in 1646 to 35.41 lakhs in 1991 and that of Metropolitan Area to 58.46 lakhs. The population growth in Chennai city and metro area is given in Table 6.1.

As may be seen from Table 6.1 that the population growth in Chennai city as well as in CMA is not at a uniform rate. The city started growing after 1921 and after Independence it touched its

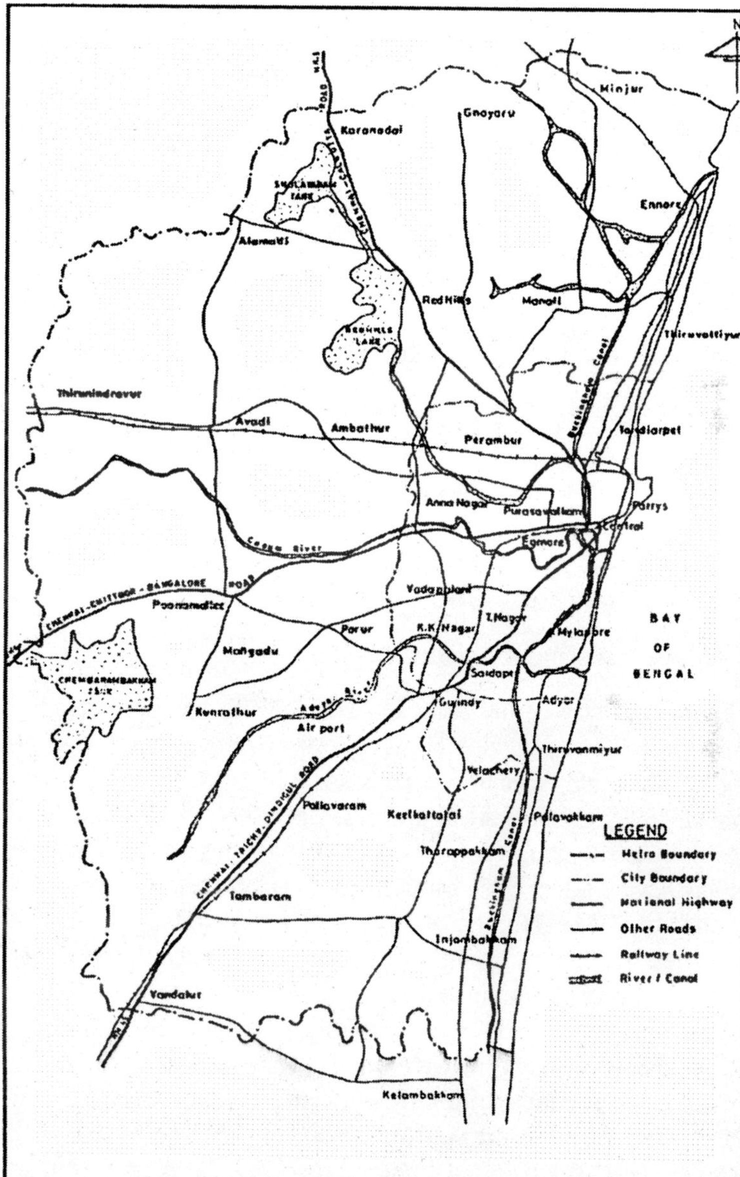

Fig. 6.1 : *Chennai Metropolitan Area*

peak. The CMDA has estimated that the city will have 60 lakhs population by 2011 while the CMA will accommodate 90 lakhs by the same year.

Table 6.1 : *Population Growth in CMA*

(Population in lakhs)

	Chennai City		CMA	
Years	Population	% variation	Population	% variation
1900	5.41	–	8.00	–
1911	5.56	2.77	8.30	3.75
1921	5.79	4.14	8.76	5.54
1931	7.13	23.14	10.56	20.55
1941	8.55	19.92	14.50	37.31
1951	14.16	65.61	17.62	21.52
1961	17.29	22.10	23.24	31.90
1971	24.67	42.68	34.75	49.53
1981	32.66	32.39	43.86	26.22
1991	38.41	17.61	58.46	33.29
2011	60.00*	56.21	90.00*	53.95

* Estimated by CMDA

Landuse Past and Presents

This capital city of the state of Tamil Nadu has been steadily growing in its importance in all aspects covering administration, higher education, specialised medicam care, trade, commerce and industry. The Municipal Corporation of Chennai city had been generally controlling the pattern of development of area till the function was taken over by Chennai Metropolitan Development Authority (CMDA). The growth of Chennai did take place in regulated manner except in a few selected areas developed under different Town Planning Schemes. It is also to be noticed that infrastructure facilities like water supply, sewerage, communication etc. did not grow along with the rapid development of the city. The thrust given for industrial development after Independence saw the city and the metro area grown in a bigger way. The uncontrolled development was brought under control by the preparation of integrated land use plan by 1969 and the same was given to CMDA for implementation in 1973. The plan mainly dealt with land development, traffic and transportation, housing and slum clearance.

By and large, the development had been taking place on the same lines as envisaged in the plan except that the satellite towns contemplated in the plan had not grown to meet the expectations. By 1995, CMDA had prepared the second master plan for CMA for the next twenty years which liberalises land use controls given in the earlier plans. The comparison of landuse at different times in Chennai city and CMA are given in Table 6.2 and Table 6.3.

Table 6.2 : *Landuse in Chennai City*

Landuse	1964		1974		1991		2011**	
	hactere	%	hactere	%	hactere	%	hactere	%
Residential	4290	33.3	5780	44.9	8138	47.7	7461	46.0
Commercial	420	3.2	760	5.9	1184	6.9	2202	13.5
Industrial	490	3.8	860	6.6	918	5.4	906	5.6
Institutional*	3700	28.8	2160	16.8	4770	28.0	4457	27.5
Open space	3980	30.9	3320	25.8	2037	12.0	1203	7.4
Total	12880	100	12880	100	17047	100	16229	100

* Include public and semi-public uses, utility services, transport and communications.

** Proposed landuse structure excluding George Town and harbour planning divisions.

Development of Transportation in CMA

The Transportation system in CMA comparises of two categories of transportaion, viz. surface transport and air transport. The surface trasnport comprising of road and rail transport serves both inter city transport while the water transport and air transport cater to inter city transport. Each category of transport has its own modes of transport ranging from mere walking to supersonic jet planes. This wide variety of modes has been developed partly through evolution and partly through design. Under pressure from successive waves of immigration and a buoyant economy, the demand for transport services has risen steadily and unlike other cities in developed countries, Chennai has notreplaced its obsolete systems but has added new systems, superimposing them on the old.

Table 6.3 : *Landuse in Chennai Metropolitan Area Excluding City*

Landuse	1964		1974		1991		2011+	
	ha	%	ha	%	ha	%	ha	%
Residential	3250	3.1	8460	8.2	20748	20.9	19278	19.4
Commercial	120	0.1	100	0.1	428	0.4	9707	9.8
Industrial	2460	2.4	2860	2.8	4705	4.7	7482	7.5
Institutional*	6990	6.7	2540	2.4	5062	5.1	6544	6.6
Open space	4420	4.3	10720	10.3	27404	27.6	332	0.3
Agricultural	86550	83.4	79120	76.2	40991	41.3	55996@	56.4
Total	103790	100	103790	100	99338	100	99338	100

* Includes public and semi, public uses, utility services, transport and communications.

** Proposed landuse structure.

@ Includes water bodies, urbanisable and non-unbanisable.

Surface Transport

The surface transport as the name goes, includes both road and rail transport. As the water transport serves only intercity transport at present, the growth of the same is described separately. In Chennai, the road transport preceeded rail transport as in other towns.

Development of Roads in CMA

The first means of surface transport was road transport where it started as a means of important lines of communication linking the old settlements. The initial roads found during 1600 were just earthen cart tracks. One of the mean roads at that time was Poonamallee High Road connecting the settlments of Egmore, Purasawalkman and Aminjikarai. The other road connected Thiruvallikeni and Mylapore.

In those days, the older settlements were along the water fronts in Chennai. The Britishers were also living along Cooum river and roads were laid to give access to them. The important thoroughfares developed during the late 1700 were Marshall road, Halls road, Monteith road, Casa Major road, Mowbrays road and

Royapettan High road. The roads which were of radial pattern before 1800, were developed on circular pattern at later stages. Mount road, the important radial road even during that time was given access through other roads like Triplicane High road, Chamiers road, Edward Elliots road, Royapettah High road etc. In addition to this, new link roads were formed to connect the newly formed residential areas. During 1845, Trunk Road Department was formed and with it began a new chapter in the history of roads in Chennai. Later on due to unsatisfactory working of the department it was abolished in 1854 and 1855 saw the birth of Public Works Department.

The Municipal Board of Chennai had more than 60 miles of roads under its jurisdiction. Heavy vehicles started using the roads during late 1800 and the main roads were wide enough for the same. The city developed with George Town and Harbour as the base and naturally all communication lines led to this centre and these were linked with each other producing a radial and ring pattern of development. In 1893, the Municipal Board had arranged for installation of Tramway in the city and it was intended for only passenger transport. The Independent India saw the development of four metro cities in all fronts.Thus Chennai was very well connected with the rest of the country by means of road, rail, seaand air. The present day road network in the study area was developed and is maintained by a number of agencies which include Highways and Rural Works Department, Corporation of Chennai, the municipalities, the panchayats,the townships, etc. The earlier earthen roads were converted into water bound macadam roads, which were then converted in cement concrete roads and bituminous roads. The roads width in Chennai city ranges from 5 m to 45 m forming single lane to 6 lane roads. The lesser carriageway widths are noticed in old settlement areas, streets and byelanes whereas most of the roads which carry heavy fast moving traffic are 2 lane roads and above. In order to facilitate uncongested traffic, the city has at present 11 pedestrian subways, 9 road underpasses, 8 road over bridges, one fly over and a number of river bridges and culverts. The road networking is primarily based on the three National Highways, leading to Calcutta (NH 5), Bangalore (NH 4) and Dindigul (NH 45). The other roads which connect Chennai to the rest of the State include East Coast road, Old Mahabalipuram road, Chennai-Thiruvallur Road, and Chennai-Ponneri road.The present day road netwok in Chennai Metropolitan Area consists of 90 km.of arteials and 468 km. of collector-distributor roads totalling to 558 km.

Development of Road Transportation

Though the road transportation started with hand cart and animal drawn carts, the advancement in technology coupled with growth of economy has brougt out different modes of transport with varying pulling powers. Both passenger transport and goods transport use the same carriageway. The vehicles that carry passengers range from cycles, cycle rickshaws to bus. The passenger transport caters to both inter city travel and intra city travel (Urban Transport) and the travel is made either in public transport or in public transport or in private transport like personalised transport (car, fast two wheelers, cycle etc.) para transit (auto-rickshaw, taxi, cycle rickshaw, jutka etc.)

Growth of Public Transport in CMA

Until 1920, there was no progress in motor transport. Due to establishment of railways, there arose a necessity for the Britishers to introduce motor vehicles for transportation. During 1890, only motor cars were seen on the roads in Chennai. M/s Simpson & Co. played crucial role in the development of transport by introducing motor cars and buses in Chennai. In 1904, Madras Electric Tramway Ltd. was formed and they proved better to cope with peak hour traffic than any other means of transport because of larger carrying capacity. However, trams had no flexibility of service and were not free to move anywhere but confined to prescribed track.

The early road transport was disorganised and individualistic in character. But the Madras Electric Tramway Ltd. operated a fleet of 50 buses during 1925-27 which was abandoned in 1928 owing to uneconomical operations. The Central Committee on post war reconstruction recommended in late 1944 that public transport should be owned and operated by the State. In pursuance of this, the State Government decided to nationalise the passenger motor transport service. In March 1947, the first stage towards nationalisation was initiated by introducing thirty buses side by side

Table 6.4 : *Growth of Bus Fleet and Person Trips in CMA*

Under Departmental set up			Under Company set up		
Years	Fleet	Passenger per day (Lakhs)	Years	Fleet	Passenger per day (Lakhs)
1947 (Mar)	6	n.a.	1972-73	1029	11.00
1947 (Dec.)	100	n.a.	1973-74	1056	13.37
1948 (Jan)	202	n.a.	1974-75	1160	15.97
1948 (Dec)	320	n.a.	1975-76	1374	18.07
1951 (Jan)	322	n.a.	1976-77	1434	18.52
1953-54	381	2.95	1977-78	1466	19.82
1954-55	402	3.04	1978-79	1576	21.48
1955-56	401	3.20	1979-80	1697	23.39
1956-57	445	3.48	1980-81	1794	24.86
1957-58	450	3.69	1981-82	1963	24.29
1958-59	470	3.80	1982-83	2103	25.59
1959-60	470	3.97	1983-84	2091	27.32
1960-61	550	5.03	1984-85	2080	29.41
1961-62	647	n.a	1985-86	2105	29.02
1962-63	658*	n.a	1986-87	2089	30.30
1963-64	771*	n.a	1987-88	2100	29.25
1964-65	890*	n.a	1988-89	2202	31.21
1965-66	1088*	n.a.	1989-90	2298	31.76
1966-67	1189*	n.a	1990-91	2296	34.38
1967-68	1391*	n.a	1991-92	2353	35.42
1968-69	1590*	n.a	1992-93	2325	36.00
1969-70	1666*	n.a	1993-94	2336	36.80
1970-71	n.a.	n.a	1994-95	2359	38.95
1971-72	n.a.	10.80	1995-96	2540	42.00
			1996-97	2849	46.00

* Includes certain long distance services.

with the exiting fleet of 188 buses run by private operators in Chennai city and by July 1948, the entire city bus service was nationalised. The closure of tramway in 1953, apart from natural growth of fleet, resulted in steady and rapid increase in fleet strength.

The Road Transport Corporation Act (RTC Act) was passed in Parliament in 1948 to enable the State Governments to form Corporations to run passenger transport services. In Tamil Nadu, the Justic Somasundaram Committee which went into the working of State Transport recommended to form autonomous Corporations either under RTC Act or undr Company's Act. This gave birth to Pallavan Transport Corporation Ltd. (PTC) in Chennai formed under Company's Act on 1st January 1972 with a fleet strength of 1029 buses and 294 routes transporting 11 lakh passengers per day on an average. From then onwards there has ben a steady growth of population, industrial activities, educational advancements, social and recreational centres and so on, which in turn has contributed to more number of trips in public transport. The economic status of the people and their insufficient disposable income to own their own vehiclees have also contributed to heavy usage of public transport. The spread of the city and the metropolitan area have forced the PTC authorities to extend the bus operation to all parts of the city and the metro area by opening new bus routes, terminals and depots. The steady growth of fleet and the person trips per day in buses are given in Table 6.4.

As years rolled, the level of service provided in buses have come down drastically because of increased cost of operation and lower fare structure. Overcrowding in buses has become order of the day. To improve the quality of service, the bus authorities have implemented various schemes in term, of operating peak hour specials, ladies specials, express and super express services, limited stop services, point to point services, industry specials, festival specials, cut trips, extended trips, diverted trips and so on. But all these measures were and are planned with the existing fleet only. Because of shortage of capital funds, the bus augmentation programmes have suffered a set back. This coupled with advancement in automobile technology and the increase in standard of living of the city people have paved way for the personalised transport.

Growth of Personalised Transport in CMA

Before the concept of public transport emerged, the personalised transport was in existence. Unlike the public transport, the personalised transport or private transport caters to the individuals providing door to door service during any time of the day. This kind of transport mainly comprises of cars, motor cycles or fast two wheelers and pedal cycles. The socio-economic status of a person/ family decides the type of vehicle that he/they own. In addition, the Government, the Corporates, the big establishments and so on own their vehicles and permit their officers to use them. From the vehicle statistics, it is seen that the growth of cars is significant in these years. There were around 9,844 cars in Chennai during 1951, which grew to 11,797 in 1961, 37,618 in 1981, 1,09,915 in 1991 and 1,63,627 in 1997. The rate of car ownership accordingly increased from 1 car per 144 persons in 1951, 1 car per 87 persons in 1981, 1 car per 35 persons in 1991 and 1 car per around 28 persons.

The inadequacy of public transport coupled with the decrease in the level of service in it and the economic status of the people have contributed to the very high growth of fast two wheelers in Chennai. The technological advancements in automobile industry and the flexible financing options have a great bearing on the increase in the number of fast two wheelers which include motor cycles, scooters and mopeds. The two wheeler population during 1951 was a meagre 1,581 and it is over 6.84 lakhs during 1997 resulting an abnormal increase in a period of around 45 years. This boom can be attributed to the Government industrial policy to give thrust to automobile industry especially to two wheeler industry.

The most predominant mode of personalised transport is bicycle. Still, it is the cheapest mode of trnasport available to all users. There were about 38,339 cycles in the city during 1952-53 which increased to 51,789 during 1956. It reached a figure of over 1.17 lakhs as early as 1971 in the city and 1.37 lakhs in the whole of metropolitan area. As there is no licensing system for cycles at present, their growth could not be ascertained. However it is estimated that there will be over 10 lakhs cycles within Chennai city alone as of now.

Growth of Para Transit in CMA

Para transit ranging from maxi cabs to cycle rickshaws, the most predominant being autorickshaw, has attracted a lot of attention in recent years. These vehicles have emerged in order to fill the gaps not catered for in overall market demand and in some cases to provide a more personalised and flexible service to section of market. The shortfall in public transport systems, difficulties in maintaining personalised transport vehicles, changing road situation with imposed constraints etc. have resulted in more demand for para transit systems in Chennai. This intermediary form of transport helps in improving the supply to match the demand in the transport system.

The first form of para transit was rickshaw pulled by human beings and this was in vogue till 1970 when it was abolished and replaced with cycle rickshaws. The rickshaws and cycle rickhshaws have limitations in their pulling power, carrying capacity, area of operation, type of goods carried etc. With the advancement in technology, the cycle rickshaws were fitted with lesser capacity petrol/diesel engines. Thus CMA continues to have both cycle rickshaws and motorised cycle rickshaws at its roads.

Once upon a time the popular mode of local transport was jutka drawn by horse. These vehicles were used even by Brithshers to move from one place to other. These vehicles had natural death when more number of motorised vehicles were put on road But still one can find jutkas operating in older parts of Chennai in very small numbers. The other mode which was popular during 1960s and 1970s was taxi. The taxi is a metered cars run on hired basis. It differs from tourist taxi which is becoming popular in recent times. The metered taxis have declined in number from around 3900 during 1970 to around 500 during 1997 (Table 6.5). The predominant reasons attributed for the same are the frequent fuel price hike and the increased cost of operation. At present the metered taxis have their presence felt at select centres like Meenambakkam Airport, Central and Egmore Railway Stations, parking lots in Star Hotels and near certain hospitals.

The most predominant form of para transit in Chennai is auto rickshaw. It started operating in late 1960s and its fleet strength was a meagre 230 during 1970 in the city. By the end of 1997, it has grown to around 33,600 (Table 6.5). There was spurt in growth of number of auto rickshaws during 1990 because of relieving them from the clutches of permits by the Government of Tamilnadu. Also the auto rickshaws which were hirtherto allowed to carry 2 persons, are permitted to carry 3 persons. This has resulted in higher usage of auto rickshaws which in turn contributed to higher growth. The auto rickshaws are there in operation not only within the city but also within CMA. It is estimated that over 42,000 auto rickshaws are in operation within CMA as of 1997. The advancement in technology has contributed to diesel engines for autorickshaws which are becoming popular in recent times.

Growth of Freight Transport in CMA

From the day it was established by East India Company, Chennai is an important place in India for commerce and trade. Because of the locational advantage, the goods produced at different parts of India are brought to Chennai and transported to other countries. For meeting the local demands, Chennai was developed as a wholesale trade market. The main carrier of fright in olden days when there were no motor vehicles was hand cart which was then supplimented with animal drawn vehicles. But these modes still continue to operate in parts of Chennai to tansport goods to shorter distances. When the motor vehicles were put on road for transportation, they were used both for carrying passengers and as well as goods. The motor vehicles which cater to the goods transportation are motorised cycle rickshaws, tempos, light commercial vehicles, trucks with and without trailers and tractor trailers.

Licensing the hand carts, animal drawn carts and cycle rickshaws were in vogue till 1980 after which it was abolished. However, the registering of motor vehicles continues to be there and permit system is also in existence for all the commercial vehicles. The goods vehicles registered under trucks category within Chennai over a period of time is given in Table 6.5. This table seen that the trucks in Chennai which was 2,770 during 1970 rose to 29,487 during 1987 but got reduced to 23,537 during 1997. But Chennai continues to experience the truck traffic not only from in and around Chennai but also from other parts of Tamil Nadu and other States.

In an earlier study conducted by CMDA, it was found that about 3,400 handcarts, 62 single bullock carts and 70 jutkas were in use in the city in 1978. The absence of the same data for the entire CMA and also for further period after 1978 have become a handicap for the planners in the city. The freight transport share the same road space along with passenger transport and hence more number of roads exceed their capacity and cause servere congestion on these roads. This problem is temporarily solved by banning or regulating the truck movement in certain important roads. Unlike the but transport, the truck/commercial vehicle transport is not in the hands of an organised sector from the beginning. There are innumerable vehicle owners ranging from a single truck operator to big operators. Also commercial vehicles do not have a fixed origin or destination nor do they follow a fixed routing with fixed timing. These have contributed to the uncontrolled growth of goods transport.

The idle parking of commercial vehicles are done at Elephant Gate (Lorry thotti), Salt Coutaurs, Perimet and Loane Square, all located within CBD of Chennai. With a view to reduce the traffic congestion because of heavey commerical vehicles, the wholesale markets of vegetables, fruits, flowers and iron and steel have been shifted from the city core to the fringes of the city (Koyambedu and Sathangadu). This has definitely relieved the city from acute traffic congestion. However, there remains more wholesale markets in the Central Business District (CBD) which attract more heavy commercial vehicles not only from other parts of Tamilnadu but also from other States.

Development of Rail Transport

Right from the beginning both metre gauge and broad gauge were in existence in Chennai. Central was first to come as a terminal as early as 1872. From then onwards, Central station has become an important railway terminal in Indian Railways. This station caters both for inter city and sub-urban traffic in broad gauge. The other terminal which serves Chennai city is Egmore station with metro gauge. It serves for both passenger and goods traffic.

To serve the city and sub-urban traffic, Chennai has four railway lines as follow :

Table 6.5 : *Growth of Motor Vehicles in Chennai City*

Year	Bus	Taxi	Trucks	Car, Van	Two wheelers	auto ric-kshaw	Total
1929	NA	NA	NA	NA	NA	NA	15931
1930	NA	NA	NA	NA	NA	NA	8096
1931	NA	NA	NA	NA	NA	NA	19741
1951	521	310	1615	9844	1581	NA	14019
1952	530	320	1745	10520	1658	136	15066
1953	554	254	1447	8770	1538	148	12891
1954	584	256	1365	8926	1604	157	13093
1955	571	304	1637	8177	1597	182	12516
1956	640	457	2131	9180	2012	308	14802
1957	656	608	2057	9666	2195	294	15553
1958	638	826	2293	10491	2700	302	17369
1959	515	901	2304	11086	2835	262	18009
1960	750	1215	2867	13666	4362	237	23222
1961	715	1252	2831	11797	4119	298	21146
1962	855	1467	3188	13538	5354	194	24700
1963	945	1686	3151	13314	5081	205	24651
1964	1077	1710	3116	14600	5597	198	26502
1965	1143	1825	3440	14809	6391	186	27948
1966	1420	2083	3607	15410	7740	140	30436
1967	1569	2479	3609	18493	9000	192	35403
1968	1768	2777	4035	18648	9882	190	37337
1969	1701	2941	4981	19103	12012	204	40980
1978	1832	2618	6933	27795	40765	1518	81461
1979	1706	2787	6681	28178	47373	2025	88750
1980	1737	2668	7527	30686	58435	791	101844
1981	1802	2264	9877	37618	62700	5856	120117
1982	1786	1890	9819	27019	74178	5834	120526
1983	2124	1599	10342	35803	72179	5640	127687
1984	2236	1514	11171	37934	86503	5548	144906
1985	2146	5731	15207	50629	122123	6595	202431
1986	2294	5028	28354	86948	201407	6152	330183

(Contd.)

(Table 6.5 *contd.*)

Year	Bus	Taxi	Trucks	Car, Van	Two wheelars	auto ric-kshaw	Total
1987	2353	5526	29847	95131	233537	5966	372360
1988	2189	6762	19051	105847	285308	5869	425026
1989	2192	6964	19534	114704	338486	5700	487580
1990	2329	5716	19826	114992	387289	6130	536272
1991	2287	5866	17305	109915	386634	10650	532657
1992	2353	4807	18445	117008	433046	16407	592066
1993	2265	3852	18530	122044	461638	18630	626959
1994	2367	4398	16639	128945	497640	20202	670191
1995	2571	5490	18122	135433	562722	21845	746183
1996	2550	5697	18528	145083	651554	25233	848645
1997*	2864	5714	23587	163627	684039	33597	936756

* Upto 30-9-97 only.

(a) Chennai Beach - Tambaram in metro gauge;

(b) Chennai Central - Trivellore in broad gauge;

(c) Chennai Central - Gummidipoondi in broad gauge; and

(d) Chennai Beach - Thirumylai in broad gauge.

The metre gauge railway line from Beach station toTambaram with a route length of 29.14 km. was electrified in 1931, which resulted in the spread of development along the railway line. This line is exclusively served by separate double tracks for suburban traffic. The trains on this corridor run between 4.00 hrs. to 23.30 hrs.The other suburban broad gauge railway lines run between Chennai Central and Tiruvellore (41.65 km) and between Chennai Central and Gummidi-poondi (46.68 km). Basically these three lines radiate from the city centre and carry about 3 lakhs trips per day. These broad guage lines were electrified in 1980.

The long awaited Mass Rapid Transit System (MRTS) was commissioned on February 1997 and it runs between Chennai Beach station and Thirumylai covering a distance of 9.0 km both on the surface and on elevated system. While the three railway lines radiate from the city centre, the MRTS traverses in north-south direction.

Development of Water Transport

During 1600s, Chennai had inland waterways in terms of Buckingham canal (running north-south along coast) Coovum river (running west-east), Otteri Nalla (running north west-east), adyar river (running south west-east) and Korattalaiyar (running west-east). The canals and rivers served mainly for the transport of agricultural produces, firewood and salt through the small country crafts. But then when the city was taking shape and the population density increased, the rivers and canals' usage was getting reduced. The silting of waterways throughout the course forced the decline of traffic.These were slowly got turned into canals of sewerage and this stage continues till today.

The flows in the rivers have become seasonal and hence they are no more suitable for navigational purposes. Added to that the formation of sand bars at the mouth of these rivers had prohibited the sea getting connected to these rivers during most of the year. Hence virtually there is no movement of water in these rivers except during the heavey rain seasons. By 1861, the British authorities realised the necessity of a harbour. A pier was constructed in 1862 and further development took place from 1868 onwards which made Chennai as one of the major ports of India. The port mainly caters to the movement of cargo, bauxite ore and crude oil to and from other countries. Containerisation gave a boost to the development of port in terms of additional berths to accommodate more ships.

Development of Air Transport

Chennai is one of the important airports in the country. It has both International and domestic. Airports catering to all clientele. Ever since its establishment, the number of flights getting operated from the terminals and the passengers using them have increased manifold. Along with that, the air cargo have also been increasing in the same trend. The result of increase in terms of air passengers and cargo is the increase of traffic in roads and rail. The roads leading to these air terminals are heavily congested and the accident rate is also considerable in these roads.

Transport and Industrial Development in CMA

As stated in the above paragraphs, the small settlements during 1600 has their own agricultural production and household industries. Hence there was very little need to transport from one settlement to another. But when the people started migrating from different places and started settling at these settlements, the requirement of agricultural production, clothes and utensils increased which necessitated transporting from neighnouring settlements. These produces were transported either through road transport or water transport or in combination.

When the population started growing further early 1900, many new industries were established inside Chennai. These industries have sprung upon the vacant or open spaces wherever available. This had contributed to the development without any order or cohesion during this period. The major development in the industrial field during the post independence era have been the establishment of a number of industries located on the outskirts in Ennore and Manali in the north, in Ambattur and Avadi on the west and in Guindy on the south-west.

The industrial development warranted the infrastructure facilities to move men and material in and out of industrial units. The raw materials and the finished products are transported either by rail or by road transport. Chennai finds a better place in the industrial map of India maily because it has better communication facilities in terms of road, rail, sea port and air port. In addition, the industrial development prodvides more direct and indirect employment opportunities thus attracting more people into the city. They in turn move around the city for work, education and the like. This has also accelerated the growth of transport in CMA.

Transport and Educational Facilities in CMA

There are six universities viz. Madras University, Anna University, Dr. M.G.R. Medical University, Dr. Ambedkar Law University, Tamil Nadu Veterinary and Animal Sciences University, Mother Teresa Women University which govern as many as six Engineering Colleges, four Medical Colleges, one Law College, one veterinery college, four dental colleges respectively within CMA. In addition there are a number of polytechnics, Industrial Training Institutes and more than 500 schools inside CMA. These educational institutions attract more person trips and vehicle trips. With the growth in population and the number of industries, the number of educational institutions have also grown up both within the city as well as within CMA.

Transport Problems

As seen earlier, Chennai is very well connected to the rest of the world by road, rail, sea and air. However each system has its own lacuna when it comes to serving the urban transport.

The road system within CMA suffers heavy congestion during almost all hours of the day and it is not able to meet the demand because –

1. The road network was not functionally developed as a system;
2. The road network is grossly inadequate in terms of length and width;
3. Different modes of vehicles occupy the same road space;
4. The growth of personalised motor vehicles at an alarming rae pose serious problems;
5. The patronage in public transport has been steadily decreasing;
6. The person trip rate and vehicle trip rate have been increasing;
7. The increase in population and the increease in income level contribute to more traffic;
8. The improper landuse planning and the existing landuse have a longlasting impact on the road system;
9. The indaquate traffic sense among the road users has direct relationships with the transport system;
10. The presence of level crossings across railway lines and the inadequacy of existing road over bridges increase the congestion;
11. The unauthorised encroachments on footpaths and even carriageways;
12. The inadequate traffic management measures;
13. Lack of proper traffic enforcement;
14. Non-availability of organised parking spaces; and

15. The financial constrains for implenting the road projects.

The public transport mainly bus could not provide better level of service to the commuters of Chennai beacuse of :-

1. Overcrowding in buses;
2. Slow augmentation of fleet;
3. Extension of operational jurisdiction;
4. Increase in cost of operation;
5. Low level of fare structure;
6. Increase in private vehicles and para transit vehicles; and
7. Slow replacement resulting in overaged buses.

The sub-urban rail network also suffers heavy burden on them because of :

1. The insufficient rolling stock of coaches;
2. The non-availability of double tracks;
3. The non-availability of exclusive tracks for sub-urban system;
4. The presence of metre gauge system in the high density corridor (Beach to Tambaram);
5. The presence of level crossings across these lines;

Transport Projects

These traffic and transport problems have been periodically studied in a number of studies and improvements suggested. These improvements range from Traffic System Management (TSM) measures to long term solutions like outer ring road, exclusive bus way, Chennai bypass etc. Important projects identified by the earlier studies are as follows :

1. Extension of MRTS from Thirumylai to Velacheri;
2. Construction of National Highway bypass connecting all three National Highways at thc fringe of city limits;
3. Constuction of outer ring road connecting again these three National Highways on the fringes of Metropolitan area;
4. Conversion of metre gauge to broad gauge in Beach Tambaram corridor;
5. Increasing the capacity of Chennai-Gummidipoondi railway line by constructing exclusive line for suburban system;

6. Construction of ICC rail corridor connecting Taramani with Villivakkam
7. Construction of bus terminals at Koyambedu, Alandur and Madhavaram;
8. Construction of flyovers at Vadapalani, Koyambedu, Thirmangalam etc on the inner ring road, at Etroms road junction, General Patters road junction and Sterling road intersection;
9. Construction of road over bridges/under passes at St. Thomas Mount, Palavanthangal, Meenambakkam, Rangarajapuram and Tambaram level crossings;
10. Construction of pedestrian subways at TVS Anna Salai, Govt. General Hospital, Eldams road junction, and Luz junction;
11. Construction of exclusive busway along Anna Salai
12. Construction of Truck Terminals at Koyamadedu, Pallikaranai and Madhavaram;
13. Fourlaning of NH 4 upto city limits.

These projects are not exhaustive but only indicative. There are other projects identified by many agencies in different order of priority. However, the above projcets if taken up for implementation will go a long way in reducing the traffic and transport problems of the citizen of Chennai.

Conclusion

Chennai city and its metropolitan area have been growing steadily in terms of population, industrial activities, educational activities, trade and commercial activities, social and other related activities in the past. Transport being the basic infrastructure to meet all the demands, its development in this region plays a vital role. Transport has not been getting its due share either from the State or Central Gevernment. The advancements in communication technology force the planners to pay more attention than so far done mainly to save the region form the transport collapse.

7

GEOGRAPHY OF TRANSPORT NETWORK IN MUMBAI METROPOLITAN REGION

Tapati Mukhopadhyay

Introduction

Transport is considered as important aspect in the crowdly and deensity population centres. The entire to and fro of people and commodity movement depend on availability of different modes and means of transport in the region. The efficienty and frequency of transport services in region can definitely help in the process of development. In addition to this quicker and safer transport may always essential to gear up the local available resources. The density of populated centres and industrial towns like Mumbai copes up with transport network. Transport is not considered only as a resource factor of region but also acts as catalyst for the developemt of various resource characters. Thus transportation survey and studies have to carried out thoroughly in order to understand spatial distribution of transport networks. With considering these aspects in mind, author has modestly attempted to evaluate the development of transport and examines networks of transport in Mumbai Metropolitan region in Maharashtra. The present study is entirely bassed on published documents available on transportation. The secondary sources was prime in this regards. Census of India, Maharashtra Series Report on Transport Development Plan, Traffic Flow Report and other relevant materials have been used in interpretting the text. The new demand cores have indentified to focus on future necessity of development of transport in Mumbai Metropolitan region.

Study Region

The Mumbai Metropolitan Region (MMR) extends over an area of 4,355 sq.km and is consisted of (1) Mumbai Corporation of Greater Mumbai (2) Thane (3) Kalyan (4) Navi Mumbai (5) Sixteen Municipal Towns (6) seven non-municipal urban centres and 995 villages. These study area includes thirteen municipalities both corporation and councils. Out of these, seven are in Thane district and six in Raigarh districts. The total length of the study region has 5016 km. (1992). Out of this 1683 km. road length (30.5 %) belong to Greater Mumbai. This shows significance of transport in the Mumbai Metropolitan Region.

History of Transport

Prior to the formulation of the regional plan, a detailed road transport plan was prepared by a American transport company, Wilber Smith in which be included elaborating network of freeways mainly for private road transport. In order to supplement the Wilbar Smith Plan, the Traffic Cell of the Town Planning Department undertook a study of mass transportation in Mumbai Metropolitan Region in 1967-68, suggesting to augmenting mass transportation capacity within the region to increasing the providing additional link road and also to provide for new rapid transit system in the form of underground railway. The regional plan in 1975 did set to do a regorous analysis of transport demand but linked all the transport projects and classified them into inter-regional, intra-regional and intra-city project. But unfortunately, most of the projects mentioned in the Regional Plan in 1975 are still being debated, thereby large part of the project remained unimplemented.

The Central Road Research Institure was commissioned to prepare a Comprehensive Road Transport Plan in 1985. The CRRI Plan prepared in 1981 and recommended a strategy comprising of : (i) Road Development Plan, (ii) Traffic Management Strategy and (iii) Landuse Planning. The road development plan under CRRI in 1985 consisted of a hierarchy of roads like free-ways, express-ways, arterial roads etc. Bombay Urban Transport Project was implemented during 1978 to 1984 with the financial assistance from the World Bank. The project was mainly for the improvement of the existing system like bus depot, signal improvement etc. In 1988 the Government of Maharashtra appointed a Committee under the Chairmanship of the then Chief Secretary, K.C. Paranjape to review various road transport projects. The Committee recommended urgent implementation of programmes of 1988-89, 1989-90. Short term stage is however, I and stage II projects were to be completed by 1994-95. However, except two fly-overs at Senapati Bapat Marg and Amar Mahal at Chembur, there is not much of progress about any of the above projects. One of the major problem was mobilization of resources.

Later on, Mumbai Metropolitan Region Development Authority with the help of World Bank, has undertaken a comprehensive study and prepared a technical reports and submitted a draft of the "Preliminary Report of Finding" in April 1994. The main objectives of this transport study were to ensure adequate level of accessibility in the expanding urban area, to assist the economic development of the region and also to improve the safety record of the transport system. These objectives were to be achieved with due consideration of from the point of view, road development strategy for the metropolitan region, there is a focus for the utilization of National Highway, Expressway connecting Mumbai with its hinterland in Maharashtra and western India and to connect various sub-regions of study area.

Existing Road System

Maharashtra is leading position for road length, having 1,80,656 km. (11.14 per cent of the total roads) in India. A further classification indicates that in Maharashtra, the predominating road network are the village roads, (59,286 km.) followed by the district roads, state highways and the urban roads. Total road in study region is 5,016 km. (1992).

In case of the study region the largest share is that of urban roads 5527.11 km. followed by village roads. This co-existence of the village roads within the metropolitan regional context indicates the need to shift the attention of future development towards these areas which are still lacking the infrastructual facilities.

Table 7.1 : *Road Length in Mumbai Metropolitan Region, Maharashtra and India (1992)*

(Length in km)

Name	National Highway	State Highway	Major District Roads	Other District Roads	Village Roads	Urban Roads	Total
India	32,446	1,15,989	n.a.	13,48,466	n.a	1,24,693	16,21,594
Maharashtra	2,955	30,548	37,224	36,403	59,286	14,240	1,80,656
Mumbai Metropolitan	228	477	327	271	878	2,835	5,016
Region	(7.7)	(1.6)	(2.7)	(1.5)	(0.7)	(2.0)	(2.8)

Source : Basic Transport and Communication facilities for Mumbai Metropolitan Region, Transport and Communication Division Mumbai Metropolitan Region Development Authority, Mumbai, March, 1994.

Note : Figures in brackets indicate percent share of study region excluding project roads maintained by various authorities like Forest, Irrigation, Electricity etc.

Types of Roads

Table 7.2 shows the classification of roads by types of surface in the Mumbai Metropolitan Region. Out of the total road while only 9.52 per cent Two words are unsurfaced roads, (527.19 km) The unsurfaced roads are mainly municipal roads and village roads. Surfaced roads are further classified into three types : (i) water bound Macada (ii) Black topped, and (iii) Concrete. Black topped is the pre-dominating type available in the metropolitan region, constituting 74.36 per cent of the total surfaced roads, while Macada contributes 18.89 per cent and concrete has just 6.75 per cent.

National and State Highways

It should be noted in this context that within the metropolitan region, municipal roads are the next important road network contributing 53 per cent of the total followed by 17 per cent of village roads. Then, importance comes of state highway (8.70%) followed by CIDCO roads (7.81%), district roads (5.91%) other district roads (5.02%) and, finally, National Highway (3.97%). The importance of the metropolitan region from the point of view of road transportation can further be seen from the origin of seven important National highways: (1) Mumbai-Agra Road, (2) Mumbai-Pune Road, (3) Mumbai-Goa Road, (4) Easterns Express road (5) Western Express, and (6) Thane-Bhiwandi by-pass road.

Table 7.2 : *Road in the Mumbai Metropolitan Region by Type of Surface*

Classification of road	Water bound	Black topped	Cement concrete	Total surface
1. National Highway	–	218.36	–	218.36
2. State Highways	23.30	475.35	0.90	418.55
3. Major district Roads	60.23	204.19	–	264.42
4. Other district Roads	104	139.55	0.25	244.20
5. Municipal Roads	148.65	2166.96	355.32	2650.92
6. CIDCO Roads	93.00	273.46	–	366.46
7. Village Roads	5125.29	258.42	0.30	774.01
Total	944.87	3718.29	336.76	4999.92
Percent	(18.89)	(74.36)	(6.75)	(100.00)

Source : Basic Transport and Communication facilities for Mumbai Metropolitan Region, Transport and Communication Division, Mumbai.

Apart from the above National Highways there are 19 State Highways in the study region, out of which nine are in Thane district and ten in Raigarh district. The major district roads covering a total of 195.97 km length of district roads in Raigarh districts, Thane district has 103 km extra roads compared to Raigarh district. But in case of minor district roads, Thane district has 26 minor districts roads with a length of 132.39 km while in case of Raigarh district number of minor district roads are 38 covering total length of 145.68 km.

Village Roads

From the point of view of village roads, Thane district has 498 village roads with a length of 625.69 km, while in Raigarh district the number of village roads, are 214 with a length of 195.46 km. A talukawise assessment of the village roads are shown in Table 7.3.

It is observed that Bhiwandi taluka has the largest mileage of surfaced roads, 44.85 km of motorable and 12.42 km of unmotorable road length. Taking into consideration only the unmotorable roads, the largest kilometerage is found in Ulhasnagar with 14.53 km followed by Bhiwandi 12.42 km. However, unsurfaced roads contribute insignificant percentage shares in case of other talukas of Thane district. In case of surfaced road, the largest kilometerage lies in Vasai taluka followed by Bhiwandi and Kalyan, Ulhasnagar and Thane. All these surfaced roads are "below standard single category". While in Raigarh district, the largest share of unsurfaced road lies in Alibag and Khalapur. The talukas, namely, Panvel, Pen and Karjat have unsurfaced road lengths are insignificant (Table 7.3). In case of surfaced roads, the largst share is constituted by Alibagh followed by Karjat, Urban, Pen and Khopoli.

Table 7.4 reveals the road length of municipal areas (urban) by types of surface. Taking into consideration individual municipalities, Kalyan Municipal Corporation consists of largest meterage of 34.85 km. unsurfaced roads followed by Thane (11.70 km). Bhayandar (9.84 km).The rest are in significant. From the point of view of surfaced road. Kalyan Municipal Corporation come as the first with 452.80 km. followed by Nizampur-Bhiwandi 218.63 km.The third in order of importance is the Thane corporation with 184.05 km and Ulhasnagar with137.98 km. A distinct difference can be indentified from the town clusters located in the central part of the metropolitan region like Kalyan, Bhiwandi, Ulhasnagar with towns located in the northern part of the region like Vasai, Virar, Bhayandar where road meterage are significantly low viz. Bhayander (60.87 km.) Vasai (36.03 km.) and Virar (11.04). In case of Raigarh district consisting of six towns with a road of mileage of 160.60 km. Out of which 131.60 km are surfaced and 29 km. are unsurfaced. This clearly indicates that the municipal corporation in Raigarh are lagging behind for road development. Out of these Municipal Corporations Panvel has the highest road length, (34.47 km) followed by Khopoli (28.77 km), Matheran (23.66), Alibagh (19.20 km) Pen (12.81 km) and Uran (12.69). With the shift of focus of development towards Raigarh district, centering Panvel, a proper attention for the development of the roads is urgently required.

Table 7.3 : *Classification of Village Roads by Types of surface (A Talukawise assessment)*

(Length in Kilmoters)

Dist.	Name of Talukas	Unclassified roads			Water Below	Standard	Bound Total Gross
		Motor	Unmotor				
Thane	Bhiwandi	44.85	12.42	121.71	—	121.71	178.98
	Vasai	1.00	—	164.98	—	164.98	165.98
	Kalyan	7.66	4.52	106.04	—	106.04	118.22
	Ulhasnagar	4.45	14.53	81.83	—	81.83	100.81
	Thane	—	—	61.70	—	61.70	61.70
	Total	57.96	31.47	536.26		536.26	625.69
Raigarh	Panvel	5.70	—	7.30	—	7.30	13.00
	Urban	12.21	—	35.03	—	35.03	47.24
	Alibag	22.10	—	102.32	—	102.32	124.42
	Khalapur	13.00	—	17.00	—	17.00	30.00
	Karjat	1.40	—	44.53	—	45.93	45.93
	Pen	5.30	—	29.57	—	29.57	34.87
	Total	59.71	—	235.75	—	235.75	235.75
Grand Total		117.67	31.47	772.01		777.01	921.15

(*Source* : Basic Transport and Communication facilities for Mumbai Metropolitan Region, Transport and Communication Division, Mumbai Metropolitan Region Development Authority, 1994)

Table 7.4 : *Road Network in Municipal Areas (Urban) By Types of Surace*

(Length in Kilometers)

District/ Municipals/ Councils	Unsurface Roads	Surface Roads			Grand Total
		Water bound macada	Black topped	Cement concrete	
Thane District					
Thane Municipal Corporation	11.70	22.90	126.95	34.20	195.75
Kalyan M.C.	34.85	39.72	376.00	37.08	487.65
Ukhasbagar M.C.	–	14.86	91.83	31.29	137.98
Bhayander M.C.	9.84	–	54.08	6.79	70.71
Vasai N.C.	–	2.62	33.41	–	36.03
Virar N.P.	4.34	–	11.04	–	15.38
Bhiwandi -Nizampur	0.38	25.76	113.54	79.33	219.01
Total	61.11	105.786	806.85	188.69	1162.51
Raigarh District					
Alibagh N.P.	–	6.80	11.13	1.27	19.20
Pen N.P.	–	0.62	12.19	–	12.81
Panvel N.P.	–	4.03	30.44	–	34.47
Uran N.P.	–	–	11.80	0.83	12.69
Matheran N.P.	29.00	23.66	–	–	52.66
Khopoli N.P.	–	7.48	16.59	4.70	28.77
Total	29.00	42.59	82.21	76.80	160.60
Grand Total	90.11	148.45	889.06	195.49	1323.11

Source : Transport and Communication facilities for Mumbai Metroplitan Region, Transport and Communication Division, Mumbai Metropolitan Region Development Authority, Mumbai, 1994.

Identification of the Gap Area

Total number of villages in each taluka are shown in Table 7.5. This table indicates that largest number of villages exist in Bhiwandi 215) followed by Alibag (202), Karjat (177), Panvel (176), Pen (165), Kalyan (114), Vasai (92), Thane (88) Uran (57). Inspite of being located in the metropolitan region which is humming with growing urbanization, these villages are without pucca road facilities. Four talukas of Raigarh district can be treated as deprive areas from the point of view of road development. This deprivation in large number of villages in these talukas is very striking. For example, Karjat taluka has 113 villages with no pucca approach roads. Similarly 96 villages in Alibagh, 87 villages in Panvel and 73 villages in Pen have no pucca approach roads. This can be further seen from the point of view of road length of villages (Table 7.5). Panvel with only 63 km of roads, top in the list with minimum road length of followed by Khalapur with a road length of 30 km. Pen with 34.87 km. Karjat with 45.93 km. and Uran with 47 km.

However, Thane district has found largest number of villages without pucca approach road facilities located at Ulhasnagar (59) followed by Bhiwandi (51). But in the rest of the talukas the number of villages without pucca approach roads are relatively less in Thane. It appears 31, Kalyan (17) and Vasai (14). The largest number of village roads is in Vasai followed by Bhiwandi. But length-wise Bhiwandi has the highest road length, followed by Vasai, Kalyan, Ulhasnagar and Thane.

Table 7.5 : *Talukwise Assessment of the Village Roads in Mumbai Metropolation Region*

(Length in Km.)

Name of Talukas	Number of Villages in each Taluka	Number of Villages in each without pucca road approach	Number of Villages roads each Taluka	Village Roads length in km.
Thane	88	18	50	61.70
Vasai	92	14	210	165.98
Bhiwandi	215	51	111	178.98
Kalyan	114	17	68	118.22
Ulhasnagar	92	54	59	100.81
Panvel	176	87	8	13.00
Karjat	177	113	25	45.93
Khalapur	122	32	22	30.00
Uran	57	26	37	47.24
Pen	165	73	37	34.87
Alibagh	202	96	84	124.42
Total	1500	581	711	921.15

Source : Basic Transport and Communication facilities for Bombay Metropolitan Region, Transport and Communication Division, Mumbai Metropolitan Region Development Authority, 1994.,

A detailed classification of village roads for each taluka has been examined by taking into consideration the distance from the taluka headquarter. Table 7.6 shows the distance and its relationship with village roads facilities. It clearly indicates that a distance of 010 km, the largest number of villages without pucca approach road are located in Alibagh taluka (54) followed by Karjat (43) Panvel (32), Pen (29). In Raigarh district, Khalapur has only four villages without pucca approach road from the headquarter, Khopoli. While Uran has fifteen villages without approach road from Uran town. In case of Thane district it is seen that Table 7.6 Ulhasnagar where thirteen villages at a distance from 0-10 km from Ulhasnagar town do not have any pucca road facility. This is followed by Bhiwandi with eleven villages and Kalyan where four villages do not have any pucca road facilities. In the second stage, at a distance from 11-20 km. from the headquarters, the largest number of villages without pucca approach road facility lie in Panvel taluka (54) followed by Alibagh (31) and Pen (28). In case of Khalapur, the number of villages without pucca road substantially increased. The figures are 5 for Khalapur, 15 for Karjat and 10 for Uran. In case of Thane district, the largest number of villages without pucca approach roads lie in Bhiwandi (14) followed by Ulhasnagar (12) and Kalyan (7).

Table 7.6 : *Villages without Pucca Approach Road According to the Distance from the Nearest Towns*

Name of Talukas	Distance from Headquarter				Name of Nearest Towns	Total
	0-10	11-20	21-30	>30		
Vasai	–	4	2	8	Vasai	14
Bhiwandi	11	14	16	10	Bhiwandi	51
Kalyan	5	7	4	1	Kalyan	17
Ulhasnagar	13	12	6	–	Ulhasnagar	31
Kulgaon	17	5	1	–	Kulgaon	23
Panvel	32	54	1	–	Panvel	87
Karjat	43	18	10	10	Karjat	81
Nerul	15	15	1	1	Nerul	32
Khopoli	4	16	–	–	Khopoli	20
Uran	15	10	1	–	Uran	26
Pen	29	28	16	–	Pen	73
Alibagh	54	31	11	–	Alibag	96
Total	244	236	71	30	Total	581

Source: District Census Handbook, Thane and Raigarh District, 1991.

In the third stage at a distance from 21-30 km. from the taluka headquarters, except Bhiwandi, there is a decline in the number of villages without pucca road facilities beyond 30 km from the headquarters, except three talukas, namely, Bhiwandi, Karjat and Vasai. The rest of the talukas do not have any village. Thus it is seen from the Table 7.6 that in the metropolitan region, the largest number of villages located at a distance per 0-10 km from the taluka headquarters without the pucca approach road facility which declined marginally in the next stage at a distance from 11-20 km. However, there is sharp fall in the number in the third stage a distance 21-30 km. beyond the distance 30 km. Except three talukas, the rest of the talukas do not have any villages.

Highlighting the Demand Core for Transport

The Fig. 7.1 shows some centres in metropolitan region which are considered as priority centres for the future growth and expansion of road transport network. Karanjade in Panvel taluka located close to Panvel municipality emerges as the centre for the highest attraction for the fast moving vehicle in the last 15 years followed by Pen Naka. These centres in future growth needs to be substantiated by the overall plan for infrastructural development of the surrounding area. Though the demand for fast moving vehicles increased by more than 3000 times in the last 14 years, Panvel taluka as a whole remained a deprived area. Panvel is emerging as an intra-regional transport total point mainly for the road transport. Hence, the regional development plan should incorporate the road transport. And, therefore, regional developmental plan should take mocro level transport planning of the taluka into consideration.

Next important centre in the southern part of the region is Pen Naka. Pen is the most rural taluka in the metropolitan region from the point of view of landuse function and human occupations. It is gaining its economic importance due to the introduction of Konkan railway and acting as a connecting point between Alibagh and Khopoli. In fact, it is the common point of the two triangle like Alibagh, Panvel, Pen and Khopoli. Panvel and Pen increase its potentiality for the future growth. However, at present, out of 165 villages, 73 villages are without pucca approach road facility. Intra-village linkages are also needed for the development so that the facility of Konkan railway reaches to the wider area. What is required is the expansion of existing roads facilities specially in the rural areas located between the growth centres like Karanjade and Pen Naka. In the eastern side along the hilly track, three urban centres are successively following each other, nemely, Matheran, Karjat and Khopoli which are connected with Panvel. However, the new emerging transport focal point is Chowk, located between Rasyani and Karjat. This will act as a subsidiary centre between Panvel and Khopoli and will provide road in the rural area establishing linkage with Panvel.

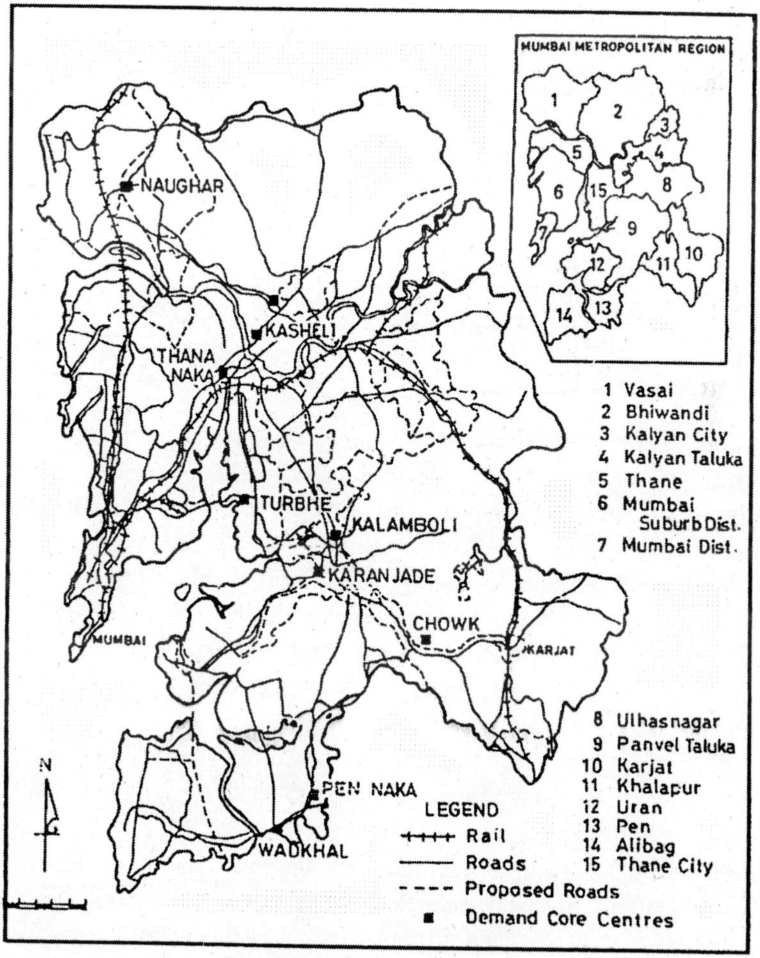

Fig. 7.1 : *Road and Rail Network*

However, in the northern part of the metropolitan region more specifically in the north of Ulhas river, Bhiwandi seems to be the only centre with no subsidiary centre developed north of Bhiwandi. Thus, the development process remains restricted within the municipal area of Bhiwandi which is surrounded by the vast rural areas. Further, development of this rural area depends on the internal accessibility. The area, therefore, needs further attention. In the north-western part of the metropolitan region emerging centre in Vasai taluka is Naughar located between Vasai and Virar on the Western express highway. Importance of Naughar increases from the point of view of both commuters and cargo movements. In Navi Mumbai, Kamboli and Turbhe are two important centres of inter-regional transport development. Turbhe due to its emergence as an agriculture product market and as well as the development of residential area, resulting to increase commuters transport and commodity movement vehicle. However, there is a natural growth of market area as it is located between Mumbai and Navi Mumbai. It is expected that Kashe between Thane and Bhiwandi will grow in demand, but due to the stagnation of Bhiwandi as a town and non-development of rural area with new type of activities, the growth of transport from the point demand is least.

Concluding Remarks

Mumbai Metropolitan Region is undergoing a transformation in its form and function that had led to the shift of focus of the future development from the old city towards the extended suburbs and peri-urban areas. Thus, it is expected that infrastructural development through systematic planning will enhance the future development of the metropolitan region. However, a review of the transport plans so far adopted by the Mumbai Metropolitan Planning 'Authority indicates that there is a lack of continuity between the plan, a large part of the plan are not implemented. As a result the benefit of the transport development could not reach a wide area in the region. This study further highlighted that though metropolitan region comprise of 13 towns but from the point of view of road network 30.5 percent of the total road length extents in the city of Mumbai. This precisely indicated the primacy of the mother city. A quantitative assessment further indicates as the distance increase from the city of Mumbai, road intensity decreases.

Barring the city of Mumbai, the metropolitan region consists of thirteen towns. From the point of surface road facilities, Kalyan tops in the list followed by Bhiwandi, Thane and Ulhasnagar. But the road meterage are significantly low in the northern part of the metropolitan region with special reference to town cluster of Bhayander, Vasai and Virar. The road facilities are comparatively better in the cental part of the region than the newly developed towns of both northern and southern part of region.

The benifit of the roads is mainly concentrated within the municipal areas beyond which there is a sharp decline of road facilities. This can be seen from Bhiwandi, Ulhasnagar as well as Alibagh, Karjat and Panvel. The existence of the village road is a great hinderance for the inter-town development of surface transport movement. Taking into consideration the distance from the taluka headquaters and the road facilities, it indicates a large number of villages do not have pucca approach road facilities, though the villages are located within a distance of 10 km. from the taluka headquarters. This phenomenon is predominant in Raigarh district particularly surounding the towns of Alibagh, Karjat and Panvel.

Panvel has already emerged as an important centre for the future development in the metropolitan region followed by Pen. The high growth of fast moving vehicles are neither seen in the coastal areas in the west nor in the hilly areas, while the northern part of the metropolitan region seems to be a deprived area specially in Bhiwandi taluka. The rural areas of Bhiwandi taluka is awaiting further development thrust. Independent economic function needs to emerge in the town which may act as a booster for the development of the surrounding areas. Mumbai Metropolitan Region being a predominating metropolitan region is suffering from the lack of uniform development of road system. What is warranted today is to focus the inter-urban road development within the region followed by the village road development. Present Metropolitan Transport Plan has given sufficient weightage for the Railway expansion followed by the development of roads corridor but it may not suffice the requirements in the deprived areas unless the transport focal points of the sub-regions are identified and planned road network in the fashion of ring roads are developed around which ultimately may facilitate the rural area to be linked with the transport network of the region. This also will ensure adequate level of accessibility in the expanding urban areas.

References

Berry, B.J. L. (1959) : "Recent Studies Concerning the Role of Transportation in Space Economy", *Annual of the Associaton of American Geographers*, No. 49, pp 328-342.

Illagovan, P. (1983) : "Transport Geography : An Analysis of relationship between Road Network Structure and Pattern of Regional Development in Tamil Nadi", Unpublished Ph.D. Dissertation, Department of Geography, Madurai University, Madurai.

Macregor, D.R. (1949) : "Town Development and Transport" *Scottish Geographical Magezine*, Vol. 65.

Mayer, H.M. (1961) : "Chicago Transportation and Metropolitan Planning", Regional Science Association, *Papers and Proceedings*, Vol. 7, pp. 241-247.

Pandey, N.P. (1990) : *Geography of Transportation : A Case Study of Madhya Pradesh*, Inter-India, Publication, New Delhi.

Ramachandran, H. (1977) : "Transportation and Urban Attributes in the Coimbatore Region : A Case Study of an Interacting System", Unpublished Ph.D. Thesis, JNU, New Delhi.

Raza, M., and Aggarwal, Y. (1986) : *Transport Geography of India :Commodity Flow and Regional Structure of the Indian Economy*, Concept Publishing Co., New Delhi.

Singh, K.N. (1984) : "Levels of Urbanisation : A Case Study of Eastern Uttar Pradesh, *National Geographical Journal of India*, Varanashi, Vo.l, 29, No.1.

8

TRANSPORT DEVELOPMENT IN DELHI METROPOLITAN AREA

M.Parida and *S.S.Jain*

Introduction

Delhi metropolitan area is facing fast urbanization, inadequacy of public transport infrastructure, phenomenal growth of personalised vehicles, increased vehicular emissions and an overall crisis in the transport system. There have been several efforts in the past to bring order into this transportation scenario but without much implications. Instead of adhoc approaches a long term transport policy can yield better results.

Delhi is the national capital of India and it is one of the metropolitan areas which has adequate mechanism for planned growth. Due to huge financial investments in the capital city every year around three lacs people are migrating into this city. To meet the transport demand of the city, Delhi Transport Corporation (DTC) buses provide the mass public transport infrastructure and they are supplemented by private operators. Eventhough Ring Railway is existing within the city since ASIAD-1982 it has failed to generate adequate patronage. The Metro Rail for Delhi may take another five years before it is operational. In the absence of a proper public transport system personalized vehicles have proliferated into the metropolitan scenario. There is a wide gap between transport demand and supply in Delhi (Jain, S.S. and Parida, M; 1996). As a result, Delhi is the second most pollutest city of the world, there are more than 2000 deaths due to road accidents each year and a number of other problems. In the light of the above issues it is essential to identify alternative solutions which can bring relief to make the living of people better.

Dimensions of Traffic and Transportation Problems

The traffic and transportation problems in Deihi have been aggravating over the years (Jain, S.S. and Parida, M., 1998) and some of its dimensions are as follows :

(i) Urbanisation trends, changing distribution of population and high density of population in certain pockets.

(ii) Increase in per capita trips and work trip lengths has resulted in longer travel time.

(iii) Congestion and delays particularly at intersections result in huge economic losses besides causing inconvenience and frustration to road users.

(iv) The road based mass transport in Delhi is not able to cope up with the fast increasing travel demands of the commuters.

(v) The ring railway remains underutilized because of its inadequate connectivity with road network.

(vi) Environmental pollution.

Environmental Pollution

The heavy volumes of traffic with predominant share of motorised vehicles on most of the city corridors is responsible for environmental degradation by way of noise and air pollution (Sharma, A. and Raichoudhari, A., 1996). About 2000 tonnes of pollutants are injected into Delhi's atmosphere every day. About 70 per cent of this is due to exhaust emission from motor vehicles. The pollutants comprise about 1,000 tonnes of carbon monoxide, 15 tonnes of suspended particalate matter, 160 tonnes of oxides of nitrogen, 250 tonnes of hydrocarbon and 10 tonnes of sulphur dioxide. The cumulative effect of these pollutants is the increasing incidence of serious health problems like tuberculosis, asthama, chronic cough and cold and irritation and other diseases of the eye which besides putting a heavy pressure on the health services and finances of the common man are lowering the overall productivity of the city. The observed noise levels on major arterials in Delhi have been found to vary between 70 to 80 dBa much above the acceptable limits.

Growth of Regional Vehicles

The motor vehicles in Delhi increased from 2.43 lacs in 1970 to 19.92 lacs in 1991 and it is expected that Delhi will be overcrowded by 4.4 million vehicles by the beginning of next millennium (Agarwal, M.K., 1996). While the population of Delhi is approximately one per cent of the country's population, the total vehicle accounts for 10 per cent (approximately) of total vehicles of the country. The total number of vehicles in Delhi exceeds the total number of vehicles in Mumbai, Calcutta and Chennai. Road accident sitution in Delhi is quite alarming. Number of people killed in road accidents in Delhi were 503 in 1972 which has increased to 2,276 in 1997 (Jain, S.S. and Parida, M, 1999).On an average everyday more than five persons are killed and twenty-two persons are injured in road accidents in Delhi. Transport demand has many dimensions (INAE, 1993)–size, characteristics, technology, finance, fiscal, institutional etc. Delhi has experienced continuous growth in urban travel demand. In Delhi, the change in mobility is conspicuous and the detailed figures are given in Table 8.1

Table 8.1 : *Vehicular Trip Rates per Person per Day in Delhi*

Purpose	Years			
	1969 (Observed)	1981 (Observed)	1991 (Modified)	2001 (Projected)
Work	0.29	0.300	0.352	0.450
Education	0.08	0.088	0.196	0.400
Others	0.12	0.012	0.196	0.100
Non Home Based	*	0.04	0.070	0.100
All Purpose	0.49	0.44	0.720	1.100

* Included in other trips.

Increase in trip length is another important factor contributing to travel demand. Segregation of residence and work areas has resulted in increased trip lengths. The average trip length in Delhi has increased from 5.4 km in 1969 to about 8 km in 1981. It is presently estimated to be about 15 km. A study has indicated the trip length of employees to a major office complex in Delhi which is given in Table 8.2.

Table 8.2 : *Trip Length Characteristics of Employees of Lodhi Place*

Mode	Trip Length (km)
Car	13
Two Wheeler	13
DTC	18
Chartered bus	18
Auto	7
Cycle	13
Walk	3
Train	26

Demand and Supply of Public Transportation

The Master Plan of Delhi 2001 forecasts that the total personal trip in 2001 is going to be 186.40 lacs (Parida, M., Gupta, A.K. and Khanna, S.K., 1994). Out of which the vehicular trip would be 116.63 lacs. In another detailed study by Central Road Research Institute, New Delhi. It has been estimated that 114 lacs passengers would require mass transit system. According to a Delhi Development Authority study the total road infrastructure can accommodate 8000 buses maximum. Currently, 5,000 bus fleet is carrying an estimated 50 lacs passenger trip per day .The growth in transport supply is very low and till MRTs project is fully implemented the scope for bridging this gap remains remote. This gap in transport demand and supply situation in the public transport environment creates a situation for opting travel by personalized vehicles.

Transportation System

The different modes of transportation in use in Delhi are as follows:

(i) Private/passenger car (personalised mode)
(ii) Scooter/motor cycle (personalised mode)
(iii) Taxi (hired mode)
(iv) Autorickshaw (hired mode)
(v) Chartered bus (hired mode)
(vi) DTC bus (public transport)
(vii) EMU train service (Rail -based public transport)
(viii) Cycle rickshaw (hired mode)
(ix) Tanga (hired mode)
(x) Bicycle (personalised mode).

Hence, it is quite obvious that Delhi has a mix of both slow and fast moving modes. The urban bus system in Delhi is already overloaded in contrast to the ring railway system which is not being used to its full capacity. Scooter, car and motorcycle are important personalised modes. The urban bus system serves as major public transport means in Delhi. It comprises of Delhi Transport Corporation and other private owners in the form of a supplementary as well as competing urban bus system (Shrivastava, P.K., 1998). The routes of DTC and private buses are fixed whereas the chartered buses have somewhat flexible routes.

Ring Railway System

The Northern Railway runs limited suburban rail services commonly called as EMU services and ring rail services called Delhi Rail Parikrama Seva. The ring railway has a route of 35 km covering 23 stations. This system was planned to carry one million passengers per day but it carries only 8000-10,000 passengers per day due to inadequate connectivity, lack of public awareness about the system, low frequency, poor parking facility at the stations etc (Pradip Kumar, 1997).

Mass Rapid Transit System (MRTS)

The MRTS of Delhi is a multicrore project being undertaken with the help of OECF, Japan. For the implementation of the prestigious Delhi MRTS Project, Delhi Metro Rail Corporation Ltd. (DMRC), a company under Indian Companies Act, 1956, set-up with equal share holding by Government of India on the one hand and the Government of National Capital Territory of Delhi on the other, has started functioning. The Detailed Project Report (DPR) for the first phase of Delhi MRTS prepared by RITES in 1995 envisages a total distance of 55.3 km including 11 kms of underground Metro corridor and 44.3 km of surface/elevated Rail corridor at a cost of 5211 corer rupees (1997 prices). 56 per cent of the cost of the project earmarked for certain specified areas of work will be met from a loan from OECF Japan. The DPR provided for an implementation schedule of 10 years from 1 April 1995 to 31 March 2005. It is proposed to take up first Shahdara-Tis Hazari six km Rail corridor on surface, which will pass through the colonies of

Shahdara, Seelampur, Gautampur, Shastri Park of East Delhi before culminating at Tis Hazari via ISBT. Proposed network of MRTS of Delhi is given in Fig. 8.1.

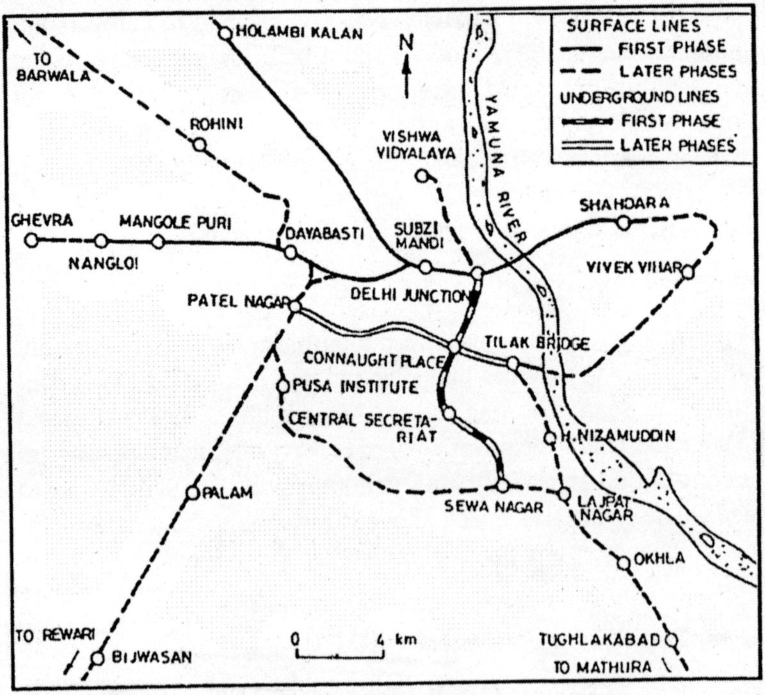

Fig. 8.1 : *MRTS For Delhi*

Traffic and Transportation Planning

To tackle the growing traffic and transportation problems a number of plans have been prepared in the past. Comprehensive traffic and transportation study for Delhi Urban Area was prepared by CRRI, New Delhi at the instance of Planning Commission, Government of India and on behalf of Delhi Administration in the year 1969. The recommendations of this study were based on a number of field studies, which included improvement of roads and intersections, construction of additional bridges across river Yamuna, road over bridges level crossings, proposals for the improvement in the CBD areas, development of off-street parking facilities and rapid transit corridors for efficient movement of passengers. Perspective plan for Delhi 2001 prepared by the Delhi Development

Authority (DDA) had proposals for transport system development which include light rail transit system, development of integrated freight terminals, new passenger terminals and other related infrastructure facilities. The study conducted by Rail India Technical and Economic Services (RITES) has been the base document for executing the MRTS project in Delhi.

Institutional Mechanism

Fragmentation of responsibility and multiplicity of authority has indeed added to the complexity of evolving a rational policy and scheme of planning and management of urban transportation (Agarwal, M.K., 1996). The national capital, Delhi, itself has to contend with about nineteen different administrative agencies. Most of these agencies function independently without any effective co-ordination strategy. At the same responsibility of individual agencies are not clearly demarcated.

Concluding Remarks

The present traffic and transportation problems of Delhi need to be viewed in an integrated manner (N.S.I.T., 1997). Eventhough, there has been continuous efforts for improving the transport system, the situation is still worsening. The following aspects needs a special cognizance in evolving a solution :

(i) A long term metropolitan transportation policy is essential to evolve an efficient public transport system, discourage growth of personalized vehicles and having least environmental impact.

(ii) The integration of landuse plan and transportation plan needs to be worked out in a co-ordinated manner. Transportation system is closely related to a landuse system and hence should not be treated in isolation.

(iii) The technology solutions to improve public transport demands extra resources and there is a need to attract private operators for projects with huge investment. The concept of road pricing also needs to be introduced.

(iv) There is a definite scope for increasing the safety and efficiency of the existing transport network through simple Transportation System Management (TSM) measures. Various enforcement agencies are already doing this job, but the only thing that is required, these agencies need to be more organised and better trained.

(v) Institutional mechanism can be worked out for improving the traffic and transportation scenario. The concept of a unified Authority can prove to be a better solution to ensure improved co-ordination amongst the multiple authorities.

References

Aggarwal, M.K. (1996) : *Urban Transportation in India*, Indian National Academy of Engineering, Allied Publishers, 1996.

I.N.A.E. (1993): "Urban Transportation" The Proceedings of the National Seminar, Indian National Academay of Engineering, New Delhi, 1993

Jain, S.S. Gupta, A.K. and Parida M. (1996) : "Evaluation of Existing Mass Transportation System and Suggestions for requirements in 2001 for Delhi Metropolitan Area," (MTSII), Final Report Submitted toUniversity Grants Commission, October, 1996.

Jain, S.S. & Parida, M. (1998) : "Multimodal Transportation System Planning and Design" AICTE/TIFAC Project under Technology Watch Scheme, Final Report, April, 1998.

Jain, S.S. & Parida, M. (1999) : "Development of a Scientific Approach to Road Safety for Indian Environment",. Project Sponsored by AICTE, New Delhi, First Progress Report, March, 1999.

N.S.I.T.(1997) : "Delhi Transportation System in 21st Century" Netaji Subhas Institute of Technology, Delhi 1997.

Parida, M., Gupta, A.K. and Khanna, S.K. (1994) : "Determinants of Mode Choice of Commuters in Delhi", Article in Edit book in *Les Transports dans les ville due sua"* CODATU/INRETS, Paris, 1994.

Pradeep Kumar, (1997) : "Analysis of a Interface between Bus and Rail Transportation System" Unpublished M.E. Thesis, Department of Civil "Engineering, University of Roorkee, 1997.

Sharma, A. and Roychoudhury, A. (1996) : "A Slow Murder - The Deadly Story of Vehicular Air Pollution in India". Centre for Science and Environment, New Delhi, 1996.

Srivastava, P.K. (1998) : "A Study on Multimodel Transport System for Delhi" Unpublished M.E. Thesis, Department of Civil Engineering, University of Roorkee, 1998.

9

MASS TRANSPORT SERVICES IN KOLKATA METROPOLITAN

Sukla Bhaduri

Introduction

The public transport system is the primary mode of mass transport within the core city of Kolkata. The transport within the area are mostly performed by surface transit modes viz, buses, both state owned and private and mini buses. The buses operate in the corrider length of 339.97 kilometres within the metro-core and carry about 6.0 million transit passengers daily. At present. 182 private bus routes operate daily within the portion of Kolkata and Howrah in the metropolitan core with an average fleet of 4,067 buses. The C.S.T.C. operates about 820 effective buses daily in 53 ordinary routes and 33 special bus routes apart from its long distance services. About 1,215 mini buses ply daily in 72 routes within Kolkata and Howrah in the metropolitan core area. The total number of routes served by CSTC. Private and Mini buses have been operating over a length of about 340 kilometres are subject to introduction with the new mode of traffic, viz., trams, vehicular traffic both fast and slow and pedestrian traffic. An attempt has therefore been made in this paper to assess the mass transport services in Kolkata with the emphasis to performances, and perspectives of state buses and routes growth of state buses and routes composition of buses.

So far the operating characteristics of state buses are concerned the majority of these are type service rather than feeder service and are subjected to considerable congestion and street interference as well as overloading. These conditions often cause delays that disrupt schedules, increase running time and generally reduce operating efficiency. Private bus services overlap existing bus services in some sections of the city, but on only three or four routes have the private buses taken over completely from the Kolkata State Transport Corporation. They operate mainly on new routes, all of which are specified in the operating permits. The private operation bus undoubtedly helped considerably to meet the growing deficiencies in mass transportation services. In 1940, when the Regional Transport Authority, constituted under the Motor Vehicle Act of 1939, came into existence, about 550 private state carriages owned by 3,222 private owners, used to ply in the following routes in Kolkata, namely, 2, 2A, 3 3A, 4 4A, 5A, 8, 8A, 10 10A, 11, 11A, 12, 12A, 12B, 13, 14, 32, 33, 35, 36 and 38. Uptil 1948, private operators had a monopoly of bus services. Since July 1948, however state buses began to appear in routes in which there was no competition from private buses.

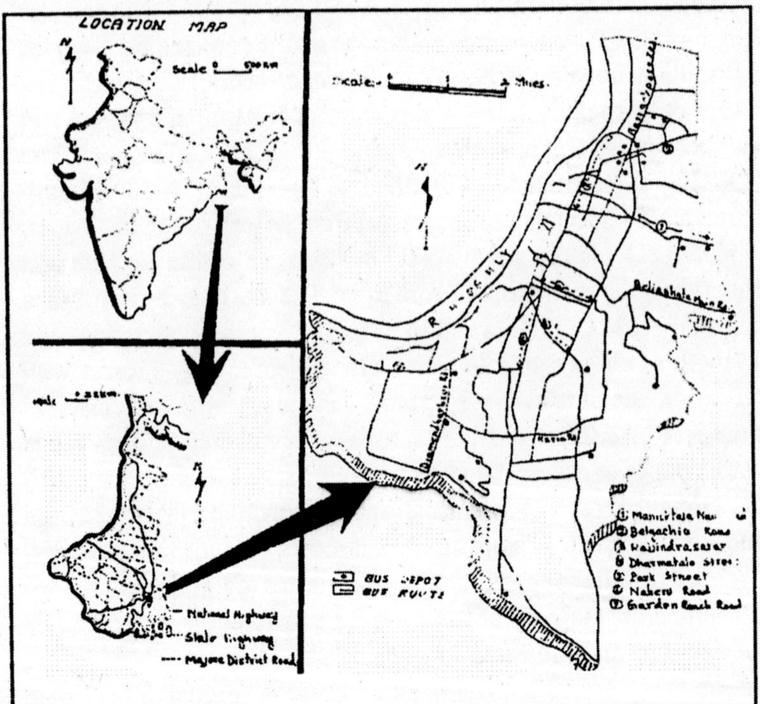

Fig. 9.1 : *Bus Route of Kolkata*

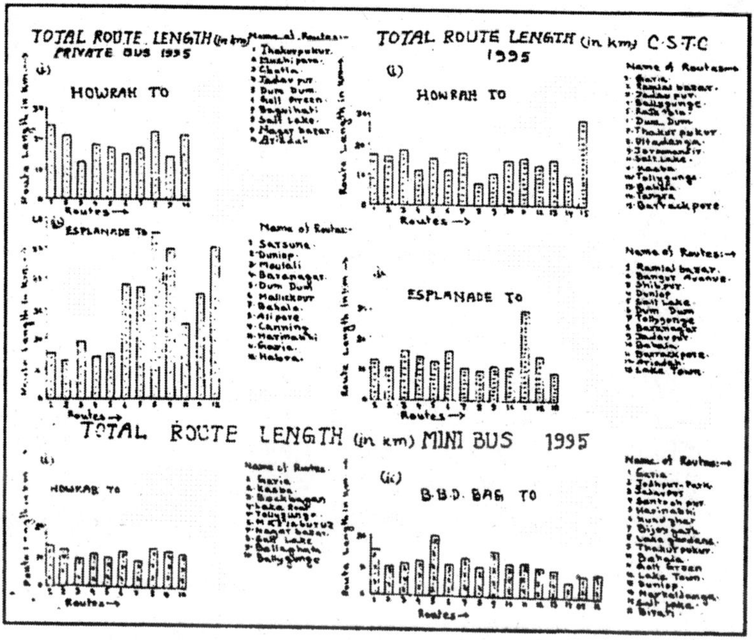

Fig. 9.2 : *Total Route Length (in Km) : Private Bus in 1995*

State Transport Corporation

The bus services in the city of Calcutta had its start in 1920. However, the State Transport Services under the Directorate of Transportation, Government of West Bengal came into existence in the form of an organised public sector on 31st July, 1948. At the outset it introduced the bus operation on six city routes with a modest fleet strength of 25 petrol driven single decker buses. Diesel era started with the acquisition of 2 double decker buses in 1949-50. The Directorate expanded its activities set up a few garages/depots and the Central Workshop at Belgharia. The Directorate was finally converted into' Kolkata State Transport Corporation' on 15th June1960 under the Road Transport Corporation Act, 1950. Till 1966-67,CSTC nationalised about 90 per cent of the city routes. However, because of acute financial constraints, the Corporation could not meet the ever increasing passengers demand and the private buses were allowed to operate in the city by the State Government. CSTC stated its first long distance services on 14th April 1968. At present, CSTC has a fleet strength of 1,205 buses. Its daily operation consists of 303 bus services along 95 mofussil routes and 850 services on 111 city routes. The function of CSTC are as below :

Function of CSTC

(i) To promote development of road transport in the areas comprising Kolkata and the districts of 24- Parganas, Haora and Hugli in West Bengal.

(ii) To Co-ordinate the road transport with railway transport in the said areas.

(iii) To extend and improve facilities for road transport in the said areas and provide an efficient and economical system of road transport services therein.

The buses are usually located in several depots and there are eleven depots altogether in Kolkata accommodating about 1,177 buses. At the same time there are also 105 other buses used for purposes other than passenger carrying.

Table 9.1 : *Depot and Carrying Vehicles*

	Name of the Depot	Actual Accommodation of buses	Non-passengers carrying vehicles
1	Belghoria	73	9
2.	Garia	73	6
3.	Howrah	148	15
4.	Kasba	132	6
5.	Lake	123	15
6.	Manicktala	139	8
7.	Nilgunje	70	2
8.	Paikpara	105	8
9.	Salt Lake	69	9
10.	Taratala	134	17
11.	Thakurpukur	111	10
	Total	1177	105

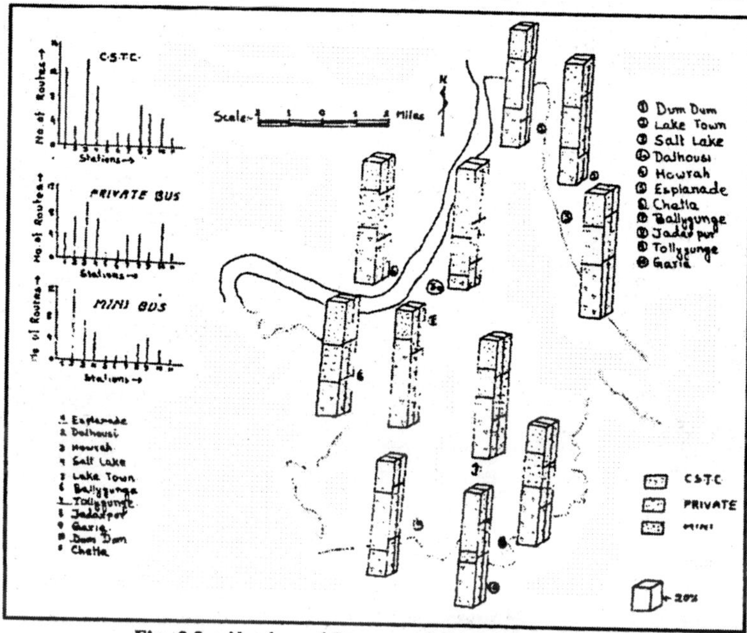

Fig. 9.3 : *Number of Routes and Kilometers of Buses (C.S.T.C./Private/Mini Buses)*

The Corporation operates 106 city routes of which 17 'limited bus services' 44 'ordinary services', 22 'special services', 5 'midi services', 4 'executive services' 4 medium services' and 5 'green line services'. It is apparent from the following table the different types of services performed by CSTC to the population.

Table 9.2 : *Type of Services and Distance Coverd*

Total No. of routes	Years of starting	Different type of services	Total distance Cover in km.
22	01.07.75	S (Special)	584.01
18	01.07.72	L (Limited)	283.89
44	31.07.48	O (Ordinary)	816.80
5	09.97.91	MD (Midi)	78.85
4	01.04.91	D (District)	152.70
4	18.02.91	E (Executive)	86.25
4	09.10.95	MS (Medium Services)	82.25
5	07.03.94	GL (Green Line)	127.30

Apart from the city services, CSTC also perform long distance services covering a maximum kilometerage of 593 km to Siliguri. This service is installed in the year of 1968 when only Kolkata-Digha routes was inaugurated covering a distance of 182 km on 14th April 1968 there are at present (1997) 118 routes. The urban transport profile has changed radically during 1990's initiating various types of buses on roads like E/MD/MS/D/E/GL, thus, bringing dynamism in urban infrastructure programme so far as transport is concerned.

Private buses

Although passengers buses started playing in the streets of Kolkata towards the end of the 19th century, it took nearly twentyfive years for the service to be regularised on an organised basis through the various routes in the around Kolkata. It is until 1925 when the Bengal bus syndicate was founded, till then individual operators run buses with scant respect for time schedule or regularity of service. Today all the private bus operators within the Kolkata Metropolitan district and organised into bus syndicates or route committees. A bus syndicate is an association of the bus owners, which looks after the interest of operation and organises the bus services. It supervises the operation of these services by forming one or more route committees, which are in turn responsible for maintaining regularity of service and are recognised by the Regional Transport Authorities. Within the Kolkata Metropolitan District, each bus route other than those operated by the Kolkata State Transport Corporation falls under the jurisdiction of route committee, which is the basic important of control. But all the routes operated by the private bus owners within the Metropolitan District do not necessarily fall under a syndicate, as some routes are supervised only by route committees.

Unlike, buses under the Kolkata State Transport Corpotation, private buses did not run in Kolkata, Haora and 24-Parganas as organised units. Buses are owned by individual route permit holders and a good deal of individualism prevails amongst them.

In the Kolkata region which includes 24-Parganas, about 4,405 buses ply in different routes. The bus have the capacity to

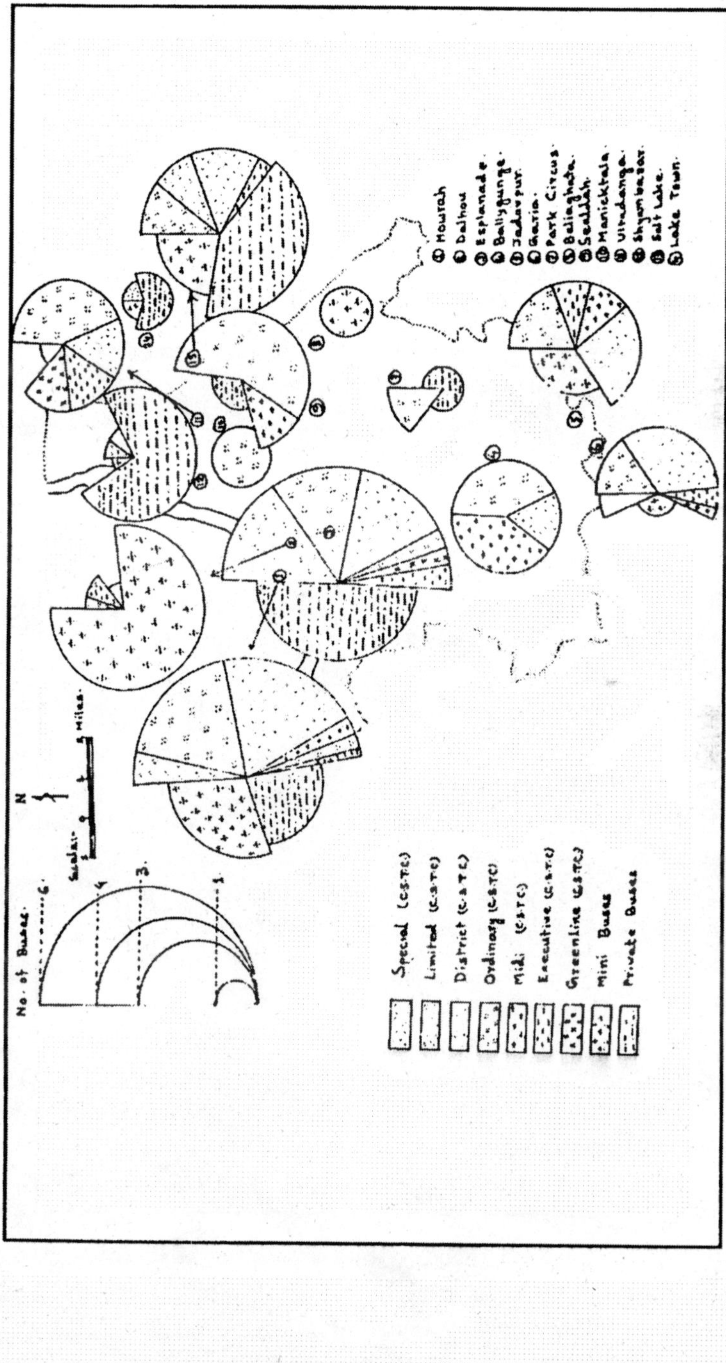

Fig. 9.4 : *Total Number of Existing Routes of Buses (Cost/Private/Mini) 1995*

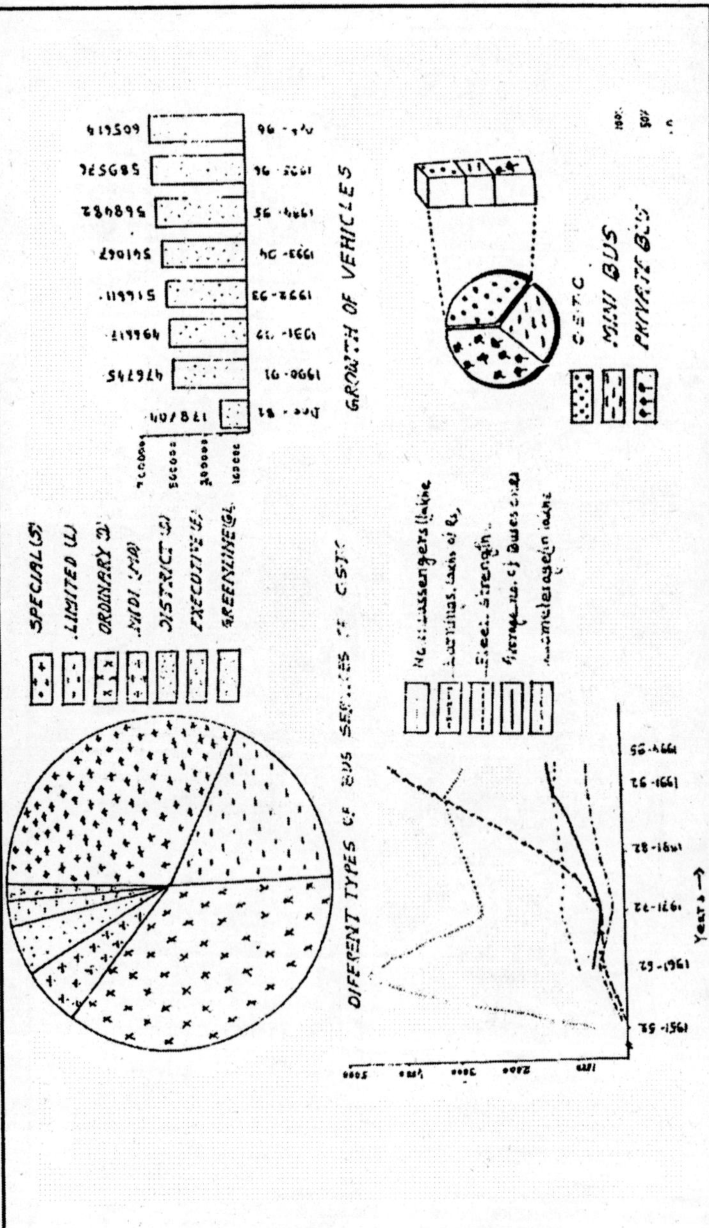

Fig. 9.5 : Growth of State Transport 1951-1995

carry 70 passengers, seated 35 and standing 35, and buses make 1 to 5 trips a day. A bus is kept from 3 to 10 hours in the run. The number of private buses plying in Kolkata city and in the sub-urban routes and also the number of private buses coming into the city are as follows:

(i) Number of private buses not plying in Kolkata city – 4405
(ii) Number of private buses plying in suburban routes – 2112
(iii) Total Number of private buses plying in the Kolkata region – 6517

In 1940, when the Regional Transport Authority, constituted under the Motor Vehicles Act of 1939, came into existence, private buses began to ply in the following routes : 2,2a, 3,3A,4, 4A, 5A, 8 8A, 10, 10A, 11, 11, 12, 12A, 12B, 13, 14, 32, 33, 35, 36 and 38. However, when the state buses began to appear in the streets of Kolkata in 1948, it drove the private buses from such routes monopolising the show for itself. But when its plan failed it again invited the private buses back to such routes under temporary licenses.

With the growth of the fleet, the State Transport Corporation faced competition from private operation. This proved still more uneconomical for the State Transport Organisation. In order to oust the private operator from certain routes, the state of West Bengal in 1954 or progressive made nationalization of routes. This became a definite and phased programme in 1955-56 when the Planning Commission recommended gradual nationalization of transport services all over the country. Between 1964 and 1967 the Kolkata State Transport Corporation enjoyed the virtual monopoly of all inter-city routes. Due to non-acquisition of sufficient number of buses for replacement of old buses, the city transport services deteriorated to a deplorable level. The State Government was, therefore, forced to move in the reverse direction of derationalization and to induct a large number of private buses in various routes of the city.

Mini Buses

The mini buses began plying in the streets of Kolkata towards the end of the 19th century. Mini buses were regularised on an organised basis under the Mini Bus Association which were run by individual operators who were members of the Association. Today all the mini buses operating within Kolkata Metropolitan District are organised under the Association. The Association of bus owners look after the interests of the operators and organises the bus services. It supervises the operation of services by forming route committees which are responsible for maintaining regularity of service and are organised by Regional Transport Authority.

Unlike the buses under the Kolkata State Transport Corporation. Mini buses mostly plying in the Kolkata are organised under one Association, suburban routes are not found. About 1215 Mini buses ply daily in Kolkata following different routes in the Kolkata and Howrah. Route length vary from 5 km in Route No. S-165 to 32 km in Route S-132. There are 72 mini bus routes in the city of Kolkata and Howrah.

State/Private/Mini Buses : A Comparison

While comparing the services provided by CSTC/Private/Mini buses, it was felt that main competition lies between state and private buses and mini buses only following that of the private buses in the line. Maximum routes are served by the state and private buses in the city of Kolkata and Howrah, almost 65 per cent is covered by these two organisations and the rest in fulfilled by mini buses. However in Park Circus, Shyambazar, Lake Town, Esplanade and Salt Lake city private buses dominated the transport services. On the other hand, Dalhousie Square, Beliaghata however is prodominated by mini bus routes.

Mini buses are well presented in serving the CBD of Kolkata City, i.e. Dalhousie Square, whereas the main outlet of Kolkata which is Howrah where CSTC played the main role in serving the passenger.

Performance and Prospect of State Buses

At first, the State Transport Organization started operating buses generally on routes which did not attract private operators on account of low volume of passenger traffic, the effect of trying only unremunerative route could not but have a bad effect on the

Table 9.3 : *A Comparison of Routes*

	CSTC	Private	Mini
Esplanade	13	4	16
Dalhousie	3	8	12
Howrah	15	10	7
Salt Lake	10	7	5
Lake Town	1	1	1
Ballygunge	2	1	1
Tollygunge	2	4	1
Jadavpur	7	4	3
Garia	4	1	6
Dum Dum	4	6	2
Chetla	1	1	1

revenue carrying capacity. Several other difficulties also affected this new state venture from its inception. There was also no suitable garage accommodation for state buses. Later on, however, two bus depots having accommodation for 150 buses each have completed – (i) at Belghoria in 1950 – (ii) at Tollygunge near Dhakuria Lack (Rebindra Sarovar) in 1951.

Still later on, four other depots also came into being at Paikpara, Manichtala, Taratals and Howrah. Lack of workshop facilities and of trained personnel was the other initial hurdle which the state undertaking had to negotiable. A central-workshop at Belgharia was completed in 1951 and arrangement for small repairs were also made at the depots. This eased the situation to some extent.

With the growth of the fleet, the State Transport Organization faced competition from private operators. This proved still more uneconomical for the State Transport Organization. In order to the private operator from certain routes, the State of West Bengal in 1954, made progressive nationalization of routes. This became a definite and proposed programme in 1955-56, when the Planning Commission recommended gradual nationalization of transport services all over the country. The process of nationalization ousted private operators from all city services in Kolkata by 1964 but the State Transportation could not take full advantage of it for reasons here in after told.

When the State Government felt that they had an effective transport organization, with garage and workshops, they decided to exercise the power under sections 3 and 4 of the Road Transport Corporation Act, 1950 and transfer the management of the State Transport Organization to a separate semi-autonomous corporation called the Kolkata State Transport Corporation. This was done in June 1960. A fleet of 670 buses, housed in several depots and the Central workshop at Belgharia were transferred to the Corporation.

The serviceable buses again were not all put in service daily. Apart from these which are said to have crossed their economic life, a large number of vehicles are not pressed into services everyday. All the outshedded buses, however, do not complete the trips assigned to them everyday. Because of low outshedding and lower effectiveness of buses, the revenue earned, it is alleged, cannot meet even the working expenses. To remedy this situation the corporation asks for a rise in the fare structure. The total earning from the fare structure have, however, been observed in Table 9.4.

Table 9.4 : *Fare structure (1996-1905)*

Years	Serviceable	Distance (km)	Rate per km	Total
1990-91	1196	2 km	1.00	2739.87
1991-92	1244	4 km	1.20	3306.86
1992-93	1233	4 km	1.20	3809.83
1993-94	1270	N.A.	1.40	3884.21
1994-95	1205	N.A.	1.70	4281.68

* Indicates buses are in running condition.

Whereas in 30th Sept.1977, the West Bengal Transport Authority has introduced a new fare structure in regard to CSTC/ Private buses, and mini buses. As per CSTC and private buses are concerned every stage increase is 50 paise upto a distance of 50 km and beyond 50 km. each stage increase is 75 paise.

Perspective

The State Transport introduced many new types of bus services

Table 9.5 : *Service Bus Types and Fare Structure*

Types of bus Services	Distance in km	CSTC/Public Buses Rate per Old fare	(in Rs.) New fare	Mini	CTC
C.S.T.C. and Private buses	2	1.25	1.25	1.60	1.75
	4	1.50	1.50	1.70	2.00
	6	1.50	1.70	*	*
	8	1.50	1.50		
	12	2.00	2.25		
	16	2.25	2.50		
	18	2.50	2.75		
	22	2.50	3.00		

* In every distance there is a rise of 35 paise and after Rs.3 rise in next stage is 60 paise.

like the special buses, the midi buses and the executive services. The double decker buses which were found to be unsuitable from the economic point of view were withdrawn from its fleet service. The C.S.T.C. also introduced other long distance services like green line service which however, was not a suitable measure. The other buses introduced in its fleet were the N.B.S.T.C. (North Bengal State Transport Corporation) and the SBSTC (South Bengal State Transport Corporation). The C.T.C. (Calcutta Tram Company) also introduced buses into its fleet in order to meet the demand of the passengers and shed some road during the peak hour. The limited stop services of the C.S.T.C. were also withdrawn from its fleet with the introduction of the new types of services, in order to stay in the competition with the private bus services the State Transport took the following measures:

(i) introduction of long distance services like operative services

(ii) introduction of NBSTC and SBSTC

(iii) withdrawal of double decker, and

(iv) withdrawal of limited bus services.

The State Transport services began as a departmentally run undertaking with only 25 petrol buses. Between 1948 and 1953, the fleet was enlarged to 481 buses consisting of 200 petrol and

281 diesel driven imported buses. The 200 petrol buses were gradually abandoned and uneconomical and more diesel buses were procured from abroad. Gradually the fleet strength further rose and by 1960 the State Transport Organization had 670 buses in service. The fleet strength gradually increased to 1,122 in 1971-72. A decrease in the fleet was noticed in 1981-82. Thereafter, it increased to 1244 in 1991-92. At present, however the fleet strength stands at 1205. A glance at the table below gives a view of the fleet strength of the CSTC from 1951-52 to 1994-95.

Table 9.6 : *Fleet strength of CSTC (1951 to 1995)*

Year	Fleet Strength
1951-52	–
1961-62	813
1971-72	1122
1981-82	1099
1991-92	1244
1994-95	1205

It will be clear from the table below the fluctuations in average serviceable fleet and in average number of buses on road together with fleet utilization ratio.

Table 9.7 : *Fluctuation in Serviceable fleet*

Years	Serviceable fleet	Average serviceable fleet	No. of All buses on road	Fleet utilization ratio (%)
1990-91	1196	1148	843	70.48
1991-92	1244	1187	914	73.47
1992-93	1233	1246	927	75.18
1993-94	1270	1228	858	67.56
1994-95	1205	1250	860	71.37

The low fleet utilization ratio can be ascribed to the increasing number of breakdowns. Failure to acquire new buses according to requirements burdened the Corporation with an age work and depleted fleet. This resulted in lesser outshedding and utilization of buses.

The number of passengers carried daily by the CSTC have remained constant over due years inspite of an increase in the fleet strength of the CSTC. The private bus and the mini buss have been responsible for shedding the ever increasing travelling passenger load. This inspite of increase in the number of daily movers there has been no reflections of the same in the passengers served daily by the CSTC which has remained at an almost same level.

Table 9.8 : *Daily Passenger (1990-1995)*

Year	Average Number of passengers carried daily (in lakhs)
1990-91	8.43
1991-92	8.68
1992-93	8.68
1993-94	8.00
1994-95	7.68

With the introduction of new routes by the Planning Commission's recommendation of nationalisation of transport services all over the country and also to eliminate private operators from certain routes the State Government began nationalisation of routes. Inspite of this, the number of routes increased from 31 in 1961 to 101 routes in 1995. This led to the increase in the total kilometerage run by the CSTC buses. The total kilometerage increased by about 6.7 times.

Table 9.9 : *Growth of State Transport at a Glance (1990-91 to 1994-95)*

Years	No. of passengers served (in lakhs)	Total No. of fleet	Total km run (in lakhs)	Carrying (in Lakhs)
1990-91	3067.41	1196.00	522.63	2739.87
1991-92	3153.76	1244.00	598.73	3306.86
1992-93	3084.65	1233.00	609.18	3809.83
1993-94	2907.22	1270.00	574.46	3884.21
1994-95	2775.24	1205.00	599.02	4281.68

Together with this urban transport system in Kolkata, a number of transport infrastructure development projects has been suggested for Kolkata Metropolitan area by the Japanese expert headed by Dr. Jura Kodera on behalf of Japan International Co-operation Agency, Tokyo. The projects includes (a) in the central city of Kolkata flyovers will be at four important intersection, on continuous flyover at two intersections and few at grade improvements (b) an elevated pedestrain plaza from Sealdah (city's main Railway Station) to C.R. Avenue (City Centre to North main road) to handle 150,000 pedestrains for 12 hours. Future extension will be made from B.B.D. Bag (City Centre) and Sealdah Station (c) traffic signalization include area traffic control (33 inter sections), arterial intersection control system (47) and isolated intersection control (36), (d) concrete pavements have been suggested in certain locations (e) two underground parking garages and proposed, one at BBD Bag (City Centre) another at Esplanade (Central Transport Terminus).

Thus, if the growing population as well as, vehicular population rise in demand for traffic, synchronises with futuristic urban transportation plan, the urban transport problem may be reduced to some extent along with the introduction of new status of state buses and new routes which will definitely give the expected multiplans effects.

Calcutta Tram Company

Calcutta Tram Company has also introduced CTC buses in different routes of Kolkata. The CTC started the services from 1995 and in 1st Sept.1996 the total number of CTC buses running in different routes of Kolkata totalled to 258 buses (estimated). The Tramways introduced their service in 1978 in 24 routes. The CTC buses introduced their services in 34 routes within a very short period of time. These buses have four depots for accommodating the buses like Tollygunge, Belgachia, Khidirpur and Rajabazar depots of which Tollygunge possess more numbers than do the others numbering to 91 buses travelling to 121 routes. Mini bus routes covered by CTC services is 186 km. between Esplanade-Gadiara and minimum distances covered by the same is Rajabazar to Howrah station a distance of 9 km.

These urban transport services has brought a radical change in the field of availability-connectivity-accessibility of city bus services. Alongwith these services Surface Transport, Metro Railway and Ferry Services also added dimention in the mass transport services in the city of Kolkata.

West Bengal Surface Transport Corporation (WBSTC)

WBSTC has also introduced the services not only for the people of Kolkata but also for West Bengal consisting 19 routes of which only two route services, nemely, Karunamoyee-Howrah and Karunamoyee-Garia-Tollygunge is for the city of Kolkata. While the ferry serives are found to be prevailing in 7 routes of which main services are found to be operative between Kachurheria-Haldia and Strand Road Howrah has the highest frequency of every day services with an interval of 30 minutes. In the year of 1995, 27th September, with a total distance of 16.45 km. between Dum Dum and Tollygunge Metro Rail has installed in the city as a underground mass transit system thereby further reducing the surface load and at the same time leading to progressive futuristic transport city. This Metro Railways started its services in 24th October, 1984 only from Esplanade to Bhowanipur covering a distance of 3.40 km only and after a decade in August 1994 it started running over a stretch of 7.64 km between Esplanade and Tollygunge. Inspite of these services the public transport system in the city is inadequate to demand, so the use of personalised vehicles has been steadily increasing.

Concluding Remarks

With the growth of the city both in terms of population and phenomenal increase in the number of vehicles and with the promise of newer expansions and investments in the industrial fronts, the role of traffic management has become extremely important. The major schemes towards improvement of traffic by way of modernization, construction of flyovers both pedestrian and vehicular, inter-section improvement there is a positive public participation and co-operation in the city. Looking into the growth of vehicular traffic, which is estimated at 1,78,704 in 1981 to 4.76,745 in 1990-91 to 6,05,614 in October, 1996, the constant rise in vehicles population in the city of Kolkata is clearly understood of which total number of buses stood at 3,185 in 1981 to 5,772 in 1990-91 to 7315 in October 1996 and mini buses stood at 2,328 in 1981 to 6,287 in 1990-91 to 7608 in 1996 including luxury taxi also.

References

Calcutta Traffic, *Annual Review*, 1996.

Halder, Dilip (1997) : *Urban Transport Problem*. Academic Publication, India.

Sarkar, Avik (Ed.) :*Anand Bazar Patrika*, 30th September, 1997 and 1.10.97

Newsletter of Centre for Build Environment : Urban Transportation in Calcutta, No. 2, April- June, 1993.

Sukla, Bhaduri (1992) : *Transport and Regional Development*, Concept Publishing, New Delhi.

The Hindu : (1993) Survey of the Environment, 1994.

10

ROAD TRANSPORT AND RESOURCES DEVELOPMENT IN MADURAI

K. Lakshmi and *N. Krishnan*

Introduction

The function of any economy requires resources and transportation. As economy develops, the production, functions and services increase and this, in turn, requires more transport facilities. Inadequate transport is one of the greatest obstacles to economic progress. Thus, transport plays a key role in the overall economic development as a conditioning factor permitting growth. In the economic sense, even transport is considered as a resource factor of a catalyst for the development of various resource characters. However, in the organization of urban areas, the relationship is more complex since a change in one sector invariably brings changes in the other with a faster rate of development at micro level.

The development process of urban centres has many dynamic components like population, manufacturing, commerce, trade, transport, communication and other multi-functional activities which are mostly segregated in and around the city nucleus. In this dynamic process of development, the city is expanding horizontally towards the rural surroundings and vertically in its nucleus centre to ease out the pressure of multi-functional activities. Such a rapid expansion leads to higher rate urbanization and transport development, especially in the million cities of various developing countries in the late sixties. It is also noticed in certain historical old cities of India. Hence in 1970, the Planning Commission of India set up a committee to study the growth of cities and transport services. Accordingly, the National Productivity. Council of India made and extensive study of transport and economic development of selected cities. It is concluded that road transport is a landmark of the urban growth and productivity. The improved transport services of the cities significantly influence the rural surrounding permitting the urban areas to penetrate the rural areas through horizontal spread of urban expansion. Hence it is understood that in Indian economy rural-urban integration is made possile only through expanding road transport. The growth and development of road providing mox interaction to people and services and vice versa. Thus the relationship between transport and local resources act in a dynamic circulatory process constantly reacting with the men and material of the city and its surroundings. But the process of development may vary in accordance with the response to interaction and on the basis of the socio-political constraints. The creation of new links and improved transport services make nodes more accessible to each others and also create changes in the economic structure of the nodes. But the degree of development of nodal growth along the major roads in certain direction is faster than the others. This is a common feature in most of the fast growing metropolitan cities in India.

In geography, the role of transport and its implication on the socio-economic development of a region at macro level were analysed by scholars like Ullman (1954), Berry (1959), Garrison (1960), Kansky (1963), Taffee (1973) and others. In urban areas problem-oriented transport studies such as travel behaviour, transport and landuse, transport and level of urbanization and the impact of transport on the metropolitan development have been attempted by Mayer (1961), Yeates (1965), Hurst (1974), Ramaswamy (1976), Ramachandran (1974, 1986), Raza (1986) and K.N.Singh (1984). In growing metropolitan cities urban expansion is more evident along the major radiating roads which leads to changes in rural-urban landscape and the resultant development of landuse and human resources. Hence, the present attempt has been made at a micro level to study the relationship between road transport and resource development by selecting the metropolitan area of the old historic city of Madurai located in south Tamil Nadu of India.

Study Area

Madurai Metropolitan area located at 77°55' East to 78°15' East longitude and 9°50' North to 10°10' North latitude with an areal extent of about 764 square kilometre comprises of Madurai city at its core, 9 urban agglomeration centres, a Class III towns and 205 villages in its surrounding. Madurai region has a hot dry climate with seasonal variation in the distribution of temperature and rainfall.

To study the relation and development of resources and road transport netowk, the revenue divisions of 21 firkas in the study area have been chosen as the basic study units for detailed analysis (Fig. 10.1). Madurai North and South talukas alone have 15 firkas. Sakkimangalam, Othakadai, Arumbanur, Kolapandi, Chathirapatti, Kulamangalam, Samayanallur and Sathamangalam are 8 firkas that constitute Madurai north taluka. In Madurai south, Keelmaduri and Melmadurai, Nagamalai, Thiruparankundram, Avaniyapuram, Valayangulam and Virathanur firkas and located. Of these Sathamangalam, Melmadurai and Keelmadurai city core firkas cover the entire Municipal Corporate area of the city. Of the remaining firkas, Sholavandan, Thenkarai and Alanganallur come under Vadipatti Taluka while Kokulam and Thirumangalam come under Thirumangalam taluka. In addition to these 20 firkas the study area comprises of Kondagai firkas which comes under Mahamadurai taluka of Sivaganga district.

The combination of rural and urban settlements in the study area produces a rural-urban complexity with varying degree of resource base and transport network development. River Vaigai which flows from northwest to southeast divides the study area into almost two equal parts as northern and southern regions. The prominent rock outcrops, terrain, drainage, distribution of rainfall, irrigation, soil cover and the crops growns in the study area produce a contrasting landuse resource base especially with the concentration of wet lands in the northern region and dry lands in the southern region. Madurai Metropolitan area is having more than a million population of which 76 per cent is concentrated in the city area and other urban centres that comprise only 15 per cent of its land area.

The remaining 24 per cent of population is spreaded in 85 per cent of its land area. This disparity in the distribution of land and human resources has resulted in an agglomeration of manufacturing and service activities in and around the city core area (Fig. 10.2). But at the same time transport allows the dispersion along the major radiating routes from the city core. The growth of road transport has further created more changes in the resource utilization. The highest concentration of non-agricultural landuses in the city core and the urban agglomeration in the study area exhibit a contrasting pattern of the concentration of net sown and fallow land. During the study period changes in the landuse pattern are predominent in the north and the south along the radiating roads. The value of land under various functional landuses have recorded a very high value in the city core than in the sourrounding peripheral firkas. Nearly 80 per cent of the small scale industrial establishments and higher order Health and Educational services are clustered in the city and the nearby urban centres. The availability of larger tract of fallow lands, cheap land value and labour in the southern region induced the new development of industries on either side of the NH-7 and other major roads.

In brief, the rural-dominated resources like net sown area, cropped land, agricultural workers and total workers are found highly concentrated in the rural areas, whereas urban-oriented resources like Iand under non-agricultural use, population density, literacy ratio, manufacturing and other workers, industrial intensity and service establishments are found more in the city core firkas. Hence, rural and urban-oriented resources have obviously established a diametrically opposite relationship.This is clearly evident from the landuse maps (Fig.10.2).

Resource development and its differential rate of growth depend on the improvement in the road transport connectivity, accessibility and service efficiency. Generally road transport connectivity and accessibility decline from the city core towards the periphery and the distance decay function is clearly evident from the Madurai city bus transport flow during the study period (Figs.10.1 and 10.2). In the study area one can notice a notable change in the landuse pattern and associated improvement in the

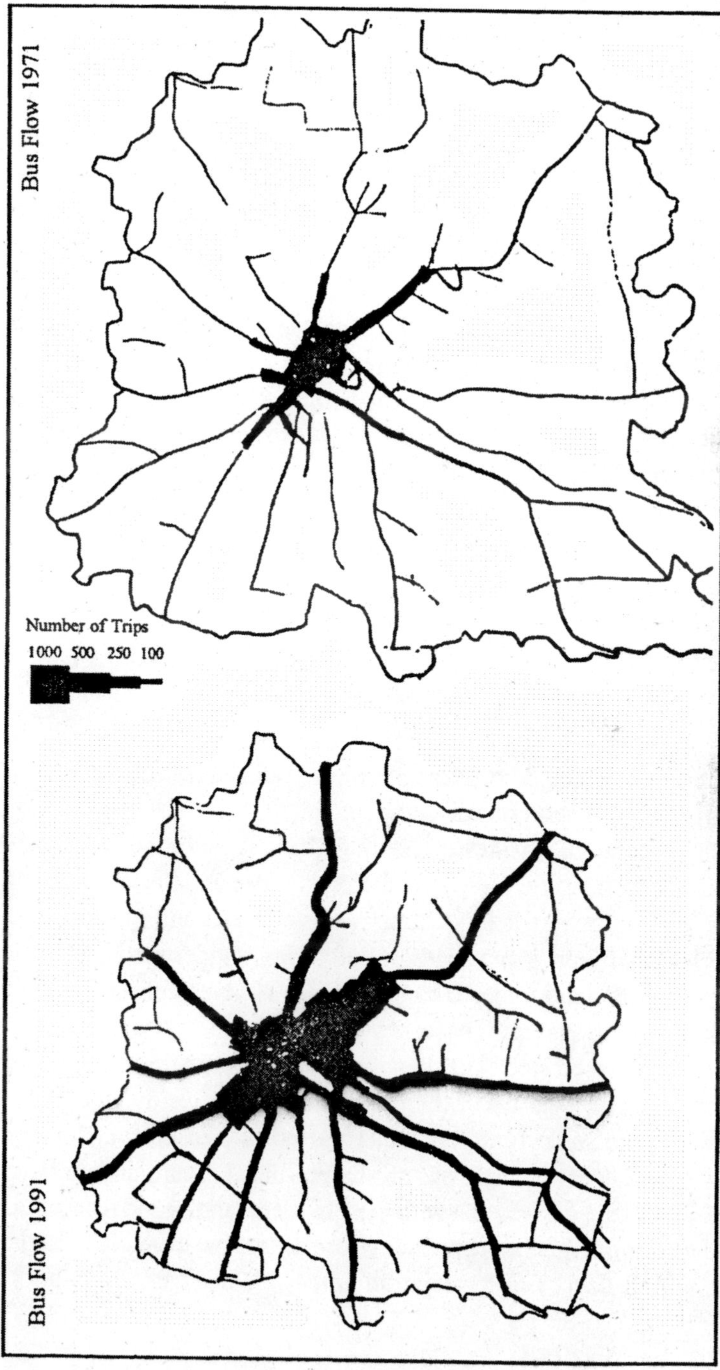

Fig 10.1 : *Madurai Metropolitan Area*

city bus service intensity of flow character. Hence, in the process of regional development, transport and resources are considered as mutually interrelated factors. At present, as a conditioning factor and permiting growth, transport is considered as dependent variable and resources are treated as indepenent variables. To evaluate the strength of transport in the regional devlopment an advanced statistical tool of multivate regression analysis is used in the study.

Variables Considered

For a detailed analysis a set of fourteen land-based resource variables another set of eight human variables are selected to assess the resources. To evaluate the road transport network characters eighteen variables of various transport connectivity, accessibility and service efficiency measures are selected. The interrelationship among all these variables and the functional interdependence and complementary character of resources and transport have been evaluated through advanced statistical tools such as multiple correlation and regression analysis with a computer aided programme.

Intercorrelation

The results of simple intercorrelation analysis have brought out the degree and direction of association among the set of fourteen land-based resource variables. Among them it is found that non-agricultural landuse, commercial crop landuse and the variables of four land value characters, namely, value of wet and irrigated use and value of land along roads, value of land under settlement use and value of land under commercial use have established a high positive relationship. These six urban-dominated variables are negatively associated with rural-oriented resource variables like net sown area, land under irrigation, area under paddy and fallow land. At the same time, the other variables like area under dry crops, market gardening crops and barren-uncultivable landuse have less significant association with most of these land-based resources. The results of eight human resource characters have clearly brought out a high degree of association among themselves. Population

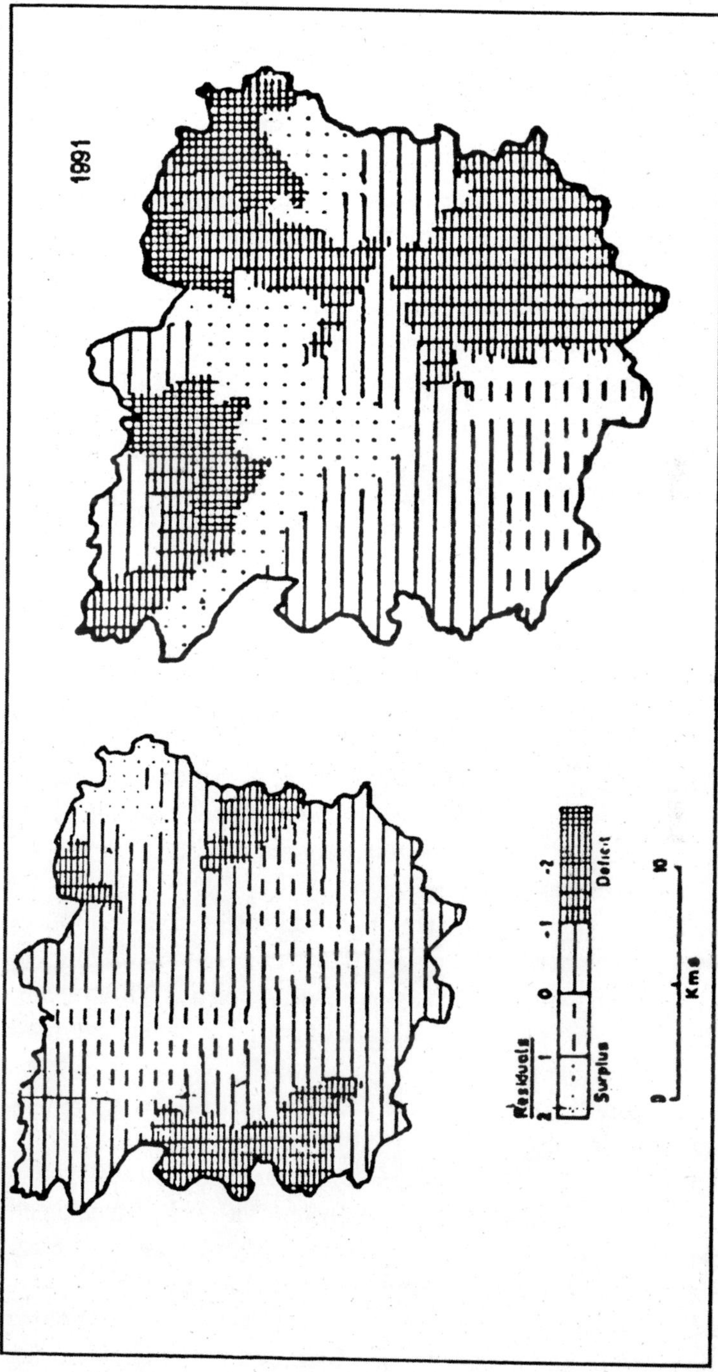

Fig.10.2 : *Madurai Metropolitan Area : Flow Intensity*

density, literacy ratio and manufacturing and other workers have positive relationship. On the other hand these three variables are negatively related with the variables of total workers and agricultural workers. The variables pertaining to industry and services (intensity index) are significantly associated among themselves and also with the other three urban-based variables of population density, literacy and manufactuing.

The urban-dominated six variables, viz. population density, literacy ratio, manufacturing and other workers, industrial intensity and service intensty are positively related with non-agricultural landuse whereas the total workers and agricultural workers establish a significant positive relationship with net sown and gross sown area. However, the four land value variables relating to both rural and urban characters possess a positive relationship with urban-dominated variable and negative relationship with rural dominated variables. The intercorrelation matrix of road network connectivity, accessibility and efficiency characters represents the varied association of eigheteen selected network variables. On the basis of existing interrelationship among these selected resource and transport variables the best associated and related variables are considered. A set of thirteen resource variables are selected as independent variables. The best related transport variable which efficiency namely, bus trip flow intensity is selected as dependent road transport network variable.

In the light of the steps involved in the computation of multiple correlation and regression statistics, the package of present date matrix consisting of twenty-one observation units (firka level), twelve independent resource variables and a dependent road transport network character is systematically processed with the help of Spss/PC plus statistical package for IBM PC Computer program. The dependent variable which explains network connectivity, accessibility and efficiency has established a statistically well significant relationship with urban-dominated population characters and land value variables relating to different functional landuse. These transport variables have an inverse relationship with rural-oriented landuse and population resource variables like population density, literacy ratio, manufacturing workers, non-agricultural landuse industrial, intensity service intensity and land value variables individually or collectively will make a notable increase or decrease in the network connectivity, accessibility and efficiency. Similarly, the increase in the rural-oriented resources like net sown area, agricultural workers and total workers will be related to the variable of network character according to the nature and magnitude of the index value (Table 10.1).

The trend of association of intercorrelation is almost the same for the two time points. Hence, to analyse and determine the complementary character of the resource development an advanced statistical method regression model has been applied.

Regression Model – Road Service and Resources

The regression model directly explains the varying degree of road network efficiency in relation to the resources in the study area. Road service efficiency is directly proportional to both rural and urban dominated resource character, viz. X_1, X_2, X_3, X_4, X_5, X_6, X_8, and X_{13}, that is any increase in these resource variables will demand for the increase of bus trip flow and it is inversely proportional to the variables of X_7, X_9, X_{10}, and X_{12}, that is any decrease in these variables will increase the frow intensity (Table 10.2). The multiple correlation coefficient and determination 'R2' (.98) shows the existing high degree of association and strength of relationship between bus trip flow intensity and the selected resource variables in the model. The standardized co-efficient of the regression model show the relative importance and magnitude of each independent resource variable in association with the dependent variable. The total variation in the bus trip intensity is highly contributed by both urban and rural-oriented resources. Population density (1.09), literacy ratio (0.31), value of land under commercial use (0.38), industrial intensity (0.23), total workers (0.1), net sown area, land value for wet and irrigated land are the potential resource variables notably contributing their share independently or collectively to tne increase of bus trip flow intensity. It is interesting to note that population variables are strongly supporting the variation of city bus trip flow intensity.

It is appropriate to state that the model developed for the year 1971 is slightly deviated from the model for 1991. In the year 1971, the resource variables like net sown area, population density and manufacturing and other workers were inversely porportional with road service efficiency. They become directly proportional and the change in the direction of relationship in the year 1991. This indicates the added contribution of transport service efficiency in the development of resources during the study period. The land value variables also have changed their direction of association in the year 1991. Others are constantly following the same direction of association in both years. This change in the direction during the study period is due to the development of road transport city bus trip flow in association with the differential rate in the growth of resources. The population resource variables are strongly supporing the variation in the bus trip flow intensity. This indicates that the concentration of population demands for more bus trip flow and vice versa. Nearly 98 percent of the road network efficiency variation is explained collectively by the eleven resource variables in both time points. (two variables in the equation is not fitted in the model due to high degree of multi colliniarity). Both models are highly significant at 01 level (Table 10.3).

The two regression models developed to explain the complementary and develpment character of road transport network efficiency with resource characters in study area highlighted the following facts :

1. The development of the resources and road network efficieny are complementary to each other in general.
2. The variation in the city bus trip flow intensity and developement issignificantly determined by a set of 11 independent resource characters.
3. Among the independent resource variables, only those of urban dominated resources are potential variables highly contributing to the development than the rural-dominated ones.

With the light of the complementary relationship of road transport and resources and to evaluate the spatial imbalance in road network efficiency for the selected resource base of the study area, regression residuals have been developed.

Table 10.1: *Intercorrelation Road Transport Efficiency and Resource Variables 1991*

	Independent Variables Resources	Dependent Variables : Bus trip flow intensity
X_1	Net sown area in hectares	0.56
X_2	Total workers in per cent	0.61
X_3	Agricultural workers to total workers	0.80
X_4	Population density	0.94
X_5	Liferacy in percent	0.75
X_6	Manufacturing and other workers	0.80
X_7	Non-agricultural lands in hectares	0.83
X_8	Manufacturing intensity	0.92
X_9	Service intensity	0.71
X_{10}	Land value for wet land Rs. 100 sq. mts.	0.67
X_{11}	Land value for lands nearer to roads Rs. 100 sq. mts.	0.90
X_{12}	Land value for lands under settlements use Rs. 100 sq.mts.	0.70
X_{13}	Land value for lands under commercial use Rs. 100 sq.mts.	0.92

Source : Author – According to 'T' test all are significant at 01 level

The residual values show the amount of deviation from the regression line compared for the dependent variable. The residual is the difference between the observed and the estimated value of variables considered in the analysis. The residual derived from the regression provides a means by which the relationship among the phenomena of dependent and independent variables can be seen in the spatial context. It shows the regional imbalance in the road network growth and the resource development in the study area.

Spatial Imbalance of Road Transport and Resource Development

The Table 10.4 and exibits the positive and negative residuals of the road transport efficiency is well explained by the bus trip flow intensity. The city core firkas record very low deficit and surplus

Table 10.2 : *Statistical Result for 1971 and 1991*

Years	Dependent Variable	Regression Models	Multiple Correlation Coefficient	Coefficient of Multiple Determination	Error	'F' Ratio
1971	Bus trip Flow Intensity	$Y = 40.0023 - 0.2041X2 - 0.1194X_4 + 0.55719X_5 - 0.02228X_6 - 0.6575X_7 + 2.4127X8X_8 - 0.07696X_9 + 0.3049X_{10} + 0.8769X_{11} + 0.0290X_{13}$	0.986	0.97	8.3180	29.6777*
1991	Bus trip Flow Intensity	$Y = -48.5736 + 0.0057X + 1.1611X_2 + 2.2765X_4 + 2.3777X_5 + 0.2174X_6 - 2.6043X_7 + 1.7280X_8 - 0.9698X_9 - 14.9385X_{10} - 0.1915X_{11} + 0.1416X_{13}$	0.990	0.98	16.8368	43.3722*

* Significant at 01 level.

Note : In the years 1971 and 1991 variables X_3 X_{14} are not included in thee equation due to high standard deviation multicollinearity and high intercorrelation among these variables.

Table 10.3 : *Standarized Regression Coefficient Values*

S.No.		Independent Variables	Bus Trip Flow Intensity 1971	Bus Trip Flow Intensity 1991
1.	X_1	Net Sown Area	-0.11538	0.00131
2.	X_2	Total Workers	0.17340*	0.09032*
3.	X_4	Population Density	-0.10322	1.0929*
4.	X_5	Literacy Ratio	0.19688*	0.31688*
5.	X_6	Manufacturing and Other workers	-0.01841	0.07321
6.	X_7	Non-Agricultural Landuse	-0.22841*	-0.37087*
7.	X_8	Industrial Intensity	0.75987*	0.23088*
8.	X_9	Service Intensity	-0.05806	-0.34624*
9.	X_{10}	Land Value for Wet and Irrigeted Land	0.02805	-0.34125*
10.	X_{11}	Land value along Roads	0.42597*	0.07275*
11.	X_{13}	Land Value for Commercial use	0.08574	0.38470*

Note : In 1971 variables X_3 (Agricultural Workers) and X_{12} (Land Value of Settlement Use) are not in the equation due to high standard deviation, multi-colliniarity and high intercorrelation.

* Significant at 01 level.

values as they have optimum level of but trip service efficiency to meet the local demand. The area surrounding the city core like Sakkimangalam and Thiruparankundram in the neighbourhood and Alanganallur and A. Kokkulam in the periphery also register a low value of surplus and deficit value. This indicates city bus trip flow intensity is concomitant with recource development. Othakadai, Nagamalai, Kulamangalam, Thenkarai, Valayankulam and Thirumangalam firkas have a medium to high surplus level of bus trip flow intensity. All these firkas have experienced significant improvement in the rate of flow during the period of last 20 years. This is due to the urban sprawl along the major arterial roads. Sholavandan, Arumbanur, Samayanallur and Chathirapatti in the north and Avaniyapura, Kondagai and Virathanur in the south exhibit medium to high deficit in bus trip flow intensity. This indicates insufficient road transport service to these firkas. Though the city infuluence is spread along the major arterial roads towards the rural sourroundings, the service efficiency is not on a par with the resource development rate.

In general two residual maps together reveal the close relationship between the development of resources and the road transport functions in the study area (Fig. 10.1). The areas registering moderate to high development of agricultural resources, urban functions, population density, land value and service characters show an increasing trend in the development of road transport service efficiency. Negative areas with poor land use resource, hilly terrain, dry lands, low population density and inadequate development service characters indicate a low degree of urban influence sustained with minimum level of road transport functions.

The regional imbalance on the service efficiency of but trip flow is highly noted in the firkas, namely, Sholavandan, Avaniyapuram, Virathanur and Kondagai, which are deficiit in terms service efficiency. In the case of Avaniyapuram, the deficit is mainly due to the greater and rapid urban expansion. On the other hand the peripheral location of Kondagai in the south and Sholavandan in the north and the poor resource base of Virathanur firkas are the reasons for deficit bus service efficiency. The immediate measure of increasing the bus trips will be a suitable solution for the removal of the problem and the overall development of the resources in these areas.

The present study is a micro level example attempt to evaluate the complementary character of road transport and resources in the urban environment. The statistical application has helped to identify the potential determinant factors of resources, which activate the transport for regional development. Apart from this in the course of time the spatial changes in the contribution of transport and resources in the overall development have also been evaluated in the study. The investigation of the research paper identified the zone of urban influence in the rural surroundings in relation to the development of resources and road transport. The

Table 10.4 : *Madurai Metropolitan Area-Rrgrssion Residuals Bus Trip Flow Intensity*

Sl No.	Name of Firkas	1971 Estimated	Residual	1991 Estimated	Residual
1.	Sakkimanagalam	15.9807	-5.1807	26.6275	-2.2275
2.	Othekadal	15.7741	-2.8741	35.9142	-20.8858
3.	Arumbanur	21.3071	16.7929	39.5687	-11.2687
4.	Koolapendi	12.7487	-4.2487	29.4404	-10.0404
5.	Chathirapatti	16.7041	-7.0041	26.0934	-13.8934
6.	Kulamangalam	16.5445	0.1555	25.1426	10.2573
7.	Samayanallur	14.1843	5.2157	45.9052	-22.2052
8.	Satnamangalam	71.3301	0.0699	123.7885	-6.3885
9.	Keelmadurai	125.1668	0.4332	303.1621	-0.2621
10.	Melmadural	114.3336	-0.6336	284.7924	4.5076
11.	Nagamalai	11.8896	5.9104	29.2854	22.0146
12.	Thirupparankundrem	25.3810	3.8190	76.6250	-4.9250
13.	Avaniyapuram	15.6858	5.6142	41.0039	-6.6039
14:	Valayankulam	8.9724	-4.7724	2.0101	7.9893
15.	Virathanur	0.8823	3.6177	13..0060	-8.5060
16.	Tirumanagalam	24.0255	-4.7255	53.8323	5.4677
17.	A Kokkulam	18.2460	-7.6460	27.4641	3.1451
18.	Sholavandan	13.2275	-0.5275	19.0704	-6.8704
19.	Thenkarai	6.3550	-0.5550	2.2454	10.2546
20.	Alanganllur	21.1204	-3.5204	34.0939	-3.6739
21.	Kondagai	7.0405	0.0595	14.9283	-6.6283

(*Source* : Computed by Author)

positive and negative residuals developed from the analysis extended the scope to identify the potential areas for future development of transport and resources. It also suggests to propose suitable plan at micro level to avoid transport problems in future.

References

Alam, Manzoor (1969). " The Hyderabad Metropolitan Research Project", prepared by Osmania University and Institute of Asian Studies, Hyderabad.

Anderson, T.W. (1958). *Introduction to Multivariate Statistical Analysis*. New York. John Wiley & Sons, Inc.

Anitha, F.P. (1952), "Transport Development in Andhra Pradesh", Andhra Chamber of Commerce, Silver Jubilee Volume, Walter.

Baliga, B.S. (1959). *Madurai District Gazetteers*. Madras : Government of Madras.

Berry, B.J.L. (1959). "Recent Studies Concerning the Role of Transportation in the Space Economy." *Annals of the Association of Amcrical Geographers*, No. 49, pp. 328-342.

Berry, B.J.L (1973) *The Human Consequences of Urganization*. Londan MacMillan and Co. Ltd., Cerdon.

Detwyler, T.R. (1972) *Urbanization and Environment : The Physical Geography of the City*. California : Duxburry Belmont.

Eliot Hurst M.E. Ed. (1974) *Transportation Geography : Comments and Readings*. Newyork : McGraw Hill

Garrison, W.L. (1960) "Connectivity of the Interstate Highway System", Papers, Regional Science Association, Vol. 6, pp. 121-139. H.W. Cilmore (1953) *Transportation and Growth of Cities*, Illinois, The Free Press.

Ilangovan, P. (1983) Transportation Geography : An Analysis of Relationship between Road Network Structure and Pattern of Regional Development in Tamil Nadu. Unpublished Ph.D. Dissertation, Department of Geography Madurai Kamaraj University.

Kansky, K.J. (1963). Structure of Transport Network: Relation between Network Geometry and Regional Characteristics Research Paper No. 84, Dept. of Geography, University of Chicago.

Kansky, K.J. (1949) "Town Development and Transport", *Scottish Geographical Magazine*, Vol. 65.

Mayer, H.M.(1961) "Chicago Transportation and Metropolitan Planning", Regional Science Association, *Papers and Proceeedings*, Vol. 7, pp. 241-247.

Owen, W.(1954) "Transport and Economic Development", *American Economic Review*, Vl. 49, pp. 179-187.

Ramachandran, H.(1977) Transportation and Urban Attributes in the Coimbatore Region : A Case Study of an Interacting System. unpublished Ph.D. dissertation, Jawaharlal Nehru.

Ramaswamy, G. (1976) "Road Network in Trichy-Thanjavur Planning Region - An Analysis", in V.L.S.P. Rao, Ed., *Golden Jubilee Volume*. Indian Geographical Society, Madras.

Ramesh, A. ed. (1984) *Resource Geography. Contribution to Indian Geography*, Vol.5. Heritage Publishers,New Delhi.

Raza, M. and Aggarwal, Y (1986) *Transport Geography of India : Commodity flow and Regional Structure of the Indian Economy*. Concept Publishing Company, New Delhi.

Singh, K.N.(1984) "Levels of Urbanization : A Case Study of Eastern Uttar Pradesh", *National Geographical Journal of India*, Vol. 29, No. 1.

Srivastava, S.K.(1964) *Transport Development in India*. S. Chand & Co, New Delhi.

Taaffe, E.J. and Gauthier, H.L. (1973) *Geography of Transportation*. Englewood Cliffs, N.J. Prentice Hall.

Ullman, E.L.and Mayer, H.M.(1954) "Transportation Geography", in P.E. James and C.L. Jones. Eds. *American Geography : Inventroy and Prospect*, Synacuse Univsersity. Press.

Wheeler, J.O.Ed. (1973) "Social and Policy Perspectives in Transportation Geography", *Economic Geography*, Vol. 49, No.2.

Yeats, M.H. (1965) "Some Factors Affecting the Spatial Distribution of Chicago Land Values" *Economic Geography*, Vol. 41, No. 1.

Reports

British Geographers (1985) *Institute of British Geographers* Vol. 10 No. 1. Review.

Govt. of India (1970) Report of the Working Group on Metropolitan Transport Services, Planning Commission, New Delhi.

National Productivity Council (1960) *Road Transport Industry in West Germany,* U.K. and U.S.A. Report No.3, New Delhi.

NCER (1967) *Regional Transport Survey Madras and Pondicherry,* National Council for Applied Economic Research, New Delhi.

NATPACT (1983) *Study of Socio-economic aspects of Rural Roads Salem District* A Case Study. Annual Report.

11

DYNAMICS OF URBAN TRAVEL AND ITS IMPACT ON URBAN ENVIRONMENT

A Case Study of Metropolitan Hyderabad

Vijaya Bhole

Introduction

The gravity of urban transportation problems before the advent of automobile was minimal. The non-availability of mechanized transport limited the urban growth to walkable distances. It contrast, the present day metropolitan cities not only have concentration of population but also spread over vast area extent as a direct consequence of development of fast means of both private and public transport. Growth in fast means of transport helped in easy movement covering long distances in short time. Here an attempt is made to understand urban transportation problems which are manifold and complex in nature. There is positive relationship between development of economic activity and transport means. As one progresses demand for more mean of transport also increase. And, thus, standard of living is adjusted with the vehicles possessed by human being. The multiple of transport system helps to determine the scope and direction of economic development in city. The development of transport system also leads to widening of the size of markets within a zone of city. Service zones of producer are determined by urban transport environment. Better the urban transport, better scope for rapid development of urban activities in city by helping rightly to gear up the modality of labour and daily workers, servicemen and laymen for their daily travelling.

Study Area

Metropolitan Hyderabad is one of the fastest growing metropolitan cities in India. Metropolitan Hyderabad with 4.28 million population (1981) has a history more than four hundred years. It is due to undergoing rapid growth of industrialization and urbanization.

It represents a blend of feudal historical and modern industrial city after 1950 showing tremendous growth in both private and public mode of transport. The figures given below speaks about the concentration of various type of vehicles (Table 11.1). It is observed that two and three-wheelers fitted with two stroke engines create maximum pollution by emitting hydrocarbons and particulate matter. The diesel engines generate pollutants like nitrous oxides and particulate matter while petrol engines generate carbon monoxide etc. Total gases emitted by vehicles every year in metropolitan Hyderabad weight 58,000 tonnes by 2000 A.D. It is expected to rise to 82,000 tonnes. About 75 per cent of air pollution is caused by automobile emission in Metropolitan Hyderabad. The pollutant like carbon monoxide is highly toxic and reduces oxygen carrying capacity of blood where as lead damages brain, hydrocarbon and nitrous oxide cause eye and skin irritation and create respiratory diseases.

Table 11.1 : *Hyderabad–Types of Vehicles, 1995*

Types	Number of Vehicles
Two Wheelers	6,50,000
Private Cars	70,000
Goods Vehicles	29,000
Taxis and Rickshaws	24,500
Buses	5,040
Others	1,000
Total	7,79,540

Source : *SPEOL*, Vol, 2, No. 33 July, 1996.

Note : On an average about 10,000 numbers of vehicles of all types are being added every month.

Data on trip generation has been obtained though sample study of household traffic zonewise like Hyderabad South, Hyderabad North, Secunderabad and Suburban areas. A sample of 400 households were chosen. They were equally distributed in all four city zones. The reason for selecting the city zones as sample areas is that in these areas firstly, the historical factors have left a deep imprint on the morphological characteristics. Secondly, each zone has got a distinctiveness in terms of spatial distribution of work centres and other economic activities. Primary data is generated through questionnaire method. The major limitation of this study is the size of the sample. Since the sample is carefully chosen to give an insight into complex transportation problems.

Trip Generation Analysis

In order to plan out the urban transportation facilities it is foremost to understand the dynamics of urban travel. Trip generation analysis presents the urban planner to determine the traffic generating capacities of the existing landuse. In the pre-mechanised transportation era, journey to work was the only important aspect of trip generation. Hence, workplace residence was a close-knit system. The proximal location of people within the urban areas was limited to walking or by transportation by horse carriages. In the post-mechanized automobile era the trip generation in urban areas became diversified due to complex and mixed nature of urban landuse pattern. Trip generation as a factor influencing the spatial fabric of the city has received scant attention in geographical work by the Indian scholars. Keeping this view in mind an attempt is made to understand the nature of trip generation, modal split analysis etc. (Table 11.2)

Workplace Residence Relationship

An understanding of workplace residence relationship is essential to appreciate the intensity and direction of daily movement of commuters. Unlike other trips in urban areas. Workplace residence trips are more stable in direction and intensity at a given point of time. It is observed from Table 11.3 that the households of Hyderabad South and suburan areas make longer trips to journey to work. Hyderabad South is the densely populated city zone and is the historical core of the city which lacks in urban activities like banks, insurance offices, hospitals, industries, educational and other institutions etc. These urban activities are major employment generating centres which are mostly located in Hyderabad North.

Table 11.2 : *Residence Relationship by City Zones, 1998*

(Percentage of Households)

Traffic Zones	Distance from Workplace (kms.)					Total
	2	4	6	8	10	
Hyderabad South	4	8	22	29	37	100
Hyderabad North	4	20	34	46	16	100
Secunderabad	5	24	30	22	19	100
Suburbs	20	18	12	26	24	100

Source : Fieldwork.

The trip length of suburban households is longer because they commute to other city zones where their work centers are located. Most of these households are migrants. Another interesting observation is that in suburban areas, considerable number of households (4%) trip length to work place is less than 4 kms. Because their workplaces are located in close proximity (mostly industrial). Households from Hyderabad North does not undertake long trip to work places because of the concentration of work centers in and around this city zone. This observation is further substantiated by the figures in Table 11.3.

The trips direction of journey to work by the sample households by city zones show that about 2/3rd of the households from different city zones make their daily trip to work centers which are located in Hyderabad North. The work place trips are focused on Hyderabad North, which is the nucleus of metropolitan activities. The origin destination survey not only highlighted the trip pattern but also the pattern of transport linkage. The transport map of Hyderabad clearly shows that Hyderabad North is well-connected and is the most accessible part of the city.

Modal Split-Travel Time Analysis

The mode of transport used is the reflective indicator of the urban economic base. Metropolitan Hyderabad has a unique mix of fast and also moving mode of transport which substantiated by the fact that nearly 1/10th of the sample households depend upon slow moving transport.

Table 11.3 : *Zonewise Traffic Trip Generation Analysis*

(Percentage of Households)

From To	Hyderabad South	Hyderabad North	Secunderabad	Suburban Areas	Remarks
Hyderabad South	5	73	12	10	TG
Hyderabad North	—	70	14	16	TG+TR
Secunderabad	—	60	18	22	TG+TR
Suburban	—	55	20	25	TG+TR

Source : Fieldwork

Note : TG = Trip Generation Area
TR = Trip Receiving Area

As seen from Table 11.4 that the half trips to journey to work is by public transport which signifies the importance of public transport as mode of transport. Majority of the sample households on account of their income level rely on public transport. Over dependence on public transport is responsible for overcrowding and more travel time. The most important factor in transport is speed and efficiency. To overcome the problem of over crowding, more travel time and for flexibility private transport is preferred mode. Use of private mode of transport depends upon level of standard of living. As seen from the Table 11.1 Hyderabad has about 6.5 lakhs scooters and an additional 70 thousands cars and as highlights in Table 11.4 that about 1/3rd of the households depends on private mode of transport. A considerable proportion of households depends on cheap and slow moving mode of transport like bicycles, rickshaws which is detrimental in the smooth and speedy urban transport. Metropolitan Hyderabad has the distinction of being one of the metros with highest concentration of bicycles.

An analysis of travel time pattern highlights spatial irregularities. The length of the trip as well as time taken by households in Hyderabad South traffic zone is longer as compared to other city traffic zone (Table 11.5).

Table 11.4 : *Modal Split to Journey to Work*

Mode of Transport	(Percentage of Households)	
A. FASTER MOVING		
1. Public Transport (R.T. C. Buses)	46	
2. Private Transport (Car/Scooters)	34	
3. Hired Transport (Auto-Rickshaws etc)	4	
4. Other (Railway)	1	
Total	85	
B. SLOW MOVING		
1. Bycycles	8	
2. Rickshaws	4	
C. BY WALK	3	
Total		15
Grand Total		100

Source : Fieldwork

Table 11.5 : *Travel Time and Journey to Work*

(Percentage of Households)

Traffic Zones	Travel time (Minutes)				
	20	20-40	40-60	60	Total
Hyderabad South	5	35	42	18	100
Hyderabad North	10	45	35	10	100
Secunderabad	8	56	28	8	100
Suburbs	20	40	24	16	100

Source :Fieldwork

Besides longer trip lengths and travel time to work place the households of Hyderabad south traffic zone undertake longer trip length to avail various urban activities like education, recreation etc. (Table 11.6)

The variation in the disparities in the trip lengths traffic zones is due to improper landuse zoning. Metropolitan function are mostly agglomerated in traffic zones others than Hyderabad South which is the historical core and static morphological zone of Hyderabad.

Table 11.6 : *Trip Length Activity Pattern*

Traffic Zones	Average Distance Travelled (kms)		
	Education	Shopping	Recreation
Hyderabad South	4.5	2.0	6.0
Hyderabad North	3.0	4.0	3.5
Secunderabad	2.0	2.5	3.0
Suburbs	3.5	4.2	6.5

Source : Fieldwork

Urban Travel Pattern and Urban Environment

The pattern of travel in metropolitan Hyderabad is the function of landuse characteristics, socio-economic set up standard of living and historical growth. These factors not only influenced the location of trip generation and trip receiving areas but also the nature and intensity of transportation forms. The manifestation of travel pattern on urban environment are complex creating a vicious circle which is explained by the schematic diagram given in Fig. 11.1.

The over consequence of urban travel pattern in metropolitan Hyderabad on the urban environment can be summarized as follows:

1. Lack of proper landuse zoning and historical factors has led to the chaotic traffic movement.
2. Mixed types of transportation i.e. slow and fast moving mode is responsible for traffic bottlenecks.
3. Increasing use of private mode of transport and traffic bottlenecks led to the emission of quantum of gases and particulate matter.
4. Convergence of transport network in Hyderabad North traffic zone led to more problem of congestion and air and sound pollution.
5. Slow movement of traffic in Hyderabad South traffic zone due to narrow roads which were laid before the advent of automobile era.
6. Packing problem in the central part of the city (Hyderabad North traffic zone) which is the hub of the metropolitan functions.

7. Longer trip lengths by the households of the Hyderabad South traffic zone which passes through Hyderabad North traffic zone generating complex transport problems.
8. Inadequate suburban rail transport i.e. suburban rail transport do not cater to the heavy traffic corridors

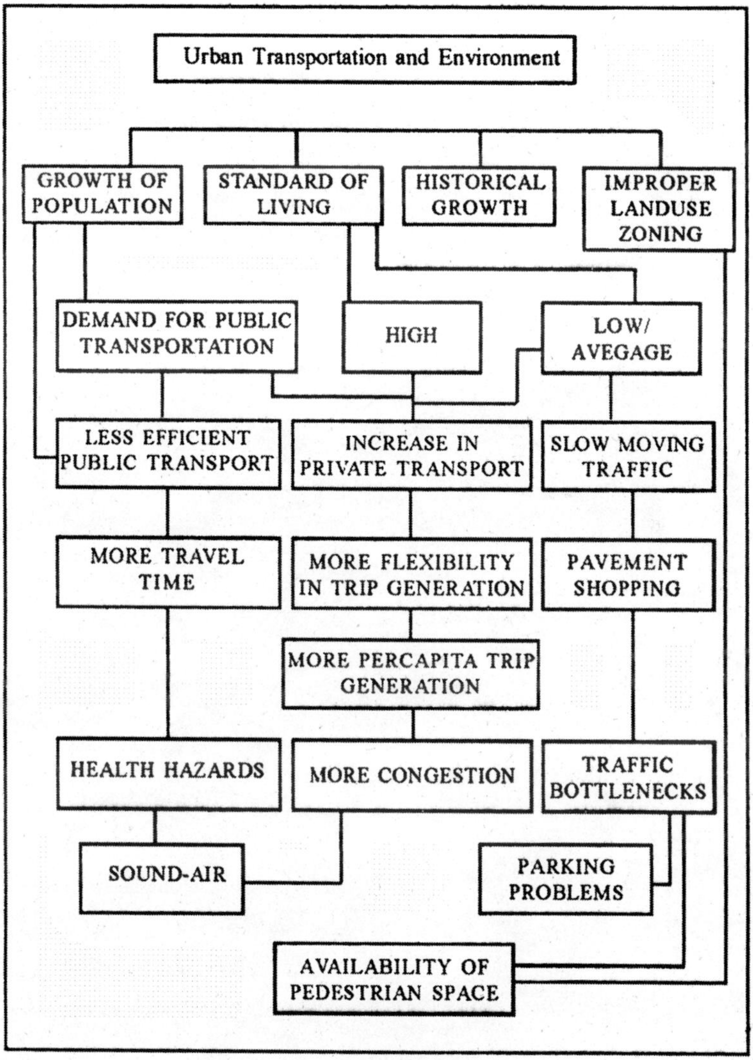

Fig. 11.1 : *Schematic diagram*

Concluding Remarks

To conclude, the Hyderabad Urban Development Authority in collaboration with various agencies is taking steps to minimize the transportation problems by implementing massive programmes like road-widening, construction of flyovers, introduction of Light Rail System etc.

References

Bhole, Vijaya (1988) : *Housing and Urban Development in India*, Classical Publishers, New Delhi.

Edward, L. Norlok (1978) : *Introduction to Transportation, Engineering and Planning*. McGraw Book Co., New York.

Hutchinson, B.G.(1974) : *Principles of Urban Transportation System and Planning*.McGraw Book Co., New York.

Vitthal Reddy, K. (1985) : *Urban Crisis in India*, Osmania University, Hyderabad.

12

TRANSPORT DEVELOPMENT IN PUNE METROPOLITAN REGION

M.V. Bagade

Introduction

Pune is considered to be an educational and cultural centre of Maharashtra. It had changed its profile to an industrial city in late 1970's. The poximity of Pune to Mumbai and the availability of infrastructure facilities, existence of transport and communication services and also the presence of skilled and unskilled manpower attracted a large number of industries to this region, especially, after a ban was imposed on Mumbai's further industrial expansion in 1960. Barring a few major industries which have been setup on the eastern side of Pune city, most industries developed around the Pune-Mumbai highway north of Kirkee cantonment in the vicinity of Pimpri-Chinchwad. The process of migration of people from different parts of the country as well as from the neighbouring areas also continued because of employment opportunities created by rapid industrialisation. It also resulted in acute housing problem and increase in slum settlements.

In early years, development of Pune was slow owing to its growth as a cultural city. The travel modes used for intra-city movement were bicycles and tongas. In fact, Pune was known to be the cycle city of India. With the spread of the city, distances and accessibility became longer and the induction of auto modes became inevitable. This traffic was expected to be served by a bus-based public transport system i.e. Pune Municipal Transport. Later, the Municipal Corporation in Pimpri-Chinchwad established a separate municipal transport corporation.

Table 12.1 : *Population Growth in PMR*

Sr. No.	Area	Area in Sq km (1991)	Population (in Lakh) 1971	1981	1991	2001 (Est.)
1.	PMC	146.11	8.56	12.03	15.67	29.47
2.	Pune Contonment	13.88	0.78	0.86	0.82	1.06
3.	Kirkee Cantonment	13.23	0.65	0.80	0.78	1.02
4.	PCMC	104.37	0.96	2-44	5.44	7.06
5.	66 Villages	538.65	1.33	2.25	2.69	3.50
6.	PMR	809.07	12.38	18.38	25.40	33.11

Source : Census Handbook, 1991

It will be observed that with a normal growth rate, the population will grow by (say) 22 per cent in a decade, while in case of Pune, it is almost 40 per cent between year 1981 to 1991 and 32 per cent till the year 1991 to 2001(Estimated)

The magnitude of commercial and industrial growth (shown in Table 12.2 and 12.3) . The growth in commercial establishment such as shops, hotels, theatres and other commercial establishment is given with their employment potential, which has almost doubled in last 10 years. The growth of industrial units with its employment, capital employed and turnover is also shown from 1978 to 1995 which is more than double in the number of units and triple in employment. With the stupendous growth in traffic generating activities and spread over of PMR from Hadapsar in the east to Nigdi in the west and Katraj in the south to Yerwada in the north, there was a huge influx of auto-vehicles. PMT's fleet which was 374 in 1977-78 was reduced to 355 buses in 1985-86. The shortfall in public transport system created a vacuum in transport supply and people had no choice but to go in for personalised modes like two wheelers and four wheelers.

Similarly, three-wheeler autorickshaw was a popular mode used to improve mobility. It had an advantage of providing end-to-end transport at a reasonable price. It wiped out the tanga as well as intra-city taxi services in PMR. Thus, two wheelers and three-wheelers almost took over the rising travel demand for the common man in PMR.

Table 12.2 : *Number of Establishments in Pune Metropolitan Region*

Type of Establishment	PMC			PCMC		
	1977	1985	1994	1983	1985	1994
Shops	31429	36280	56440	2365	622	12043
Commercial Establishments	7679	14301	33071	1183	2556	4663
Hotels	2289	2845	3818	334	410	918
Theatres	36	55	55	5	5	4
Total	41433	53481	93344	3887	9191	17628
Employment (Estimated)		67300	139000		15000	25000

Source: Directory of Maratha Chamber of Commerce and Industries

Table 12.3 : *Industrial Units*

Years	Units (No.)	Emp. (No.)	Capital Emp. (Rs. in Crores)	Turnover (Rs. in Crores)
1978	2515	111965	407	670 (est.)
1985	4529	148470	1563	2749
1990	6766	197830	3204	5445
1995	5838	337417	12423	26093

Source : Directory of Maratha Chamber of Commerce and Industries.

When a small car like Maruti appeared on the personal transport scene and became affordable to the middle income group families, the use of four-wheelers shot up adding a large number of vehicles on the roads causing further congestion.

When all these vehicles were pouring in the road network in PMR almost remained stagnant. Naturally, the resultant road congestion created major problems :

(1) Mobility on the roads became poor even though hi-tech, high speed vehicles were being bought in greater numbers.

(2) Due to poor speed, noise pollution and air pollution increased tremendously.

(3) Rate of accidents went up due to traffic congestion roads.

On the whole the complete traffic environment and mobility on roads was badly affected due to a shortfall in the public transport system. The following statistics will highlight the growth of auto-vehicles as well as increase in congestion. During the last 30 years, the growth of passenger vehicles was almost twenty-four times and to serve one lakh population, the auto vehicles increased nine times. This is because of the growth in small capacity vehicles owing to the absence of an adequate and efficient mas transit system.

Table 12.4 : *Growth of Auto Vehicles* *(Passengers)*

Type of vehicles	1970	1987	1995	1999	1970	1999	No. of times
Two-wheelers	15048	153664	320398	500706	1225	15172	12.4
Three-wheelers	2660	12185	24986	41519	208	1258	6.0
Four-wheelers (light)	7565	21988	43086	69000	616	2091	3.4
Private buses	386	1817	4729	7484	31	227	7.3
PMT+PCMT	280	588	1000	1000	23	30	1.3
Total	25539	190242	39242	619709	2104	18779	9.0

1. *Source* : Road Transport Authority, Pune.
2. Vehicles per lakhs Population.

Parking Facilities

Hardly any attention was paid to the development of proper parking areas for vehicles. Road are mainly being used as parking lots, which has affected the availability of road space for actual movement of traffic, PMR being an industrially and commercially developed centre attracts traffic from various neighbouring areas on the national highways : Mumbai Road, Satara/Bangalore Road, Nasik Road, Ahmednagar Road and Solapur Road. The floating population of incoming and outgoing vehicles and parking of these vehicles, particularly on the road has severely affected the traffic environment in PMR. It is also worthwhile to highlight the safety aspect on the roads of PMR. Since 1995 there has been a significant increase in road accidents and fatalities (barring 1996 which is exceptional).

It is also worthwhile to identify accident prone vehicles causing maximum mishaps on roads.

Table 12.5 : *Road Accident from 1991-1999*

Year	Road Accidents	Persons killed
1991	1949	283
1992	2757	284
1993	2282	286
1994	2184	282
1995	2562	308
1996	2279	271
1997	2687	302
1998	2644	334
1999 (Jan. to June)	1353	156

Source : Road Transport Authority, Pune.

As far as passenger vehicles are concerned, personalised modes i.e. two-wheelers have maximum accidents followed by those of cars. The root cause for rise in the rate of accidents as well as fatalities is inadequate and inefficient public transport system which had almost stagnated when the whole of PMR showed increasing traffic demand due to high economic growth.

Table 12.6 : *Type of Vehicles and Accidents (January-June, 1999)*

Types	Total Accidents %	Fatalities (%)
Two wheelers	323	26
3-seater	120	1
6-seater	49	3
Cars	232	20
MGV	114	12
HGV	236	42
PMT+PCMT	58	14
MSRTC	25	5
Pvt. Buses	28	8
Unknown	118	23
Others	13	2
Total	1316	156

Air Pollution due to Automobile Traffic

According to the air quality standards adopted by CPCB (Central Pollution Central Board) the limits prescribed for the concentration of SPM in the ambient air is 500 micrograms/cu.mt. It will be interesting to note the actual SPM concentration at certain locations in Pune.

Table 12.7 : *SPM Concentration Levels*

Sl No.	Location	SPM Concentration (Micrograms/cu.mt)
1	Yerwada Chowk	1010
2	Bund Garden Road	1616
3	City Survey Office Chowk	570
4	Paud Phata	2222
5	University Circle	1068
6	Mhatre bridge chowk	370
7	Sassoon Hospital Chowk	629

Source : Environmental Status Reports 1997-98 for Pune.

Noise Pollution

The acceptable outdoor noise level for city area is 45 to 50 decibel level (dBCA). It is interesting to note that noise pollution on sample locations exceed the acceptable limits substantially.

Table 12.8 : *Sound levels (in dBCA)*

Sl No.	Location	Morning 10 a.m.	Evening 8 p.m.
1.	Alka Talkies	86.0	87.0
2.	Hadapsar Gadital	81.2	83.4
3.	Swargate	82.2	83.4
4.	Sambhaji bridge	84.0	78.5
5.	Deccan Bus Terminus	88.6	87.8
6.	Simla Office	81.0	87.1
7.	University Chowk	89.5	89.2
8.	RTO Welleslary Road	83.8	87.6
9.	Sassoon Hospital	79.0	91.2

Source : Environmental Status Report (1997-98) for Pune.

Transport Environment of Pune Metropolitan Region

Status report on the transport environment of Pune reveals that :

(a) Migration and growth of population in PMR is three times the normal annual growth rate.

(b) Road capacities and network have almost stagnated for over 15 years.

(c) Over thirty years, the vehicle population has increased twenty-four times, as against a three time increased in the public transport system (of bus fleet).

(d) Road congestion has increased over 15 times (measured as PCUs per road km) resulting in poor speed on roads and higher consumption of fuel, inspite of, hitech fuel efficient vehicles.

(e) There is a significant rise in road accidents particularly 2 wheelers and cars. The rate of fatalities has also increased.

(f) Air and noise pollution have significantly increased. The survey undertaken for various locations have conclusively proved this fact.

The diagnostic analysis indicates that when the population was increasing at a galloping pace due to rise in industrial and commercial establishments spread over the whole of PMR, there was bound to be a stupendous rise in travel demand. The vacuum created by the shortfall in public transport system caused a tremendous inflows of personalised as well as para transit vehicles affecting mobility, safety, traffic flow and increased air pollution and noise pollution damaging the normal health of the city.

Role of Public Transport System as a Mass Transit

The role of public transport system in developing a safe and healthy urban environment is through satisfying growing travel needs of masses adequately and efficiently at an affordable price so that citizens may not prefer or switch over to personalised modes out of compulsion. One bus is capable of carrying 60 passengers which is almost equivalent to 40 scooters, 30 autorickshaws or 20 cars. These personalised modes not only occupy more road space causing congestion on roads but also result in higher consumption of diesel and petrol. PMT and PCMT did not realize the need for growth as one of the important public utilities serving the mobility needs of PMR. As far as the State Government is concerned, it was concerned about collection of passenger tax, while PMC and PCMI sometimes provided some temporary support which is not a

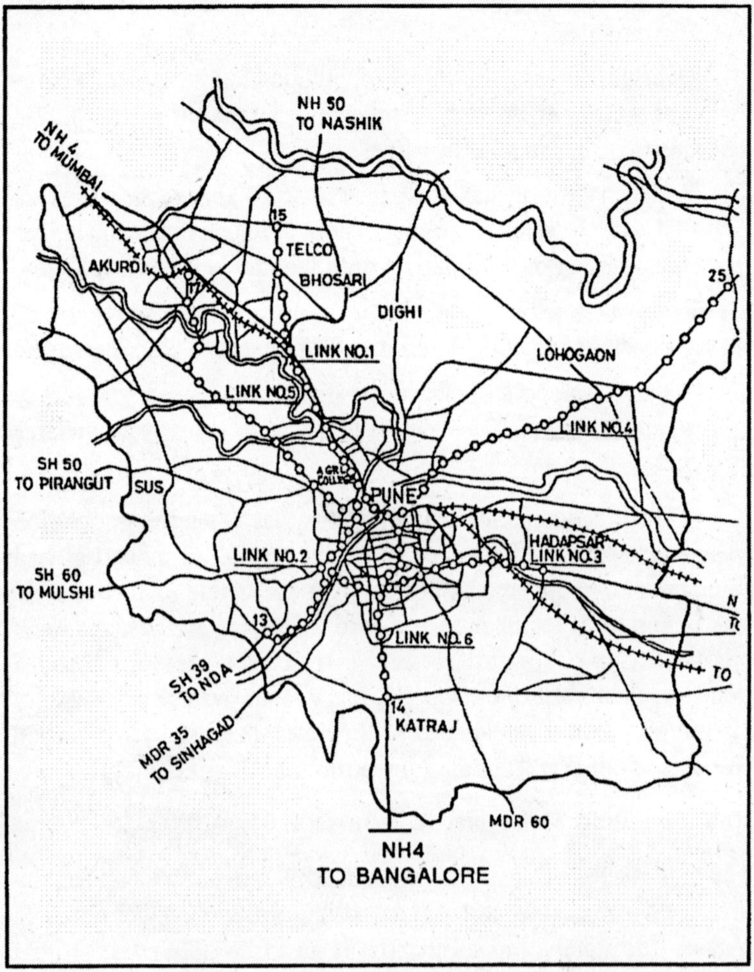

Fig. 27.1 : *MRTS for Pune Metropolitan Area*

permanent solution. In most of cities in the country, public bus transport system is nobody's baby.

In order to specifically identify the role, one can assume that the growth of auto-vehicles on the roads is a function of growing travel demand. On the basis of various studies, the total travel demand can be estimated for PMR. These parameters are average occupancy of a travel mode, daily trips and the percentage vehicles on road. This data helps to develop a multiplying factors which converts the vehicle data into person trips. The estimated trip rate per person by auto-vehicles was 0.30 in 1971 which doubled in 1991 to 0.63 and for 1999 it was 1.30

Thus for PMR, total person trips are 41.30 lakhs for a population of 31.6 lakhs. The problem of traffic environment in an urban agglomeration is that of peak hour demand. Various studies have indicated that 50 per cent of the total trips are operated during peak hours only. A fair estimate of traffic carried by company buses, school buses and private buses at peak hours is 10.48 lakh passenger trips.

The analysis goes as follows :

(a) 41.30 lakh person trips (Est.) per day (1999)

(b) 20.65 lakh person trips are operating by other buses

(c) 10.48 lakh person trips are operating by other buses

(d) Thus, 10.17 lakh person trips are served by public transport, intermediate public transport and personalised vehicles.

If 75 per cent of these trips are catered by the bus system, the expected level of mobility standards can be achieved which will definitely improve the traffic environment. It means that approximately 7.63 lakh person trips should be operated by the bus system. On converting these trips into number of buses, the requirement is of 1192 buses or 37 buses per lakh population for PMR only. Thus, considering the population growth in PMR, PMT and PCMT should increase their fleet to serve the travel demand.

If PMT and PCMT want to serve neighbouring towns like Alandi, Saswad, Urulikanchan, Rajgurunagar , Chakan, Talegaon (Dabhade), Talegaon (Dhandhare), Shikrapur, Khed Shivapur, Nasarapur etc., they will have to procure additional buses for the purpose. In retrospect at the cost of overlooking the travel demand in PMR, PMT and PCMT have ventured to serve neighbouring towns by starving the citizens in PMR for which they were originally created. This starvation has created 6 seater autorickshaws as a powerful competitor on city roads. In the future, it is essential that PMT and PCMT should concentrate mostly on the route network of PMR to ease out various environment problems in PMR and operate beyond its boundaries only if adequate fleet is available and the services financially viable.

References

Govt. of Maharashtra (1975) : Town and Country Planning Department, Traffic Department Report, Pune.

Govt. of India (1985) : Transportation Study of Pune Metropolitan Region, Unpublished Report Undertaken by CIRT, Pune.

Govt. of India (1985-96) : Road Transport Statistics, Published by Transport Commissioner for 1985-1996.

Project Report (1993) : Public Transport Studies conducted by Pune Municipal Corporation and Pimpri-Chinchwad Corporation, Chinchwad during 1993.

SECTION-III

REGIONAL TRANSPORT SYSTEM

13

RAILWAY AND ROAD TRANSPORT IN RAJASTHAN STATE

B.R.Meena

Introduction

Transport is function of various activities. Means of transportation play a key role in world as the circulatory system of the body. It is important to the individual carrying life blood of commerce, supplies necessary for substance of population groups (Davis, 1984). Transportation is the movement and exchange of people, goods and ideas. It is ubiquitous and there is no escape from it. It is difficult in conceive the situation where transport does not play a major role in regional development. Transportation is an important absorber of scarce resources, enormous wrong directed or misplaced investment and has a serious impact upon the whole economy. Poor transport is an obstacle in making effective use of regional resources. Most of the regional resources lie idle or are partially utilized due to lack of access as it is the function of transport to bridge the geographical gap between produces and consumer so that goods and services may be exchanged to their mutual benefit. Transportation by definition implies a movement and every individual from the earliest times possessed his own building ability to move, albeit within a limited areas.

Historically and culturally, Rajasthan has been considered as unique and important region of the northern India. Transport connects the eastern plain, the rugged and scattered hills of the central part and sandy terrain of the Thar desert in the west. This perplexity in transportation has naturally mean to economic separation of the region and tendency towards isolation. The economic development is associated with the introduction of modern means of transportation. There are areas which are otherwise potential but due to lack of essential infrastructure, they are still underdeveloped. Natural and cultural diversities of Rajasthan provide an ideal ground to study the role of transportation in regional development. In view of the above fact, Rajasthan has been selected to in-depth study of road and rail transport network.

Evolution of Transport Network

Human culture shows two outstanding development, namely, rapid and advance technology and fast growth of cities. The role of transport in these development have been the most significant in organising and utilizing the human efforts and resources in changing landscape through better spatial linkages. However, the nomadic movements have preceded to rise to trade routes. The trails may go to track and ultimately to roads. Nomads discovered the most convenient traffic routes in adjustment with the nature. It generally followed the natural line of drainage that afforded certain advantages along their courses. Thus, the routes were evolved rather than created, the geographers call them as "Natural routes". But the settled people have paradoxically enough created such important construction, both in their nature and the number of vehicles using them. After the advent of specialized wheels in carriages, the roads intended for bullock carts and reflecting their local needs to came to be gradually modified for the use not only by pedestrians, animals, animal driven carts, bicycle but also by automobiles. They also began to cater the needs of more extensive areas. In the evolution processes, the routes and roads in the Rajasthan were following this pattern. Though numerous trade, carven, and army route existed prior to present century, but no systematic account is available about their nature, structure and length etc. According to Manu the trade internal or external brings a good amount of money to the state exchequer in the shape of taxes levied on the movement, production and sale of different kind of commercial goods. Besides this, surplus goods are sent to the deficit areas to meet the local demands which adds to the prosperity and happiness of the state. The king should, therefore, adopt such trade policy as offers of his state.

In study region, trade routes existed sincerely medieval period connecting various towns of Marwar and Mewar through Pali, Jalore, Merta, Barmer, Nadulai and June. Early epigraphic records of 10th to 13th centuries mentioned that Madhya Pradesh, the country lying between the Himalayas on the north, the Vindhyas on the south and Sindhu on the west were connected with various parts of Rajputana. The route from Delhi to Gujarat passed through Rajputana via Ajmer, Nadulai and Pali. Their trade routes maintained trade between different Rajputana as well as the other parts of India. The situation of Pali in the trade route made it a central place from where merchandise moved to Kashmir, China, Europe, Africa, Persia and Arabia in the 10th century. Jaisalmer enjoyed the position of transaction between the goods of eastern countries, the valley of Indus and those beyond that stream. Ajmer, Pushkar and Chittor continued to enjoy the position of big commercial centre. Akbar established political unity, stability and peace which had helped trade and commerce.

The Mughal rulers monopolized certain trade such as gun powers, lead, saltpetre etc. and gave monopoly to certain persons to trade in certain commodities like Indigo etc. The princes and queens were too engaged themselves in trade and earned immense profit. The emperor fixed the prices of market sales and monopolized the banking and, thus, earned a lot by issuing permit to banking business in the country. No horse could be sold without the knowledge of the king or their agents. The emperor could permit or check a traders from doing particular business by issuing a Forman (Sharma, 1948). No outsider could sell in the empire to make any business without the prior permission of the Emperor. Taverrier says "The King receives royalty from the merchants for permission. Another aspect of their trade policy was to encourage trade by giving facilities and privileges to European and Indian traders like remission in taxes, granting loan from Royal mints, sarais to give shelter and protection along the highways providing safety and security on the trade routes and instructing the officers of different administrative divisions not to harass the traders. The Rajput ruler encouraged trade by providing many facilities and privileges to traders in their states by they neither monopolized nor engaged themselves in any trade. Their policy was free trade policy i.e. open to all. Some of the important trade routes either passing through Rajputana or connecting it with different parts of India during this periods in Rajas than.

Agra to Ahamedabad Road

This is shortest and the easiest trade route from the Mughal capital of Agra to reach manufacturing city of Ahmedabad and the busy port of Cambay lay through Marwar. From Agra to west, the road passing through the loyal state of Jaipur to the Imperial city, Ajmer was safe. Thereafter, it crossed a gap in the Aravali barrier and runs through the Marwar cities of Pali and Jalore to Ahmedabad. Another route from Agra to Gujarat passed through Shambhar, Ajmer and Mandal Mewar. A third route from Agra to Gujarat Passed through Dausa Muilabad, Ajmer and Pali. There was yet another route which passed through Sikar, Fatehpur, Lindme, Chenigo (Chandangaon), Mugalsari, Nonnigong, Gamgram, Cherriat, Berat, Landancy (Ladana), Munsalde (Mozabad), Bralandany, Bramerande (Bonde Sindre) to Ajmer. From Ajmer the road led to Merta to Pipar Jogesgong (Jogikagaon), Severrange, Candempe (Khende), Joloure, Murtra (Merta) to Ahmedabad.

Delhi to Ahmedabad Road

This route is from Delhi to Ahmedabad passed Alwar through Jaipur, Ajmer, Udaipur and Rikhabdev. This route was taken by Ulughkhan in his attack on Somnath. It branched off at Ajmer and passed through Nadol and Pali and terminated at Ahmedabad. Ujjain to Marwar route : Maharaja Jaswant Singh left the battle field near Ujjain and reached Banswara, Ghatol, Salumber, Udaipur, Marwar via Sojat. Delhi to Dakhim route is from Delhi to Dakhim via Ajmer was commonly used by Maghal Emperors. Gujarat to Udaipur route: According to Abdul Fazal, Man Singh of Amber came from Gujarat by way of India to Dungarpur and then to Udaipur.

Agra to Chittor and Jodhpur Road

According to Abbas Khan these roads were constructed by Sher Shah. Jaipur and Ajmer was connected with road leading to Mewar. Amber, Pali, Merta, Siwara, Dunnara, Jodhpur and Pali, Sande Rav, Nathdwara and Chittor. Delhi to Bikaner route runs from Bhiwani, Hissar to Rajgarh in Bikaner. It forked into two, one leading to Bikaner via Rini and the other to Pali via Churu, Ratangarh, Nagaur, Phalodi, Osain, Mandore, Jodhpur. *Bikaner to Jaisalmer route.*: Maharaja Gaj Singh of Bikaner went to Jaisalmer for his marriage following the route through Kolayat Nokhera, Sirod, Bap, Khara, Ramdevra, Odhani, Lathi and Bhatdratalab as stated by Banke Dass and returned by another route which passed through Lathi, Pokaran, Phalodi, Jambha, Nokha, Chaknand and Bidasar. Other routes were from Bikaner to Sirsa via Rajgarh from Bikaner to Mewar via Pushkar and Nathdwara from Jaisalmer to Bhawalpur, Multan and Sindh via Barsalpur and Jaisalmer. Sindh, Gujarat, Hansi, Nagaur Bikaner, Krishnagarh, Udaipur, Kota via Chittor, Begun and Bhaisroadgarh were also linked. *Delhi to Jodhpur route* : This route passed through Sambhar, Ajmer and Merta. *Agra to Nagaur route*: According to Gulbadan Begum Humayan while going to Marwar took the road leading to Nagaur and hence he proceeded to Phalodi and stayed at Kul-i-Jogi a nearby pond. *Sirohi to Pokaran route*: This inter-state route passed via Phaloond, Gow, Balotra, Ramsin. Sironj to Bharatpur and Agra route :It was situated at equal distance from both Agra and Bharatpur. *Nagaur to Narnaul route* : Nagaur was connected via Didwana, Sikar, Udaipur, Mukundgarh, Jhunjhunu, Khetri within Narnaul in Haryana. *Ajmer to Sindh route* : This route runs via Nagaur, Jodhpur and Jaisalmer. *Pali to Jaipur route :* There were two routes running from Pali to Jaipur one via Parvatsar and Nairana Kuchaman, Sambhar. The other via Maroth and Jhadond Jodhpur was linked with Udaipur, Kota, Jaisalmer, Bikaner, Ajmer, Jaipur, Bundi and Sirohi, *Bikaner to Kota route* : Horse traders from Kabul and Kandhar passed through Bikaner and Marwar while going to Kota. Jaipur was connected with Bikaner via Maroth and Nagaur in Marwar and via Maroth with Amberkot. (Fig.13.1).

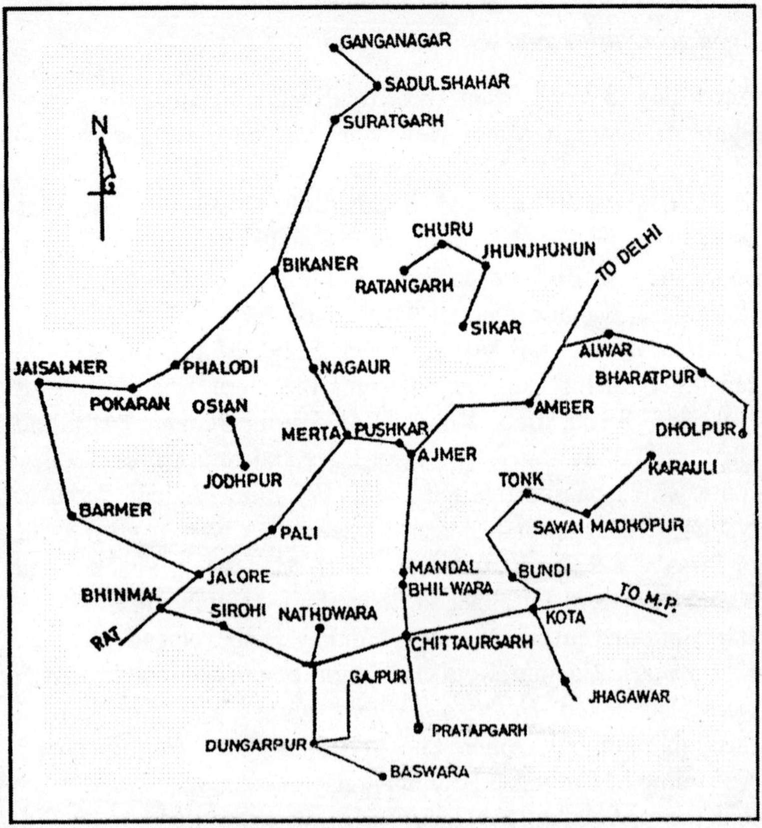

Fig.13.1 : *Rajasthan (Rajputana) Trade Routes of 10th Century.*

Besides these routes and Inter-State trade routes. There were other trade routes connected the capital of that particular princely state. Such as Jodhpur to Phalodi via Chainpur. Derijhar, Balharva, Biraj, Bhedri Tatal and Jolivado, Jodhpur to Sojat and Merta, Jodhpur to Pali, Pali to Jalore, Sanchore and Bhinmal. Bikaner to Rajgarh via Rini Churu, Bikaner to Anupgarh and Deshnok. Bikaner to Nohar via Bhather. The main routes in Kota were from Shahbad to Kota via Kishanganj, Baran, Anta and Bigod and from Shergarh to Kota via Raipur, Bilwara, Mandalgarh, Bijolia, Bundi, Kota, Darrah and Jhalawar. It is quite evident that the various cities of Rajputana were linked with Gujarat, Uttar Pradesh, Sindh, Multan, Kabul, Kandhar, Panjab, Haryana, Agra, Malwa, Mondu and Dakhin. (Fig. 13.2). The cities which were

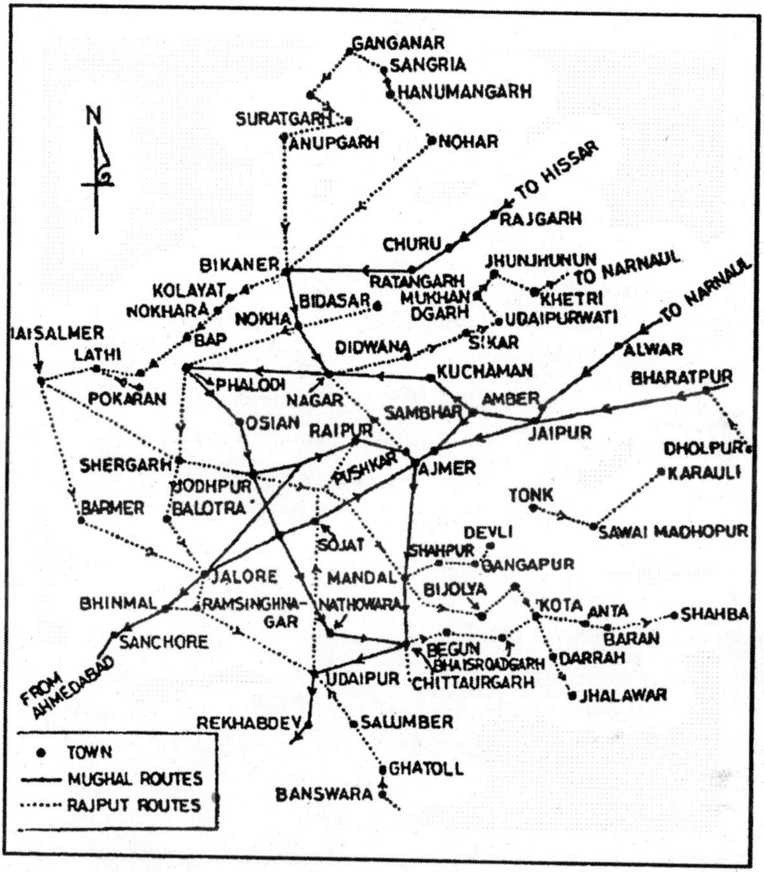

Fig.13.2 : *Rajasthan (Rajputana) Trade Routes of 18th Century.*

situated on the main routes or throught which many routes passed flourished as business centres in Medieval period. These routes or roads were simple tracks or paths demarcated either by tree or thorny bushes. There was no improved surface. The camels and oxen connected the various commercial markets of India and Rajputana. The lack of good road and means of transport were in hindrance to trade but the trade encouraging policy of Rajput rules more than compensated these drawbacks and resulted in the development of old commercial cities and growth of new commercial manis in different Rajput State. And henceforth, the construction of the roads was being undertaken by the imperial or State Government in a systematic manner. A motivation for construction of the roads in the study regions as conceived and interpreted by the author are either any one or a combination of the following :

1. To connect administrative centres e.g. Jaipur, Udaipur, Jodhpur, Bikaner, Kota, Ajmer etc.
2. To connect urban settlements e.g. Sujangarh, Khairthal, Dausa, Jhalrapatan, Amet, Bhinmal.
3. To connect places of archaeological and historical importance e.g. Peelibanga, Deeg, Amber, Chittorgarh and Kishangarh.
4. To connect places to tourist interest e.g. Mount Abu, Udaipur, Jaisalmer, Jaipur, Sawaimadhopur, Chittorgarh.
5. To connect cultural centres i.e. social, religious, education etc.e.g. Udaipur, Ajmer, Nathdwara, Jaipur.
6. Exploitation of Agricultural resources e.g. Sriganganagar, Alwar, Ramganjmandi, Bharatpur, Laxmangarh, Jhunjhunu etc.
7. Exploitation of minerals and industries and stones for construction purposes e.g. Khetri, Bhilwara, Kota, Jodhpur, Didwana, Makarana and Alwar.

Road Transport

On August 15, 1947 India achieved Independence with emerging the new outlook for the construction of roads. The State Government took over the charge of Nagpur Plan from British companies. India launched the Five-Year Development Plans but due to paucity of capital and underdeveloped economy of the state very few roads were added to the earlier network. The partition of the country also somewhat affected the progress of roads in Rajasthan (Table 13.1).

Roads in the Period of 1950-60

It may be regarded as the period of scattered nodes and the beginning of penetration lines (Taffee, Morill and Gould, 1963). During this phase, there were small scattered towns like Bhinmal, Sardrshahar, Osion, Raniwara, Hindaun, Lalsot, Anupgarh, Nohar,

Table 13.1 : *Road Network in Rajasthan*

Sr. No.	Periods	Road Length (kms)	Increase of Road Length	Percent of increase over the Proceeding period
1.	1950-60	3467	—	—
2.	1960-70	7809	4342	55.00
3.	1970-80	11777	3968	33.00
4.	1980-90	13671	1884	13.00

Source : P.W.D. Government of Rajasthan, Annual Progress Report, Jaipur.

Bhadra, Rajgarh, Banswara on the periphery of the state and short roads were connecting Jaipur-Focal point of transport on Delhi-Ajmer road, with administrative and the religious town both within the state and adjacent province to mark the beginning of the penetration lines (Table 13.1). The principal roads during this period as compiled from the record and an account of Public Work Department office and annual progress report of Rajasthan.

Roads in the Period of 1960-70

The total length of metalled roads was 3467 km. which comes to 79.84 per cent of the present road network. Besides these, there were other good kachha (unmetalled) roads e.g. the roads connecting Delhi with Jaipur passing through Alwar and running for 199 km within this State, Jaispur to Bassi, Jhunjhunu to Churu etc. (Table 13.1).

Period of Extension (1970-90)

This may termed as the period of extension of penetration lines. During this period the existing roads were extended. This period shows a slow progress in the construction of metalled roads. There has been an increase of 50.81 per cent (3968 km) road length over 1960-70 in Rajasthan State. The roads added in this period is showing in Table 13.1.

Period of Moderate Growth after 1980

The period is marked by development of new penetration and feeder lines connecting the important roads to towns mainly so far in accessible cultured nodes or the settlements in the agriculture regions. Such connection has given rise to unique phenomena of the development of end nodes at such advanced stage of the sequence of road development. It is evident that more than 50.81 percent (3968 km) roads are connecting only terminal nodes. The road length was increased 16.08 per cent (1894 km) during this period (Table 13.1)

Type of Roads in Rajasthan

The State has three major type of roads viz., national highway, state highway and major district roads. Among them, state high shares the maximum percentage having 10,006 km. on the other hand national highway has the minimum percentage (15.00) while major district roads has 31.00 per cent (Table 13.2). Therefore, the state badly needed the development of national highway at rapid rate. The detail of road is shown Table 13.2.

Table 13.2 : *Road types and length in Rajasthan*

Sr.No.	Roads	Length in km	Percentage	Name of Districts
1.	National highway	2998	15.00	Dholpur M.H.-3) Alwar, Jaipur, Ajmer, Udaipur (N.H-8), Bharatpur, Sawaimadhopur, Sikar, Churu, Bikaner, Tonk, Bundi, Kota, Jhalawar, Ganganagar, Jalore, Jodhpur, Jaisalmer, and Barmer
2.	State Highways	10,006	52.00	All Districts in State
3.	Major District Roads	6010	31.00	All Districts in State
	Total	19014	100.00	

Source : P.W.D., Government of Rajasthan Annual Report, Jaipur

The State has an average roads density of 3.99 which is much lower than the country average. Among all the districts 12 districts enjoy the average more than 5.00 sq. km. Ajmer has the maximum density (7.68) in the state, while Jaisalmer has the minimum density of the roads followed by Bikaner (1.80), Barer (2.67), Ganganagar (2.62) and Jodhpur (4.08). The density of metalled road in Rajasthan is shown in Fig.13.3).

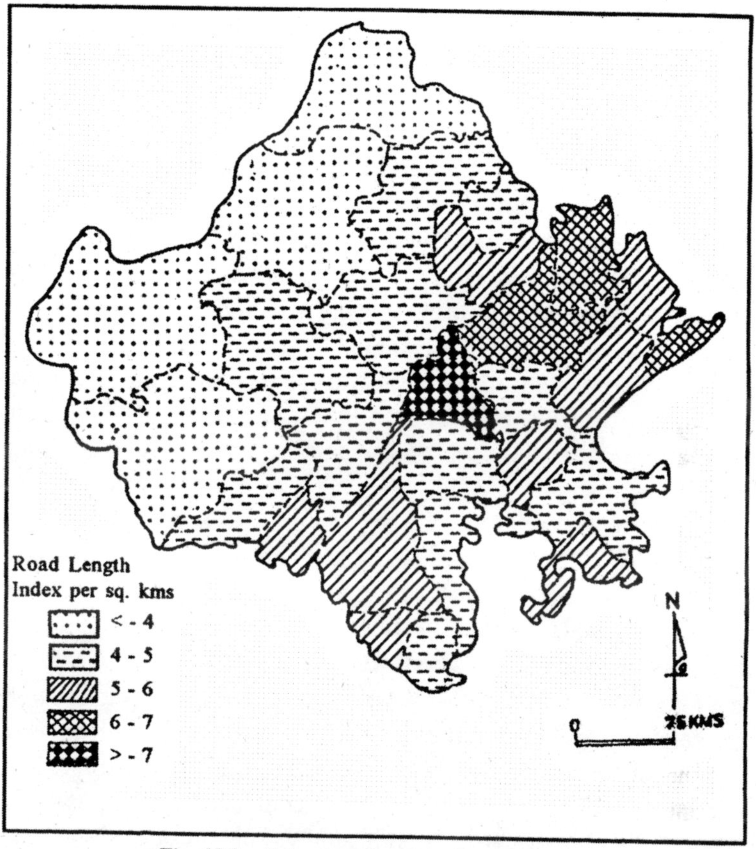

Fig. 13.3 : *Rajasthan Density of Metalled Road*

Railway Transport

The evolution of railway network of Rajasthan cannot be studied in isolation. It is closely linked with evolution of railway network of the country as a whole. Britishers introduced railway to our country in the middle of the 19th century. They planned the railway routes mainly to serve the commercial and political interest of the British Empire. Commercially the aim was to link agriculture and mineral rich areas with the ports so as to facilitate the export of raw materials and to import and distribute. British manufacturing and finished goods to the consumption areas. And therefore, those places were connected by railway which were suitable from the point of view or raw materials and also vulnerable for the finished goods imported from England. In this route system no care was taken to connect the continuous trade points nor were feeder lines constructed to promote the internal trade.

Thus the trunks lines and the main lines in Rajasthan are part of the National Railway network. The construction of these lines was entrusted to private foreign companies by British Government. The princely states also granted liberal concessions to these companies to construct railway in their own states wherever these concessions were not granted by the princely states, they remained neglected and the railway routes were not designed Benswara, Dungarpur, Tonk and Bundi in the eastern Rajasthan and Jaisalmer in Thar desert of western Rajasthan bear testimony to this neglect.

The construction of the railway prior to Independence was a three-tier effort and arrangement. The Central Government (British) was constructing penetration lines, states (Princely States) were responsible to construct lines in their own areas and the combined efforts of both constructed some feeder lines to establish inter-connections. After 1947 the construction of the railway became the sole responsibility of the Central Government. No railway line whether it be a trunk, main or feeder line can be constructed in continuation. These railwaylines were controlled by the physical, economic and political factors. The construction of bridges over rivers tunnels and road cutting in hilly areas, filling in low lands require much time and resources and thereby inhibit continuous construction. Similarly, the economic availability and the political control of different Government in a region also impeded the continuous construction of railway tracks. Hence, the chronological order of the evolution of the railway network is an inconsistent phenomenon in the evolution of transport networks. Till 1873, there was not railway line in the whole of Rajasthan (then known as Rajputana). Bharatpur holds a pioneer place in the advent of railways in Rajasthan. It was in August 1873 that the first railway line was opened between Agra and Bharatpur (Table 13.4).

The first railway line in Rajasthan was opened for traffic in August 1873 from Bharatpur to Agra, thereafter the skeleton of present railway network was initiated as below :

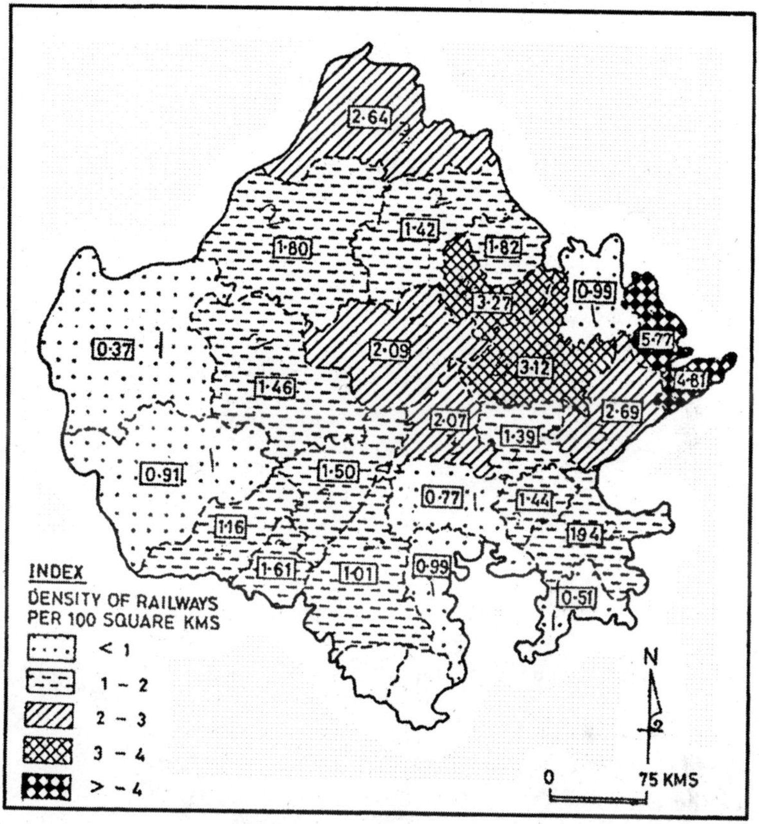

Fig. 13.4 : *Rajasthan : Density of Railways Route*

1. Delhi-Ahmedabad via Alwar, Bandikul, Dausa, Jaipur, Kishangarh, Ajmer, Marwar Junction.
2 Delhi-Mumbai (broad guage) via Bharatpur and Kota.
3. Agra-Karachi and Hyderabad (Sindh) via Bharatpur-Bandikul, Dausa, Jaipur, Phulera, Merta, Jodhpur, Luni and Barmer.
4. Great Indian Peninsular Railway passing through Dholpur.
5. The main lines from Ajmer to Khandawa via Nasirabad, Bhilwara, Chittorgarh, Nimbahera, Jodhpur-Biakner via Merta was constructed.

Distribution of Railway Transport

Rajasthan with an area of 3,42,239 sq km. served by 6424.3 km of railways. Out of this 2778.2 km are meterguage, 3532 km broadguage and 114.3 km narrow guage. The State is served by the western, eastern and central zones of the Indian Railway (Table 13.3).

Table 13.3 : *Distribution of Railway lines*

Sr No. Groups	Length in Kms	Percentage	District served
1. Meter Guage	2778	43.79	All districts except Balsara, Jhalawar and Bundi
2. Boad Guage	3532	54.61	Bundi, Bhilwara, Chittorgarh, Swaimadhopur Kota, Bharatpur, Dholpur, Bikaner, Sriganganagar
3. Narrow Guage	114.3	1.00	Dholpur
Total	6424.3	100.00	

Source : Ministry of Railways, Govt. of India, New Delhi.

Rajasthan has an average of 16.4 km of railway per 1000 sq km of territory accounting an average for India is nearly 2 km for the same territory. As the population in the State is sparse the average railway kilometerage in respect of population is better than the average in India. These average whether in Rajasthan or in India, do not give a clear account of the nature of the railway unless compared with other countries. When the average is compared with other developed and industrialized countries. It is realized that along with India this state is poorly served by a network of railway. At persent out of twenty seven districts in the state there are five districts headquarters which are not connected by rail and have railway stations at a distance ranging from 21 km to 85 km. The density of railway in Rajasthan is shown in Fig. 13.4.

During the period of quarter century (1900-1925) the existing lines were extended to complete moist of the trunk and main lines. Ahmedabad-Delhi trunk line via Phulera. Reengus was completed, Jodhpur-Bikaner track was extended to Ganganagar via Suratgarh and Hanumangarh, Degana-Sadulpur line was constructed in Nagpur and Churu districts to provide linkage with Jodhpur and Jaipur. Similarly, Jaipur was linked to Loharu (via Reengus, Sikar and Jhunjhunu. A bridge lies between Bikaner-Ratangarh was also constructed. Some feeder lines were also emerging e.g. Jodhpur-Phalodi; Anupgarh-Suratgarh, Kota-Barmer Dholpur-Bari and Sarmathura, Ratangarh-Sardarshahar etc. A total of 2061.82 km (33.1%) rail network track were added to the earlier railway network which was largely constructed in the north-central Rajasthan.

The period of 1925-1947 is known as slow growth period. The development of feeder line and beginning of the interconnection is marked by indifference on the part of British Government largely due to the momentum gained by the freedom struggle of the nation, thereby the progress in construction of the railways was slowed down. Several small new feeder lines viz. Bolotra-Pachparda-Salt, Tilwara-Tilwara fair, Makarana-Parbatsar, Sikar-Fatehpur, Pipar Road-Bilara, Samadri-Raniwara etc. were developed. The existing feeder lines and penetration lines were extended to establish inter-connections. i.g. Mauli junction-Marwar Junction across the Aravali ranges), Hanumangarh-Sadulpur (along Rajasthan Punjab border). Sawaimadhopur, Jaipur with the extension of the Sawaimadhopur-Sheodaspur line to Jaipur, Suratgarh-Ganganagar with the extension of Suratgarh, Raisinghnagar link to Ganganagar via Kishringpur. Thus, the construction of new railway lines during this period were mainly confined to Ganganagar, Churu, Jodhpur, Jalore and Udaipur districts. A total of 828.77 km railway track was constructed during this period which was 13.30 per cent during this period.

After Independence railway in Rajasthan emerged as sector for its construction . The Central Government took over the charge of Indian Railways from British Companies and princely states. India launched the Five Years development plans, but due to paucity of capital and underdeveloped economy of the states very few new railway lines were added to the existing network. The partition of the country into India and Pakistan also affected the progress of railways in Rajasthan. The Barmer Frontier Railway connecting Hyderabad (Sindh) and Karachi and Ganganagar-Bhawalpur (Pakistan) were terminated at Godra Road and Hindumalkot. During this period, only two significant extra regional linkages with Ahmedabad from Udaipur via Himmatnagar and from Rooiwara via Bhildi were established and two new feeder lines vix. Movli-Barisadri and Sanganer-Todarasingh were constructed. Table 13.3 reveals the new addition of railway during this period.

The railway trunks are unevenly opened over Rajasthan. Banaswara district has not rail line, Dungarpur, Jhalawar, Tonk, Sirohi and Jaisalpur find one railway line and few kilometer age. The main reason for this is that these areas either lie in the most arid and sandy part (e.g. Jaisalmer) or are located in hilly and undulating (e.g. Panswara, Dungarpur) terrains (Table 13.4) Such areas have sparse population and have been considered by Hartshorne as "Static and Kinetic" barriers in the development of transport network.

Table 13.4 : *Districts Headquarters Unconnected by Rail*

Sr. No.	Districts	Distance from nearest Railway stations in kms
1.	Banaswara	86
2.	Jhalawar	29
3.	Sirohi	22
4.	Tonk	32

Source : Railway Report, Ministry of Railway, New Delhi.

Conclusion

The above discussion presents the study of road and rail transport in Rajasthan state. It is safely said that routes are the prime source in the development of new roads in the study area. The state highways accounts highest percentage (52%) than National highways. In 18th century the roads in study area were linked only those nodes which had been unconnected by construction of roads (Table 13.2) The eastern and southern parts have highest density of roads shared by Ajmer and Udaipur districts. The state has an average road density of 3.99. The railway in Rajasthan served by better commercial centres to mobilize the available resources. It is obvious that broad guage has largest share of 54.61 crossing; nine districts in state, while narrow guage is shown its layout only in Dholpur districts accounting only (one per cent). It is seen that both roads and railway route in Rajasthan has developed in eastern and southern parts. This needs to be replanned with taking into consideration of latest development of commercial centres in these areas and more attention be given to west part in state to exploit the materials.

References

Appleton, J.H. (1962) : *The Geography of Communication in Great Britain*, Oxford University Press, London.

Avondo Bodino, G. (1962) : *Economic Application of the Theory of Graphs*, Gordon and Breach, Science Publishers Inc., New York.

Benson, D. and Whitehead, G. (1975) : *Transportation and Distribution*, W.H. Allen, London.

Briggs K. (1972), *Introducing, Transportation Network*, University of London Press, London.

Chand, R. (1980) : Measurement of Transport Connectivity and Economic Development of India, Unpublished Dissertation, Kumaun University, Nainital

Haggett, P. and Chorley, R.J. (1969) : *Network Analysis in Geography*, Edward Arnold, London.

Jain K.C. (1972) : *Ancient Cities and Towns of Rajasthan : A Study of Culture and Civilization*, Motilal Banarasi Das, New Delhi.

Joharapurkar, B.P. (1986): "The Growth of Road Network System in Pune District, "T.C. College, Baramati, District Pune.

Misra, V.C. (1962) : *Geography of Rajasthan*, National Book Trust of India, New Delhi.

Sexena, H.M., *Geography of Transport and Market Centres : A Case Study of Hodoti Plateau,* S. Chand & Co., New Delhi.

Sharma, K.K, (1975) : *Motor Transport : Its Development and Planning in Rajasthan*, Sterling Publishers Pvt. Ltd., New Delhi.

Tripathi, P.C. (1972) : *Rural Transport and Economic Development,* S. Chand &, Co., New Delhi.

SPATIAL VARIATION IN ROAD TRANSPORT IN ORISSA STATE

L.V. Singh

Introduction

The regional development of any country and more so in India is dependent on three important networks, such as irrigation, power, and transport; of which transport plays a vital and basic role in the process of regional-cum-economic development. Transport network keeps relation and makes integration among all other economic activities over space. It is presumed and observed that in any developed nation; the higher the economic development, the higher is the transport network development and there is always high positive correlation between these two. This particular analogy emphasized the role of road transport network in the context of overall economic development in the districts of Orissa. In a State like Orissa, where other means of transport are yet to be developed, naturally road networks play important role in connecting the inter and intra-district settlements. The road density alone cannot give the true picture of transport development of any particular region or district. On the other hand, the connectivity and accessibility are the two important aspects of road network analysis in addition to the shape indices such as *Alpha, Beta* and *Pi* etc.

An attempt is being made here to highlight the analytical framework to study the road network development; (preparing composite index out of some key indices of road network like road density; *Alpha, Beta, Gamma, Pi* indices) taking into consideration of the connectivity and accessibility of intra-district settlements. There are 50,845 settlements in Orissa. The study of connectivity and accessibility of each settlement is not at all possible as well as desirable. The settlements which have been most interacted for various functional and service purposes are obviously selected out of those settlements for the study of connectivity and accessibility indices. The centres or nodes related for road transport network study have been identified on the basis of the following characteristics :

(i) Urban and Industrial Centres;
(ii) Administrative Headquarters;
(iii) Regional Marketing Centres;
(iv) Important Tourist Centres;
(v) Centres for Higher Education;
(vi) Centres for the collection of mineral and forest produce; and
(vii) Port towns and fish collecting centres along the coast.

On the basis of above seven characteristics, there are 328 small and large centres which have been listed in Table 14.1 for each district. The settlements, chosen, for the study are distributed over the entire State. For the study of connectivity and accessibility, each area has been taken into consideration through these selected settlements. The level of development of a region (district) can be found out on the basis of the following indicators :

(i) Road density
(ii) Degree of connectivity between the settlements inside the districts
(iii) Beta Index
(iv) Alpha Index
(v) Gamma Index
(vi) Pi Index
(vii) Degree of Circuity
(viii) Detour Index.

The idea behind each of the analytical techniques is to apply some kind of measurement to a network, in order to provide a numerical value which can be used (i) to describe that network quantitatively, and (ii) to compare exactly the network with another. At the present stage of the development of the network analysis, however, there are two points to this. The first is that some indices give a high number for a well developed network while other indices give a low number for a well developed network.

(a) High value index indicates well developed network.

(i) Road density
(ii) Alpha Index
(iii) Beta Index
(iv) Gamma Index
(v) Pi Index

(b) Low value index indicates well developed network.

(i) Degree of circuity (ii) Detour index

Table 14.1 : *Districtwise of Indentified Transport Centres in Orissa*

1. Balasore District	Balasore , Bhadrak , Jaleswar Chandbali, Batgar, Huguli, Balipal, Singla, Basta, Rupesa, Remuna, Chandipur, Nilagiri, Soro, Anantpur, Basudevpur, Khaira, Benta, Tihidi, Dhamanagar
2. Bolangir District	Sonepur, Terbha Patnagarh Kentabandhi, Bolangir, Titlagarh, Belgaon, Bengomunda, Belpara, Lathor, Khapakhol, Saintala, Deogaon , Loisingha, Dugripali, Biramahorajpuir, Sindol Binke, Rampur.
3. Cuttack District	Cuttack, Jajpura, Road, Jajpur, Kendrapa Paradip, Chowdwar, Banki, Athgarh, Jagatsinghpur, Daitari, Sakinda, Panikoili, Gurudijhatia, Dharmasala, Chhtia, Barchana, Salipur, Indupur, Patamundai, Rajnagar, Aul,Rajkanika, Marshaghai, Machgaon, Ersama Balikuda, Tirtol, Niali, Kujang, Patkura, Binjharpur, Bari, Kabirpur, Mangalpur,Tigiria, Baramba, Narasinghpur, Damaparagarh
4. Dhenkanal District	Bhuban, Talcher, Angul, Dhenkanal, Athmallick, Kamakhyanagar, Pallahara, Hindol, Khamar,Condia, Kankadahad, Joranda, Tarava, Bhanpur, Motanga, Parjang, Mahidharpur, Bantala, Tikarapara, Handapa, Purnagarh, Purnakot, Thakurgarh.
5. Ganjam District	Gopalpur, Kavisuryanagar, Buguda, Sorada, Chhatrapur, Rambha, Polasora, Aska,Khalikote, Bhanjanagar, Bellaguntha, Tarsingi, Paralakhemundi, Purusottampur, Berhampur, Jagannathprasad, Ballipadar, Gadulabadi, Bedgadu, Adaba, Mohana, Chandiput, R.Udayagiri, Guma, Raygad, Ramgiti, Chikati, Digpahandi, Sharagada, Narendrapur, Taptapani, Sharagada, Kodala.
6. Kalahandi District	Bhawanipatna, Kesinga, Junagarh, Khariar Road, Khariar, Dharmagarh, Nowpara, Sinapali, Utkela, Golamunda, Santpur, Koksara, Ampani, Jayapatna, Thualmal Rampur, Gurupur, Bijepur, Lanjigarh, Korolu, Omat, Rampur.
7. Keonjhar District	Anandapur, Barbil, Joda, Keonjhar, Hatodihi, Patna,

	Ramachandrapur, Ghasipara, Ghatgaon, Dehnkikot, Kanjipani, Talkoi, Narainpur,Bainspar, Jhompra, Parsore, Bemebari, Bhadrasahi, Ghampus.
8. Koraput District	Jeypore, Sunabeda, Rayagoda, Koraput, Gudari, Nowrangpur, Gunupur, Kotpad, Malkangiri, Motu, Umarkot, Raigarh, Chandahundi, Dabugaon, Pappadhandi, Boipariguda, Ranigedda, Boriguma, Govindepalle, Balimela, Chittarkundu, Machkunda, Nandapur, Pottangi, Similiguda, Kakkiriguma, Narayanpatnam, Dasamantpur, Lakshmipur, Khumbhikota, Ramannaguda, Kolnar, Padampuram, Kashipur, Kalyansing puram, Bisam, Katak, Ambodella, Therubali.
9. Mayurbhanj District	Baripada, Rairangpur, Koptipada, Tiring Bahalda, Bodo Raruan, Barakhunta, Bisoi, Thakurmunda, Udala, Baisinga, Betnoti, Dantiamuha, Muruda, Chitarda, Similipalgarh, Kuliana, Deoli,Bangiriposi, Deoli.
10. Boudh-Khandmal District	Boudh, Phulbani, Saliguda, Manmunda, Kentamal, Gantapara, Tumudiband, Kotagad, Baminigam, Daringbadi, Niagaon, G.Udayagiri, Tikaballi, Phiringia,Madhpur, Gadimunda, Ranipathar.
11. Puri District	Puri, Bhubaneswar, Jatni, Khurda, Nayagarh, Gop, Astrang, Nimapara, Konark,Satyabadi, Delang, Pipli, Brahmagiri, Kanas, Phulnakhara Balugaon, Garh Krishnaprasad, Banapur, Odgaon, Nuagaon, Desapalla, Bhapur, Khandapara, Ganie Begunia, Bolgarh, Rampur, Kantilo Pipli.
12. Sambalpur District	Sambalpur, Brajarajnagar, Jharsuguda, Bargarh, Hirakud, Burla, Barpalli, Padampur, Kuchinda, Govindpur, Deogarh,Rairkhol, Paikmal, Jagadalpur, Melchamunda, Sohela, Bhatli Ambabhona, Dungri, Attabira, Gurbhaga, Dhama, Jujomura, Sason, Rengali, Janukira, Pravasuni, Barkot, Naikul, Raimal, Naktideula, Bagadihi, Laikera, Kolabira, Mahulpalli.
13. Sundargarh District	Rourkela, Biramitrapur, Sundargarh, Rajgangpur, Panposh, Koira, Barsuan, Bonaigarh, Nonipara, Banki, Kukurbhuka, Bargaon, Sabadeg, Talasara, Balisankara, Lifreepara, Himgiri.

The second disadvantage is that some of the ranges of values are not absolute; they have no fixed highest and lowest values, and the range values known at the moment results simply from examples which have been measured of the indices listed above. Only the Alpha index gives absolute values between 0 and 100.

Alpha, Beta, and Gamma Indices

One of the simplest application of graph theory has been its use in comparing sets of networks. By reducing the network to a topological graph, by selecting three basic measures and by manipulating the three measures in a series of ingenious indices, we can form some useful common yardsticks for network comparison. The three basic parameters on which the indices are based on (i) the number of separate (non-connecting) subgraphs in the network (G), (ii) the number of edges (links) in the network (E), (iii) the number of vertices (nodes) in the network.

Alpha Index

The alpha index is a ratio measure of the number of actual circuits, to the maximum number of possible in a given network. A circuit is defined as a finite, closed path in which the initial node of the linkage sequence coincides with the terminal node. In practical terms the existence of circuity means the establishment of additional or alternative paths between nodes in the network. The number of alternative path is determined by the number of linkages added to a minimally connected network. The range of the index is from a value of 0 for a minimally connected network to a value of 1 for a maximally connected one. For convenience, the numerical value may be expressed as a percentage of circuity in the network.

Beta Index

The simplest description is to relate the number of edges and Certices in a ratio E/V. This Beta index differentiates simple topological structures with low Beta values from complicated structures with high Beta values. The lower section of the Beta scale (from 0 to 1) differentiates between different types of branching networks and disconnected networks values of 1 and above differentiates circuit network.

Gamma Index

The Gamma Index is simply the ratio of the number of edges in a network to the maximum number possible in that network. Network. connectivity as measured by the gamma index is expressed in terms of a graph theoretic range that varies from a set of nodes having no inter-connections of one centre to a set of nodes in which every node has an edge connecting it to every other node in the graph. The connectivity of the network is evaluated in terms of degree to which the network deviates from an unconnected graph and approximates a maximally connected one. The numerical range for

Table 14.2 : *Computation of Alpha, Beta, Gamma and Degree of connectivity Indices*

Name of the District	No. of edges *(E)*	No. of vertices *(V)*	No. of subgraphs *(G)*	Characte-ristics of the graph	Degree of connectivity	Beta Index	Alpha Index	Gamma Index
Balasore	25	21	1	N.P.	11.90	1.19	2.63	43.85
Bolangir	20	18	2	-do-	13.01	1.11	2.94	41.66
Cuttack	43	38	3	-do-	06.11	1.10	1.20	39.81
Dhenkanal	29	24	1	-do-	10.50	1.20	2.37	43.93
Ganjam	53	33	1	-do-	11.03	1.60	4.63	56.98
Keonjhar	31	24	1	-do-	10.00	0.95	0.58	35.18
Koraput	18	20	2	-do-	05.93	1.12	0.85	79.27
Kalahandi	44	39	1	-do-	11.23	1.29	2.89	46.96
Mayurbhanj	37	24	1	-do-	14.40	1.54	5.53	56.06
Phulbani	23	19	1	-do-	13.45	1.21	3.26	45.09
Puri	43	31	1	-do-	09.24	1.38	2.98	52.43
Sambalpur	39	35	1	-do-	06.55	1.11	0.89	39.39
Sundargarh	21	20	1	-do-	11.05	1.05	1.16	38.88

Table 14.3 : *Computation of Pi, Detour and Degree of Circuity Indices*

Name of the District	Total length of the Network in kms.	Longest(shortest) path distance - distance between two points		Pi Index	Degree of Circuity	Detour Index
		Name of the Place	Distance in Kms.			
Balasore	1290.96	Chandbali toHuguli	199	06.48	0431.33	128.87
Bolangir	1470.48	Bengmunda to Sindol	162	09.07	0501.95	126.33
Cuttack	2744.29	Paradip to Narasinghpur	194	14.14	1239.02	138.12
Dhenkanal	2253.49	Palladip to Athmallik	163	13.82	1170.55	149.00
Ganjam	3188.93	Rambha to Ramgiri	217	14.69	1577.02	148.53
Keonjhar	1276.78	Hotadihi to Bamebari	172	07.42	0679.03	105.89
Koraput	3097.57	Khariar Road to Ampani	385	07.99	2916.31	144.97
Kalahandi	2097.57	Khariar Road to Ampani	178	11.78	0470.05	126.65
Mayurbhanj	2779.51	Tiring to Similipalgarh	159	17.48	0500.45	128.36
Phulbani	2002.13	Boudh to Kotagarh	169	11.84	1148.66	145.99
Puri	2651.22	Puri to Astrang	211	12.56	1055.21	138.31
Sambalpur	3715.48	Govindpur to Paikmal	271	13.71	2003.40	148.06
Sundargarh	2308.65	Barsuan to Himgir	226	10.21	0961.14	140.95

the gamma index is between 0 and 1. For convenience, the numerical value may be expressed as a percentage of connectivity.

Maximal Connectivity

The relationship between the number of nodes and the maximum number of linkages in a planner graph is always e = 3(V-3). The inclusion of the additional node to a network of more than 2 nodes increases the number of possible linkages by a value of 3.

Minimal Connectivity

Minimally connected network is one in which there are no isolated nodes and the number of connectivity linkages is less than the number of nodes. If the linkage is removed the network is divided into two completely separate parts. On the other hand, if one or more linkages is added to the work, the connectivity is increased beyond the minimal configuration in which only a single and unique path can be identified between all pairs of nodes. Additional linkage in a network create circuity.

Numerical representation of the indices are given below:

$$\text{Beta index} = \frac{E}{V}$$

$$\text{Alpha index} = \left(\frac{E-V+G}{2V-5}\right) \times 100 \text{ For planner graph}$$

$$\text{Gamma index} = \left(\frac{E}{\frac{V\,V-1}{2}}\right) \times 100 \text{ For planner graph}$$

$$\text{Alpha index} = \frac{E\,V+0 - (V-1)}{\frac{VV-1}{2}} \times 100 \text{ For non-planner graph}$$

$$\text{Gamma index} = \frac{E}{3(V-2)} \times 100 \text{ For non-planner graph}$$

It may be noted here that planner graph are characterised by edges which have no inter-sections or common points except at vertices while in non-planner graphs the condition does not hold.

Degree of Circuity (Route Factor)

One of the simplest measures proposed to describe path wandering is to relate the length of the observed path (O_L) to the length of the expected path (E_L), where this is measured as the direct "desire line" distance between the two ends of the path. For transport lines the terminal points are usually clearly defined and Kansky has proposed the gemeral formula for a connected network of lines.

$$\sum_{i=1}^{n}(O_L - E_L)^2/V$$ where V is the number of vertices connected.

For the purpose the formula has been adopted as –

$$\sum_{i=1}^{n}\sum_{j=1}^{n}(O_L - E_L)^2/V^2$$ because the district is taken as the areal unit for comparison.

Detour index : $\frac{\text{Shortest distance on network}}{\text{Direct distance}} \times 100$

$$= \frac{O}{E} \times 100$$

Degree of connectivity = $\frac{\text{Observed no. of routes}}{\text{Maximum possible links}} \times 100$

Maximum possible links = $\frac{n(n-1)}{2}$ where $n = v$

Pi index : Degree of development = $\frac{0}{d}$

$$= \frac{\text{The total length of the network}}{\text{The longest (shortest possible) path distance between two points}}$$

The computed values based on the above considerations are given vide Tables 14.2 and 14.3 for further analysis.

Level of Transport Development

The final selection of indicators has been done from the intercorrelation matrix. The indicators are : Beta index *(B)* , Alpha index *(A)*, Gamma index *(G)*, Pi index *(P)* and Detour index *(D)* Degree of connectivity and degree of circuity are neglested because of negative correlation with other indices. The inverse relation of these two variables with respect to others gave the same picture as before. Road density has been taken into consideration for the preparation of composite index along with above set of variables inspite of negative correlation only with gamma index. The reason is quite obvious as road density is the most important indicator to measure the transport development.

For constructing composite index, the technique of principle component has been used. This technique justifies its application as it does explicitly take account of the problem of malticollinearity among original indicators by orthogonalisation of the whole set of variables and also gives objective mathematical weightages for constructing a composite index by linear aggregation. The computational technique for finding out first principal component and composite index has been given in Table 14.4. Inter-correlation matric of the selected variables are as given in Table 14.4.

Table 14.4 : *Composition Technique for finding out first principal componant and composite Index*

Calculate the intercorrelation matrix ($n \times n$), where n is the number of development indicators taken in account. To get the values of composite index (CI) is the objective.

Step I : Calculate correlation coefficient matrix(rij)

Step II : Calculate (ni) = $\sum_{j=1}^{n} (rij)$

Step III : Find out the greatest number R*i* in the (R*i*) *q* vector and divide (R_i) vector by R*i* to get (Vi) vector.

i.e., $\frac{1}{R_i} \times (R_i) = (V_i)$

Step IV : Multiply r_{ij} by Vi to get New R*i*

i.e., (r_{ij}) X Vi = New Ri

Step V : Continue this process until a new Vi talleys with its just precedent Vi upto 3 places after decimal.

Thus (Vi) (New) is the stabilized characteristic vector.(Hotelling's method of iteration).

Step VI : Truncate the (Vi) after 4 or 5 decimal places, and denote it by (K*i*), The last iteration which has yielded find (V_i) has the relation (V_i) = $R_i(R_i)$. This constituent Ri is the largest principal component or the largest root or largest eigen value, denoted by *i*, that is final iteration $R_i = \lambda_i$. The total variation explained by this principal components equals $\lambda = \lambda i/N$, where N is the order of the inter correlation matrix.

Step VII : Calculate $\alpha = \frac{\lambda i}{\Sigma k_i^2}$, i = l, n

Step VIII : Calculate Mean and standard deviation of the indicators and denoted them by X_1, X_2 --X_n and SD_1, SD_2 --SD_n etc.

Step IX : Calculate α E *i*, *i* = 1, *n*

Calculate λi SD, *i* = 1, *n*

Calculate α K*i* X*i*, *i* = 1,n

Calculate α (K*i* X*i*)/(λi SD*i*),*i* = 1, *n*

Calculate $\sum_{i=1}^{n}$ (α K*i* X*i*)/(λi SD*i*) =L

Calculate $L\lambda i$ SD*i*,*i* = 1, *n*

Step X : (α Ki Xi)/(Lλi SD*i*) = (a_1), =1, *n*

a_i is the weightages of 1th indiviudal indicator.

Let the original set of variables be (X_{ij}), then assign weightage that is to say, multiply the matrix by the vector (a_i) and obtain the composite index(C_i).

$$(X_{ij})(a_i) = (C_i)$$

Sometimes two very important set of indications do not have positive correlation. Then for adding them up the following formula is suggested by Dr. M.N. Paul.

$$Ci = \frac{(Ai \times SDBi) + (Bi \times SDAi)}{(SDAi \times Bi) + (SDBi \times Ai)}$$

Where Ci is the composite index

A*i* and B*i* are the 1st and 2nd sub-composite indices of two groups of variables respectively.

SDA*i* and SDB*i* = Standard deviation of A*i* and B*i* vectors; A*i*, B*i* are mean of A*i* and B*i* vectors.

Table 14.5 : *Correlation Matrix*

	B	A	G	P	D
D	1.00000	0.89141	0.42634	0.66747	0.27707
A		1.00000	0.23699	0.54847	0.07070
G			1.00000	0.04724	0.33307
P				1.00000	0.41855
D					1.00000
Mean	1.22100	2.45400	47.65300	11.63000	136.16000
SD	0.18000	1.45000	11.14100	3.11300	12.03000
Eigen Vector	1.0000	0.88525	0.49883	0.82118	0.50311
Weightage	0.35371	0.07612	0.11125	0.15997	0.29693

Note : The largest root of the metrix (eigen value) is 2.68945. So total variation explained by it is 53.78%.

Multiplying these weightage values with respective indicators and adding up together we get a composite index, designated by CI. The equation is represented as : CI = 0.35371B+0.07812A+ 0.1112582+0.15997P+ 0.29693D. The final composite index CI*i* is derived (Table 14.6) by combining road density and respective CI as suggested by Pal

Table 4.6 : *Road Density and Compesite Index*

Name of the District	Road Density (R.D.)	CI	CI*i*
Balasore	20.19	44.80663	0.94688
Bolangair	16.51	44.23387	0.88583
Cuttack	24.47	48.19729	1.05791
Dhenkanal	20.81	51.95015	1.06067
Ganjam	25.45	59.72020	1.15304
Keonjhar	17.72	36.92397	0.79578
Koraput	15.49	53.60543	1.00888
Kalahandi	11.39	45.39696	0.82967
Mayurbhanj	26.69	47.81078	1.08386
Phulbani	18.08	48.66179	0.97340
Puri	26.09	49.63193	1.10215
Sambalpur	21.14	51.00089	1.05145
Sindargarh	23.86	48.27297	1.05029
Mean	20.60692	48.01678	0.99999
SD	4.46497	4.34865	0.10395

The classification of groups of development is done (Table 14.7 and Fig. 14.1) on the basis of mean and standard deviation technique.

Table 14.7 : *Level of Developmant*

Class Interval	Level of Development	Name of the District
0.79200-0.89604	IV Order	Keonjhar, Kalahandi, Boalangir
0.89604-0.99999	III Order	Balasore, Phulbani.
0.99999-1.10394	II Order	Koraput, Sundargarh, Sambalpur, Cuttack, Dhenkanal, Mayurbhanj, Puri.
1.10394-1.20789	I Order	Ganjam

From the analysis it is observed that the transport development does not keep relation with economic development in case of Orissa. The spatial variation in the context of transport development is due to topography and obstacles created by the presence of many rivers with large many tributaries and distributaries. The following example will illustrate in a better way. There are only three points of crossing (bridges) over the river Mahanadi at Sambalpur, at Sonepur over the Tel (tributary to Mahanadi) and the conjest bridge at Cuttack not sufficient to establish a better accessibility land connectivity for inter and intra district settlements. The coastal district which is identified as developed region having a very poor picture of transport development because of the fact that, this plain is interacted by a large number of distributaries of the rivers like, the Baitarani, the Brahmani and the Mahanadi. At many places in coastal district inland water transport is the important means of transport and road transport is yet to be developed. It can be safely concluded as it is at present that the spatial variations in road transport development does not affect directly much in the spatial variation of the economic development of the state perhaps due to being neglected for so long.

However, it should not be inferred from the above analysis that transport networks do not play significant role in rapid

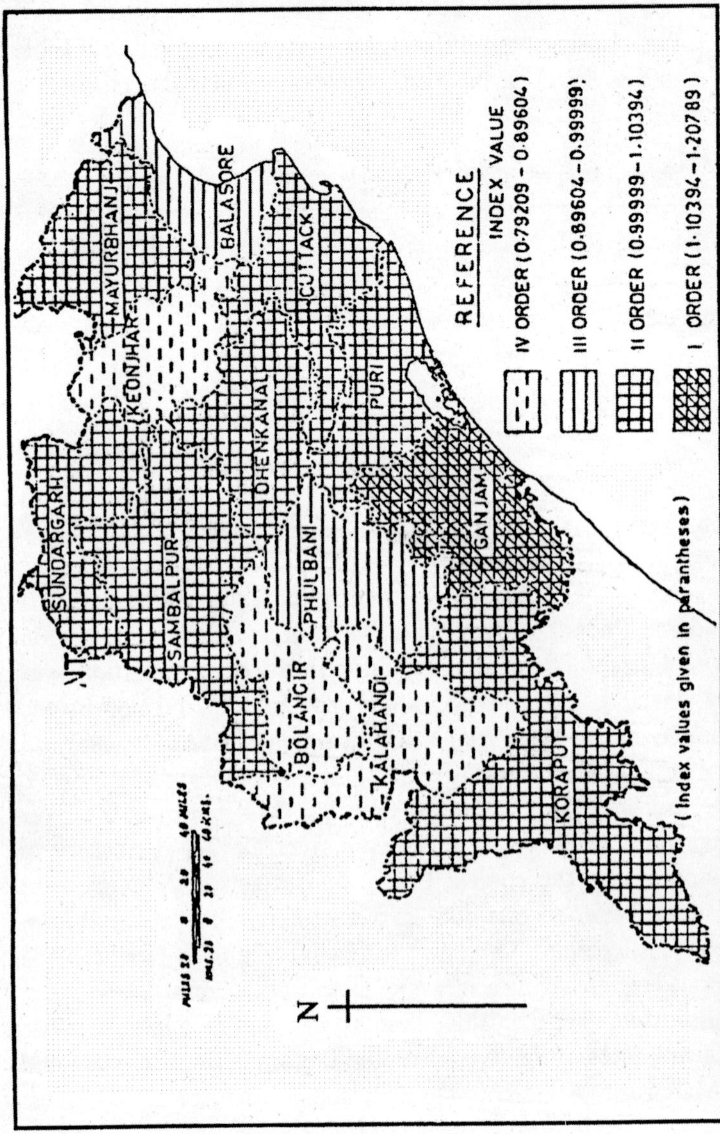

Fig. 14.1 : *Orissa Leavles of Development in Road Network*

development. The regional development and road transport net are highly and intimately connected. In a properly planned developmental strategy infrastructure plays a vital part and acts as a mode for mobility and functionality of economic operation both for public and private purposes. But the road transport network development in Orissa could not be much effective becasue of the present stage of agrarian economy (primary stage of development) of the state. The detail and efficient planning of networks required in the context of industrial-cum-urban phase (secondary) of development to have a better economic development in the state, for which Orissa is highly rich in scarce as well as abundant mineral resources, forest resources and also power resources of the country. The future development can be foreseen through road network planning taking into consideration of the facts discussed above.

References

Chorley, R.J. & Haggett, P., (1967) (ed.), *J :Integrated Models in Geography*, Methuen, London.

Gauthier, H.L., (1970) : "Geography, Transportation and Regional Development", *Economic Geography*, Vol. 46, October.

Haggett, P. & Chorley, R.J. (1969) : *Network analysis in Geography*, Edward Arnold, London.

Mandal, R.B. & V.N.P. Sinha (ed) (1980) : *Recent Trends and Concepts in Geography*, Voll. III, Concept Publishing Company, New Delhi.

Pal, M.N.(1975) :"Regional disparities in the Levels of Development in India", *Indian Journal of Regional Science*,Vol.7, No. 1,IIT, Kharagpur.

Patnaik, S. & Chattopadhyay, R.N.(1975) :"Spatial Variations in Levels of Development in Orissa", *Indian Journal of Regional Science*, Vol.7, No.1, IIT, Kharagpur.

Singh, D.K. Routray, B. & Jena, S. (1976) : "Spatial Analysis of Rural Settlement Pattern in Orissa", in *Geographic Dimension of Rural Settlements*, edited by R.L. Singh, K.N. Singh & R.P.B. Singh, National Geographical Society of India, Varanasi.

15

ROAD AND RAILWAY TRANSPORT DEVELOPMENT IN MAHARASHTRA

B.C. Vaidya

Introduction

Transport modes and means act as a circulatory system in any region. It provides basis for economic, social and commercial progress as well. And, therefore, facility of transport is considered as an index of gearing up local and regional resources at the best. The access in part either taluka or village level is easily possible by means of transport. In Maharashtra roads and railways plays a vital role in the developmental process. This is also true in reverse that the region may remain backward and less developed if there is lack of transport facilities. Thus both ways availability of transport reflect on the scenario of region. The pace of progress is entirely depended on transport facility for real development. Maharashtra state has extensively expanded road and railway routes which attracted people for their daily commuting to urban centres and good transport to countryside along with movement of people. With considering main trust of movement, author has attemped to assess and examine the trend and development of road and railway transport in Maharashtra State.

The present study is primarily based on data collected from Central and State Government published from 1951 to 1991. Author has sincerely attempted and assessed the trend of transport development in Maharashtra State. The focus is given on road and rail transport network. The road density pettern for per 100 km and per 100,000 persons have been computed and this later on, depicted in maps with suitable cartographic representation showing spatial dimension of transport in Maharashtra State.

Study Area

Maharashtra State lies in central west part in India extending from 16^0 40' to 22^0 10' North latitude and $17^0$6' to $50^0$9' East longitude. The areal spread of study region is 3,07,690 sq. kms. The study region is physically divided into different divisions and are closely related with transport modes. The region has three modes, namely, land, water and air. Land transport is extensively extended over the region except western part where Sahyadri mountain does not allow to easily spread for construction of roads and railway routes. Even then, railway and roads are well-connected with market centres, capital, district headquarters and taluka places. Not only this, but also route of roads have been reached to village level and plying buses for transport service. Such vast extension of means and modes of transport has to be studied intensively. And therefore, present treatise has been undertaken to assess transport development in Maharashtra State.

Transport Development in Maharashtra

The inter-regional circulation of people and commodity is operated by railways, buses aind different types of vehicles and to some extent by waterways and airways. Irregularity in surface and numerious physical barrier like relief features have given the hinder the lying of railway and road transport than others. And these have extended mostly in plain region in Maharashtra State.

Road Transport

The study region has a length of 1,76,446 km roads in 1991. This includes national highways, state highways, major district roads, other district roads and village roads. The road length in state has therefore increased during the three decades (Table 15.1). In 1961 it was 37,957 kms. which increased significantly to 1,76,446 kms in 1991. Table 15.1 reveals the types of roads development during study period. This table also exhibits notable changes occurred in layout of road transport in Maharashtra State.

Table 15.1 : *Road Length in Maharashtra*

Roads	Length of Road (kms.)				Percent
Years	1961	1971	1981	1991	
National Highways	2309	2445	2945	2979	1.68
State Highways	9539	14203	18949	30737	17.45
Major District Roads	10692	17684	25233	38527	21.83
Other District Roads	6664	11012	25404	38040	33.88
Village Roads	8753	20020	28105	59782	3.61
Total	37957	65364	51131	176446	100.00

Source: Statistical Abstract (1961, 1971, 1981 and 1991), Directorate of Economics and Statistics, Mumbai.

There are ten national highways acrossing the state territory. Out of these, one national highway No. 50 connecting Pune-Nasik of 192 kms length is entirely located in Maharashtra state (Table 15.3). In terms of spatial distribution, highest occupying areal extent of National Highways in western part (Fig. 15.1) appears to be radiating toward Mumbai metropolitan city namely, National Highways No. 3, 4 and 8 (Fig.15.1) These routes of national highways in study region connect major cities and market centres. It is also observed that only seven districts in state did traverse national highways, namely, Aurangabad, Beed, Nanded, Parbhani, Chandrapur Jalna and Latur while few districts like Kolhapur and Sangli are either marginally touched by one or other national highway. The present national highways act as lifeline of industrial development of the state. It is noted that the industries of Pune, Kolhapur are entirely depended on road transport than railway shipment. Dhule-Kolkata highway connects with district centres like Jalgaon, Amravati, Nagpur and Bhandara traversing of 686 km while Mumbai-Bangalore and Panvel-Mangalore highways are traversing to 371 km. and 482 km. respectively linking Konkan.

In study region, state highways and major district roads provides basic internal cirulation for transport. It is vital shipment of goods for industries. The raw material, capital and manufactured goods are transported in state primarily by road transport from major sources of production to market places. It is found that state highways and major district roads have constructed with considering

Table 15.2 : *Divisionwise Road Length in Maharashtra*

Sr. No.	Divisions	Road Types / Years	Road Length in kms. 1961	1971	1981	1991
1.	Mumbai Division	National Highways	757	851	1344	1303
		State Highways	2463	3682	4623	8108
		Major District Roads	3020	4696	5976	8371
		Other District Roads	2904	4185	6556	8879
		Village Roads	5149	8260	9123	15837
		Unclassifed Roads	Nil	Nil	10151	4606
		Total	14296	21674	37725	47104
2.	Pune Division	National Roads	731	829	761	787
		State Highways	2471	3262	4821	7246
		Major District Roads	5063	7026	8286	12910
		Other District Roads	2706	4232	8700	13415
		Village Roads	2678	8148	8236	19054
		Unclassified Roads	Nil	Nil	17973	178
		Total	13650	23497	48777	53590
3.	Auranga-bad Division	National Highways	79	79	79	79
		State Highways	1943	3122	4923	6134
		District Roads	674	2293	5419	7584
		Other District Roads	96	1019	5959	5980
		Village Roads	194	1786	5945	10196
		Unclassified Roads	Nil	Nil	4073	70
		Total	2987	8299	26396	30675
4.	Nagpur Division	National Highways	740	686	761	810
		State Highways	2660	4137	4577	9249
		Major District Roads	1933	3669	5552	9662
		Other District Roads	757	15576	4189	9766
		Village Roads	730	1826	4801	14695
		Unclassified Roads	Nil	Nil	8298	895
		Total	6822	11894	28180	45077
5.	Mahara-shtra State	National Highways	2307	2445	2945	2979
		State Highways	9539	14203	18949	30737
		Major District Roads	10691	17634	25233	38527
		Other District Roads	6464	11012	25404	38040
		Village Roads	8752	20020	28105	59782
		Unclassified Roads	Nil	Nil	40495	63381
		Total State	37756	65364	141131	176446

Source : Statistical Abstracts, Maharashtra State, Mumbai

gravity and handling goods and movement of people in study region. In addition to this other district roads and village roads provide better linkages between talukas and villages in state for better movement of local people and goods. Table 15.2 reveals road development (1961-91) in state where divisionwise national highway, state highways, major district roads, other districts roads and village roads are computed. It is also computed the trend of roads length in study region. It is seen from Table 15.2 that route length of National highways in Mumbai division (1303 km.) followed by Nagpur division (810 km.) whe as lowest length of national highways observed in Aurangabad a vision (79 km.) traversing only in Osmanabad district. (Table 15.3 and Fig. 15.2).

Table 15.3 : *National Highways and their Length (1990-91)*

Sr. No.	National Highway Number	Routes	Length (kms.)
1	3	Mumbai-Nasik-Agra	391
2	4	Mumbai-Bangalore-Chennai	371
3	4b	Navasheva-Kalamboli-Palspe	27
4	6	Dhule-Nagpur-Kolkata	686
5	7	Varanasi-Nagpur-Hyderabad	232
6	8	Delhi-Jaipur-Ahmedabad	128
7	9	Pune-Solapur-Hyderabad-Vijaywada	336
8	13	Solapur-Vijapur-Chitradurg	43
9	17	Panvel-Goa-Mangalore	482
10	50	Pune-Nasik	192
		Total	2888

Source : Statistical Abstract, Govt. of Maharashtra, Mumbai.

The length of road is one of the reliable indicator of transport development in any area as density per unit area provides the more basic variable to describe actural existence on the ground of specialized routes whether those routes be intensively or lightly used. (Ginsburg, 1961, Hogget, 1979). Taffee, Morrill and Gould (1963) while, investigating the distribution or road density of Ghana and Nigeria have also supported strongly suggesting density of route as major variable. In study region, value of this indicator ranges from 26 km length per 100 sq. km in Gadchiroli district to 84 km in

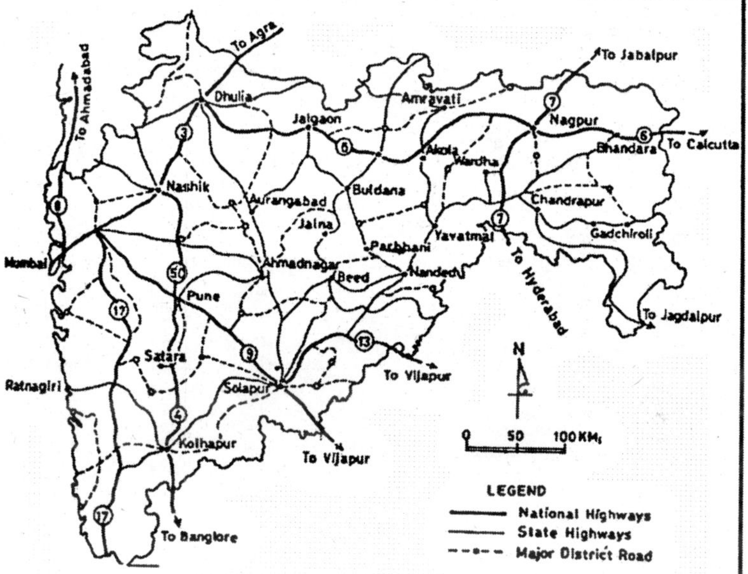

Road

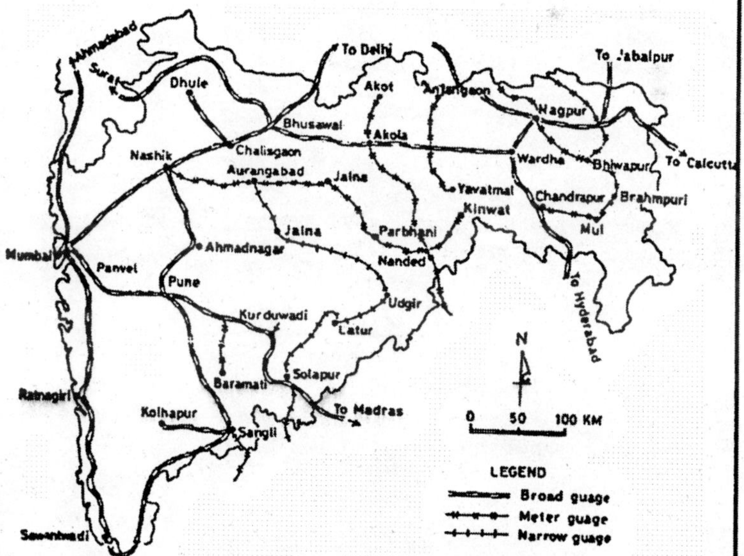

Rail

Fig. 15.1 : *Road and Rail Transport in Maharashtra*

Table 15.4 : *Length and Density of Roads in 1991*

		Road Length in km.				
Sr. No.	Districts	Metalled Roads	Unmetalled Roads	Total Length	Road Length per 100 per square kms	Road Length per 1,00,000 persons
1.	Greater Mumbai	49	—	—	—	—
2.	Thane	2020	3331	5351	56	102
3.	Raigarh	1849	2202	4051	57	225
4.	Ratnagiri	1653	3249	4902	60	326
5.	Sindhudurgh	1254	2058	3312	63	414
6.	Nashik	2408	10219	12628	81	322
7.	Dhule	1593	6895	8498	64	339
8.	Jalgaon	1618	6711	8329	71	260
9.	Ahmednagar	2550	9361	11911	49	350
10.	Pune	2519	7039	9558	61	173
11.	Satara	2235	6489	8724	84	363
12.	Sangli	1664	6201	7865	92	357
13.	Solapur	2136	7732	9868	67	308
14.	Kolhapur	1718	3946	5664	75	188
15.	Aurangabad	1102	4120	5222	51	237
16.	Jalna	732	2425	3157	41	225
17.	Parbhani	1221	4550	5771	52	274
18.	Beed	1603	3902	5505	51	305
19.	Nanded	1423	5534	6957	66	302
20.	Osmanabad	1329	2734	4063	54	312
21.	Latur	1401	2728	4129	58	242
22.	Buldana	1297	2445	3742	39	196
23.	Amravati	1326	3425	4751	39	215
24.	Yeotmal	1428	3758	5186	38	259
25.	Wardha	668	1965	2633	41	263
26.	Nagpur	1157	4374	5531	56	167
27.	Bhandara	797	4595	5392	58	256
28.	Chandrapur	925	4629	5554	49	308
29.	Gadchiroli	658	3139	3797	26	474
	Total	43846	132600	176446	57	221
	Percent	(24.87)	(75.15)	(100.00)		

Source: Statistical Abstracts, Maharashtra State, Mumbai.

Satara district. The average length of road/100 sq. km measures about 75 km. within twelve districts above the averages and remaining districts lie below on the basis of the indicator the district road length has been identified in their categories, number of districts and name of distric in Maharashtra state. Table 15.5 shows that there are four districts, namely, Nashik, Satara, Sangli and Greater Mumbai which have more than 80 km. of road/100 square km. The main reason behind this is that one of four is state caiptal and other three have comparatively greater number of market and service centres and therefore enhancing more transport connectivity. The 60 to 80 range of value lies in seven districts located in western and central parts in state causing development of roads (Fig 15.2). The reverse is found in case of medium and low range of value indicating less development.The lowest range is observed in four districts, namely, Amravati, Buldhhana, Gadchiroli and Yeotmal, except Buldana district, other three districts are partially tribal and backward and hence is less developed in the state showing low range of value of less length of road/100 square km. in the study region.

The Fig. 15.2 reveals the spatial distribution of length of road per 100,000 persons in Maharashtra. The density of road measured in terms of population is also significant and one of the most reliable measures of transport in the region (Berry, 1960, Taffee, Morrill and Gould, 1963). On the basis of the length of road/one lakh population, the region has identified into four category. It is evident from Table 15.6 that two districts have highest scores on this indicator, namely, Sindhudurgh and Gadchiroli due to backward nature of the districts. The high scores is identified in eleven districts located in western and central parts in the study region. Ten districts ranges 200-300 value labelling as medium. The very interesting and more explicitly note that 100-200 range value appears to seven districts in Maharashtra which have larger population in number either by industrially well developed like Greater Mumbai or rapidly increase in urban sprawl around the cities and, therefore, this district has lowest value of range scores more than in the study region. Other six districts lie near the big cities or districts headquarter showing high range of value. These districts are namely, Thane, Pune, Kolhapur, Buldhana, Akola and Nagpur.

Table 15.5 : *Districts and Road Length*

Category	Length of Roads/per 100 square km		Name of Districts
	Range of Value	No. of Districts	
Low	20-40	4	Amravati, Buldhana, Godchirali, Yoetmal
Medium	40-60	15	Thane, Raigarh, Ratnagiri Ahmednagar, Aruangabad, Jalna Osmanabad, Latur, Akola, Wardha, Nagpur, Bhandara, Chandrapur, Parbhani, Beed
High	60-80	7	Sindhudurga, Dhule, Jalgaon, Pune, Solapur, Kolhapur, Nanded
Very High	>80	4	Nasik, Satara, Sangli, Greater Mumbai

Source : Author.

Table 15.6 : *Districts and Road Length*

Category	Length of Road per 100,000 persons		Name of Disticts
	Range of Value	No.of Districts	
Low	100-200	7	Thane, Pune, Kolhapur, Buldhana Akola, Nagpur, Greater Mumbai
Medium	200-300	10	Raigarh, Jalgaon, Aruangabad Jalna, Parbhani, Latur, Wardha, Amravati, Yeotmal, Bhandara
High	300-400	11	Ratnagiri, Nashik, Dhule Ahmadnagar, Satara, Sangli, Solapur, Beed, Nanded, Osmanabad, Chandrapur.
Very High	>400	2	Sindhudurgh, Gadchiroli

Source: Author.

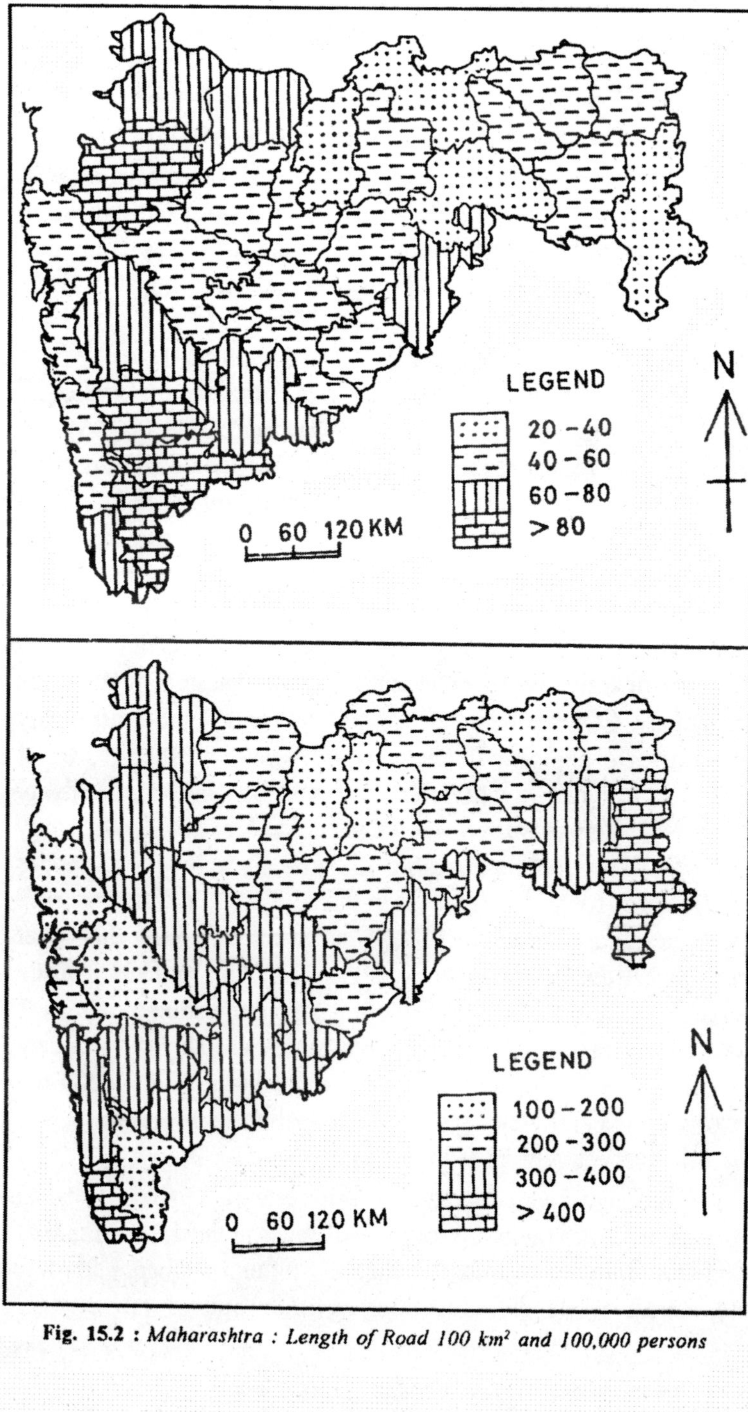

Fig. 15.2 : *Maharashtra : Length of Road 100 km² and 100,000 persons*

Railway Transport

Railway transport is imperative for handling shipments goods and movement of people. It is seen that there are four railway division traversing areal extent in Maharashtra, namely, shoutheast, central, south central and western railway. The total route length of rail in state accounted for 5,440 km. in 1991. It is historically noted that this state is recipient honour to start rail in 1853 carrying 400 people covering distances of 21 miles between Mumbai to Thane in about 75 minutes. The train took off with a regal 21 gun salute to celebrate this occasion. It is observed from Table 15.8 that state has speedily expanded during last fourty years. It was noted that in 1961 rail route was 5,057 km, has increased to 5,442 km in 1991. The notable changes has been occurred in case of broad guage length from 2,767 to 3,345 km. in 1991. This was mainly increased due to conversion of metre guage into broad guage and consequently metre guage and narrow guage have been decreasing later on (Table 15.8). There are four railway divisions in Maharashtra owned by Ministry of Railway, Government of India. These four divisions are namely, Central, Western, Centralwest and Southeast railway divisions. The highest percentage of railway routes lies (53.67%) under central railway division followed by southeast railway (27.31%), whereas western railway (Table 15.7) division has only 6.72 per cent route length. Table 15.8 reveals railway route types and trends of route development in the study region. It is observed that there is rapid growth of broad guage railway route that other increasing from 2,767 km in 1961 to 3345 km. in 1991 sharing 61.18 per cent in the study region. It is also interesting to note here that narrow as well as metre guage routes are slightly reducing due to conversion into broad guage route in study region. Both narrow guage and metre guage are contributing 18.18 and 19.43 per cent respectively in Maharashtra.

The significant increase in rail network appears to be in Mumbai as centre radiating routes to all parts in India, i.e. Mumbai-Bangalore railway, Mumbai-Delhi and Mumbai-Ahmedabad route and, therefore, Mumbai is working as trunk route of railway.

Table 15.7 : *Railway Length in Maharashtra*

Sr. No.	Railway Divisions	Types of Route in km. Broad Guage	Meter Guage		Narrow Guage	Percent of Total Route
1.	Western Railway	366	Nil	Nil	366	6.72
2.	Central Railway	2250	Nil	670	2920	53.67
3.	Southern Central Railway	760	1000	Nil	1486	27.31
4.	Southeast Railway	254	Nil	414	668	12.20
	Total	3356	1000	1084	5440	100.00
	Per cent	(61.7)	(18.4)	19.9)	100	

Source : Author

Table 15.8 : *Railway Routes in Maharashtra*

Sr. No.	Types of Routes	Route Length in kms Years 1960-61	1970-71	1980-81	1990-91	Per cent
1.	Broad Guage	2767	2752	3131	3345	61.18
2.	Meter Guage	1197	1308	994	1024	18.88
3.	Narrow Guage	1093	1102	1100	1053	19.43
	Total	5057	5162	5225	5422	100.00

Source : Statistical Abstracts, Directorate of Economics & Statistics, Govt. of Maharashtra, Mumbai, 1961, 1971, 1981 and 1991.

Mumbai's railway acts to link to almost all big cities and market centres in India and it serves both as national and international centre for movement of people and commodities. Not only other state capital places are linked by rail route to Mumbai but also district places within Maharashtra linked with Mumbai, namely, Aurangabad, Pune, Nagpur, Kolhapur, Nanded, Jalgaon, Wardha, Amravati, Akola. These centres are also working as regional links with state territory. The new Konkan railway network of 392 kms. in state started in 1996 could be used for exploitation of resources of entire western Maharashtra giving opportunity to opened new industries based on local resources available in this region. It is significant to note that railway existence in Konkan in further would carry unmistakable imprint on industrial landscape in Konkan area in study region. Mumbai-Pune and Kolhapur and Nagpur clusters of Maharashtra state are located on railway lines. These railway junctions and routes proved easy access to different urban and industrial development and thus Maharashtra state is oriented toward railway lines. This is comparatively visible and better in study region than elsewhere.

Conclusion

The present attempt has been made to assess and examine trend and development of road and railway transport in Maharashtra. It is clear from indepth discussion that existensive routes have been constructed in three-fourth part in study region except the natural barrier area of Sahyadri mountain. Both road and railway provide better facilities of the study region as they connect major cities and district headquarters. This is more true in case of national highways and state highways (Table 15.2 and Fig. 15.1). The total length of road route is accounted for 1,76,446 km. sharing 79.19 per cent transport. Railway route in study region is also well connected with market centres and big cities and it is radiated toward Mumbai, state capital. Such linkage with state capital by railway give access to speedy movement of people and commodities. The total lengh of railway routes in Maharashtra is 5443 km. Four divisions of railway, namely, western, central, southcentral and southeast are available to people. It is observed that the conversion of narrow guages to broad guage is in progress and, therefore, it reduces in length rapidly. Thus it can be safely said that both road and railway are useful to movement of people and commodities in study region.

References

Bhagwati, A.K. and Sharma, B. (1984) : "Urban Centres and Spatial Pattern of their Road Accessibility in Assam", *Geographical Review of India*, Vol.13, No. 3, pp. 14-18.

Govt. of India (1993) : India, Ministry of Information and Broadcasting, New Delhi

Hibbs, J. (1970) : *Transport Studies : An Introduction*, John Baker, London.

Pandey, N. P. (1990) : *Geography of Transportation*, Inter-India Publication, New Delhi.

Ray, S.K. (1995) : *Transport Planning for Developing Countries*, Prentice Hall, New Delhi.

Rao, P.D. (1988) : *Trend in Indian Transport System*, Inter-India Publication, New Delhi.

Saxena, H.M.(1975) : *Geography of Transport and Market Centres*, S. Chand and Co. Pvt. Ltd., New Delhi. pp.25-211.

Singh, R.P. (1960) : Transport Evolution of Punjab, *National Journal of India*, pp. 12.

Singh, D.N. (1989): Role of Transportation in Rural Development : A Case Study of Basti district. *Geo Science Journal*, Vol. pp. 19-39.

Singh, D.N.(1991) : Transportation and Agricultural Development : A Case study of Gorakhpur Division, *Geo Science Journal*, Vol, 6, pp. 71-79.

Singh, D.N.(1997) : Prespective on the Role of Transport in RegionalDevelopment in book edited by Dr.B.C. Vaidya, Reading in *Transport Geography : A Regional Prespective*, Devika Publication, Delhi.

Singh, J.(1968): Transport as a Factor in Regional Planning with particular Reference to India in R.L. Singh (Ed.) *Applied Geography*, National Geographical Society of India, Varanasi.

Thorner,D.H. (1963) : Pattern of Railway Development in India. *The Far Eastern Quarterly*, New Delhi, pp. 212-304.

Vaidya, B.C. (1998) : Transport Network and Planning Strategy for Backward Region in Maharashtra state. A Case study of Gadchiroli district in edited book on *Readings in Transport Geography – A Regional Perspective* edited by Dr. B. C. Vaidya, Devika Publication, Delhi.

White, H.P. and Senior, M.L. (1983) : *Transport Geography*, Longman Group Limited, London.

16

IMPACT OF TRANSPORT DEVELOPMENT ON INDUSTRIALIZATION IN GOA

G.S. Badigar and *K.R. Badigar*

Introduction

This study deals with the development of road transport system and its impact on changing industrial structure in Goa state especially after liberation (1961). The period from 1961 to 1991 has been taken as the study period with the objective to understand the changes in the transport characteristics and the nature of relationship between the transportation and industrialization and consequent spatial disparity in Goa. It is the development of transport system in particulars, has played a pivotal role as an artery for moving men and material betwen the region for the recovery of economic health of Goa from devastated state of economy, specially at the end of 1961, when portuguese retreated from Goa. Presently, Goa represents as one of the fast industrialising states with the second highest per capita income in India. This state of economy cannot be thought of in isolation without the efficient development of transport.

Geographically, Goa is small with an area of 3702 sq. km. It is bounded by Western Ghats in the east, north and south and by Arabian sea in west. Drained by perennial rivers originating from Western Ghats. Generally, the area is hilly, except in parts of Bardez taluka and Salcete taluka. It has rich mineral resources such as iron, manganese, bauxite etc. Goa state experiences tropical monsoon and equable climatic condition. Receives maximum rainfall from summer rains between June and September, average annual amount of rainfall is 300 cms.

Administratively, Goa is divided into two districts Namely, North

Goa and South Goa. It comprises in all 11 Talukas six in North Goa and five in South Goa. The total population of the state in 1991 was 11,69,793. The state comprises 41 per cent of urban population and 59 per cent of rural population. It enjoys high literacy and the second highest per capital income in India. Panaji is the capital city of Goa situated on the left bank of river Mandavi.

Data Sources and Methodology

The present study is relying on secondary data. Major sources are, Statistical Pocket Books, District Census Handbooks from 1961 to 1991. The *Gazetteer of India*, Union Territory, Goa Daman and Diu (Part-I). A lot of data have also been collected by conducting survey especially data concerned to industrial characteristics. Industrial and commercial directory of Goa is another secondary source of date regarding industry. This study is divided into three parts. The first part is the analysis of development of road transport. To understand the growth trend of transport and its efficiency, structural indices of transport network such as : Alpha, Beta, Gamma and Cyclomatic number were calculated. In order to understand accessibility surface and nodal accessibility, the shimbel index dispersion index method had been employed in the second part, degree of industrialization has been calculated with reference to the number of work force per 1000 inhabitats.

The third part assesses the growth of transport and industry for that network efficiency indices have been composited, that is: "mean value has been calculated for each indicator the values of all the indicators for each year have been divided by their respective mean-values thus arrived at scale free figures. These scale free value were added for each year which gave composite index". Same method is repeated to calculate composite indices for industrial components in order to establish the association between transport accessibility and industrial characters. The coefficient of correlation has been employed. Maps and diagrams are also used to give the visual impression of spatial data.

Development of Transport

The transport system was quite inefficient and far from being satisfactory upto 1961. The earstwhile Portuguse Government was interested just in harnessing the local resources, so for that convenience some segments of loads were developed. But orientation was not for social convenience. In pre-liberation period more attention was given only to water transport, since, it was less expensive and locally advantageous.

The advent of road transport in modernsense is quite late and is a characteristic of post-liberation in Goa (Fig. 16.1). After liberation Goa has become an integral part of India and is linked to the neighbouring states and important cities of India by highways and other modes of transport. The development of roads and vehicular traffic broke the comparative isolation of the area. Locations of various elements on geographical space have relatively been altered and reorganized because of increased efficiency of the overall transport system.

Geographically, Goa state is very small and in such a micro region, the road transport system (alone) plays a pivotal role in handling both passengers and goods traffic. The Table 16.1 gives an idea about the pace of road transport development in Goa through various components such as length of road, number of vehicles and bridges (1960-1995). The demand for transport has increased rapidly after the liberation. In response to this, provisions were made through different plan periods to meet the demand by improving, straightening, re-aligning, widening, bridging, etc. of the existing transport network and constructing new roads as well.

The length of the road in the region has increased from 1,655 km. in 1962 to 3,865 km. by 1971 and rose further to 6,230 km. by 1985. In the next eight years time that is by 1993 the length has stretched to the ever higher figure of 7543 km. It is an increase, slightly more than 4.5 times for the period of 30 years and is a remarkable growth. The construction of bridges through the study period has also registered a substantial growth. In the year 1962, there were only 13 major bridges, which rose to 38 by 1971, the figure, nearly trippled within nine years of time. By the years 1981 the total number of bridges has risen to 42. Further it rose to 47 by 1995. Apart from all these, four more major bridges are under construction and will be opened shortly.

In case of vehicle traffic, it is the teaming number of vehicles

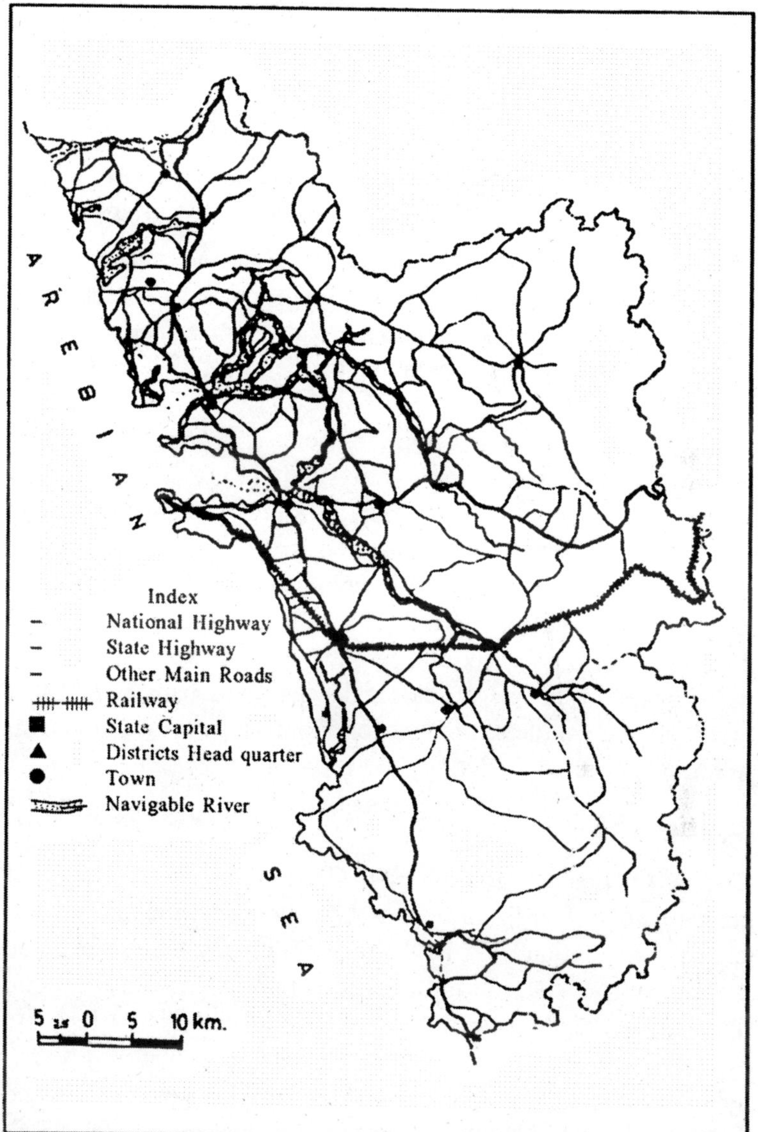

Fig. 16.1 : *Goa Transport (1990)*

with revolutionary trend. During pre-liberation period, vehicle number was just 8,531 which dwindled down to 7,923 by 1965, the reasons, however, are not known. From the 1971 the real pace of road transport is marked. In the year 1971, on Goa's roads, there were 11,710 vehicles which have increased to 32,799 by 1979, it was nearly the four fold growth; which continues further and reaches to 65,808 by 1985 and in the next 8 years of time the number of vehicles rose remarkably to the 1,74,656, that was by 1995. The addition in terms of percentage was 1947 over the 1960 figures (Table 16.1). The tremendous peace with which the vehicle number increased is unprecedented and improportionate as compared to the growth of fixed assets; such as roads. As per the local news paper report the vehicle population, by the end of 1996, has grown to 2.30 lakh.

In all, the region comprises some 225 kms of National Highways, 232 kms of State Highways and 815 kms of major district roads and rest of the length out of 7543 kms accounts for village roads both of mettled and unmettled roads. The road transport industry, in the year 1993-94, commands a total of 1,74,656 commercial vehicles of which 2337 were stage carriers (Mini buses/ buses) both of state owned and private. Another 16,361 were goods vehicles and 21,284 private cars and jeeps but as many as 1,26,040 were the two wheelers, account for about 70 per cent of the total vehicles registered in Goa.

Density of Roads

The distribution of road network in Goa is very uneven. Density of major roads for each taluka has been derived with reference to area and population units. High densities of roads, that is more than state's average, occur in the talukas of Bardes, Pernem Bicholim, Ponda, Tiswadi Marmugao and Salcete (Table 16.2) which are relatively developed and densely populated area with fairly low relief. Sanguem (33 km), Canacona (35 km). Quepem (37 km) and Sattari (38 km) are having the density of roads less than the state's average (45.51 km/100 sq km) reflecting the hilly terrain, sparse population and less development.

Coastal talukas with large size of population present a little length of road mileage for a unit of 10,000 population. The density of main road per 10,000 population records a minimum of six kms in Marmugao taluka and maximum of 16 kms in Bardez. As against

Table. 16.1 (a) : *Growth of Components of Industry and Transport in Goa, 1961-1995*

Sr. No.	Item	1961	1971	1981	1991	1995
1.	Population	5,89,997	7,95,120	10,07,749	11,67,793	12,50,815*
2.	Workforce	2,45,221	2,54,475	3,11,247	3,82,369	4,25,277*
3.	Industrial workers	—	5,940	22,100	39,193	47.060
4.	No. of Industries (S. S. T. ,L.,M. S.,)	46	621	2,229	4,763	5,638
5.	Per Capita Income	434	916	4,931	7,634	7,634@
6.	No. of Goods Vehicles	2,500	2,613	7,198	12,470	16,361
7.	No. of Vehicles	8,531	9,601	32,799	1,18,518	1,74,656
8.	Length of Road (km)	1,655	3,834	4,067	7,202	7,543
9.	Density of Road (km/100 sq km)	45	103	110	194	203
10.	No. of Bridges	13	38	42	44	47

* = Estimated population
@ =1991 figure.

Table. 16.1 (b)

Sr. No.	Item	1961-71	1971-81	1981-1991	1991-95	1961-95
1	Population	34.76	26.74	16.08	6.93	112.00
2.	Workforce	3.77	22.31	22.85	11.22	73.43
3.	Industrial workers	—	272.05	77.34	20.07	792.25@
4.	No. of Industries	1250.00	258.90	113.68	18.37	12156.50
5.	Per capita income	111.00	438.38	54.81	—	1659.00
6.	No. of Goods Vehicles	4.52	175.47	73.24	31.20	554.44
7.	No. of Vehicles	12.54	241.62	261.35	47.37	1947.31
8.	Length of Road (km)	131.66	6.08	77.08	2.87	347.60
9	Density of Road (km/100 sq km)	129.00	5.82	4.76	6.82	344.44
10	No. of Bridges	192.31	1052	4.76	6.82	261.53

@ = 1971-95.

these, the hilly talukas with small size of population reflect comparatively higher density ranging between 21 km in Quepem and 51 km in Sanguem. The present pattern of density of roads per unit area and population unit shows the general unevenness in economic and resource development.

Similarly the road density may also be derived in term of

Table 16.2 : *Talukwise Distribution of Factors of Industry and Transport in Goa.*

Sr. No	Taluka	Population (1991)	Workforce (1991)	Industrial work-force (1994) x	No. of. Industrial (1. M. SSI) y	Industrial Intensity $\frac{x+y}{2}$	Road Length (km.)	Road Density /10000 popn.	Road Density /100 km²	Accessibility Shimble Index
1.	Tiswadi	1,46,443	49,806	6604	897	16.07	138	11	70	31
2.	Bardez	1,89,383	60,770	5516	1041	16.80	242	16	95	37
3.	Pernem	66,689	21,305	770	175	2.48	190	32	79	50
4.	Dicholim	84,352	27,305	3355	387	7.41	167	23	72	33
5.	Satari	49,530	17,633	1509	113	2.79	190	47	38	33
6.	Ponda	1,28,227	43,269	7041	788	15.21	204	19	71	27
7.	Sanguem	59,355	22,623	608	104	1.61	287	51	33	44
8.	Canacona	40,716	13,267	453	110	1.51	122	34	35	50
9.	Quepem	64,518	22,836	2672	262	5.49	119	21	37	39
10.	Salcete	2,19,897	64,817	7571	1207	19.80	208	11	71	33
11.	Marmugao	1,20,503	38,691	6667	409	11.51	60	06	57	44
	Total	11,69,793	3,82,369	42,766	65493		1685	$\overline{D}$=17	$\overline{D}$=45.51	SI=38

length of road for a convenient unit of vehicle population. The density of vehicles per road km has increased from 3.26 per km in 1965 to 23.15 per km by 1993. Whereas, the road density per 100 sq km has increased 45 km in 1960 to 204 km by 1993. Obviously, the vehicle density per road km has registered faster growth than that of road density per unit area. Uneven pattern of road density, thus, has a far reaching effect in the rapid growth of population and economic activities but only in some pockets of the state while others remain backward.

Structural Characteristics of Transport

The efficiency of the transport network plays a pivotal role in the changing trend of regional and economic development. Accessibility is such a phenomenon depends highly on the efficiency of the network, that can be expressed through the intra-regional or inter-regional contrasts in economic development or the development of the region through different time periods.

The Table 16.3 shows the increasing complexity of the road network through time (1951 to 1991). The efficiency indices namely, Alpha, Beta, Gama and Cyclomatic numbers portray the ever increasing trend of efficiency and complexity of the network for instance, the road network of Goa has Registered the Beta index of 1.20 in the year 1951 which rose to 1.65 by the year 1991. This represents consistent increase in the complexity of the network. The values of 1991 compared to the aggregate index values of 1981 network show a considerable decline, thus, it can be stated that the accessibility has declined. However, the network still maintains complexity and offers a better connectivity. Decline in accessibility is due to improportionate growth of towns (vertices) compared to the development of roads (edges).

Nodal Accessibility

The relative location of an urban centre or of an area in the network of transport offers a particular degree of accessibility to a given centre. Centrally located vertices are usually having high accessibility as compared to the peripheral ones. Such variation in

Table 16.3 : *Index of Network Efficiency*

Years	No. of Vertices	No. of edges	Alpha	Beta	Gamma	Cyclomatic Number
1951	5	6	40.00	1.20	66.66	2
1961	11	14	23.52	1.27	52.00	4
1971	11	16	35.30	1.45	59.25	6
1981	16	27	44.00	1.66	64.00	12
1991	26	43	38.30	1.65	60.00	18
			181.12	7.23	301.91	42
			A= 36.22	C 1.44	C=60.38	CH=8.4

the degree of accessibility over the surface can be indentified with the help of Simbel index.

Shimbel index : It is the improved and simple method to show the varying importance of nodes in a network, by filtering the redundancies, i.e., indirect linkages give an idea about the relative position and respective accessibilty of a node in term of number of arcs required by a given node to reach the rest. Smaller the number of arcs or edges required to reach all the other nodes of the network, from a given node, greater the accessibility. This enumerates larger values to the nodes located further and peripheral and smaller values to central and nearer nodes.

A comparison of the Shimbel index (dispersion index) values for the topological graphs of road network for the year 1981 and 1991 yields a meaningful idea about the changing accessibility character of the nodes in the network through time. It is discernible from the matrices (Table 16.4 (a) and (b)) that Phonda (V8) the highly accessible centre (1981), retains its position intact even in 1991. The Shimbel index value rose from 27 in 1981 to 55 in the year 1991 but such change has not affected the centrality of Phonda in the network. This centre is followed by other accessible centres viz; Panaji (60), Valpoi (63), Sanvordem (63), Cortalim (65), Raja (67) and Pale (68). It is conspicuous that the relative accessibility of Panaji, Valpoi, Sanvordem has improved compared to Cortalim (V 16) in the year 1991. This is mainly due to emergence of new towns in the near vicinity and added direct connectivity between

Table 16.4 (a) : *A Shortest Paths (Dispersion Matrix, 1991 Network*

	1	2	3	4	5	6	7	8	9	10	11	12	13	14	15	16	17	18	19	20	21	22	23	24	25	26	SI	Rank	AN
1	0	2	1	1	2	3	6	4	3	6	4	5	5	6	6	5	4	4	2	3	2	3	3	5	7	7	99	25	7
2	2	0	2	1	2	3	4	2	1	4	2	3	3	4	4	3	4	3	2	3	2	3	1	3	5	5	71	10	5
3	1	2	0	1	3	6	4	3	6	4	5	5	7	5	6	5	4	4	2	2	2	2	3	5	7	7	97	23	7
4	1	1	1	0	1	2	5	3	2	5	3	4	4	6	5	4	3	3	1	2	1	2	2	4	6	6	77	16	6
5	2	2	1	1	0	2	5	3	3	5	4	5	5	6	5	4	3	3	2	1	2	1	3	4	6	6	84	19	6
6	3	3	3	2	2	0	3	1	2	3	2	3	3	4	4	2	1	1	1	1	1	3	2	2	4	4	60	2	4
7	6	4	6	5	5	3	0	2	3	1	3	4	2	3	3	1	2	3	4	4	4	6	3	2	1	2	82	17	6
8	4	2	4	2	3	1	2	0	1	2	1	2	2	3	3	1	2	1	2	2	2	4	1	1	3	3	55	1	4
9	3	1	3	2	3	2	3	1	0	3	1	2	2	3	3	2	3	2	3	3	3	4	1	2	4	4	63	3	4
10	6	4	6	5	5	3	1	2	3	0	2	3	1	2	2	1	2	3	4	4	4	6	3	1	1	1	75	14	6
11	4	2	4	3	4	2	3	1	1	2	0	1	1	2	2	2	3	2	3	3	3	5	2	2	3	3	63	3	5
12	5	3	5	4	5	3	4	2	3	2	1	0	2	2	1	3	4	3	4	4	4	6	3	3	4	3	83	18	6
13	5	3	5	4	5	3	2	2	2	1	1	0	0	1	1	2	3	3	4	4	4	6	3	2	2	2	72	11	6
14	6	4	7	6	6	4	3	3	3	2	2	2	1	0	1	3	4	4	5	5	5	7	4	3	3	1	94	21	7

Contd.

Table 16.4 (a) (*Contd.*)

15	6	4	6	5	5	4	3	3	3	2	2	1	1	1	0	3	4	4	5	5	5	7	4	3	3	2	91	20	7
16	5	3	5	4	4	2	1	1	2	1	3	2	2	3	3	0	1	2	3	3	3	5	2	1	2	2	65	5	5
17	4	4	4	3	3	1	2	2	3	2	3	4	3	4	4	1	0	2	2	2	2	4	3	2	3	3	70	9	4
18	4	3	4	3	3	1	3	1	2	3	2	3	3	4	4	2	2	0	2	2	2	4	2	2	4	4	69	8	4
19	2	2	2	1	2	1	4	2	3	4	3	4	4	5	5	3	2	2	0	2	2	3	3	3	5	5	74	12	5
20	3	3	2	2	1	1	4	2	3	4	3	4	4	5	5	3	2	2	2	0	2	3	3	3	5	5	75	14	5
21	2	2	2	1	2	1	4	2	3	4	3	4	4	5	5	3	2	2	2	2	0	3	3	3	5	5	74	12	5
22	3	3	2	2	1	3	6	4	4	6	5	6	6	7	7	5	4	4	3	2	3	0	4	5	7	7	109	25	7
23	3	1	3	2	3	2	3	1	1	3	2	3	3	4	4	2	3	2	3	3	3	4	0	2	4	4	68	7	4
24	5	3	5	4	4	2	2	1	2	1	2	3	2	3	3	1	2	2	3	3	3	5	2	0	2	2	67	6	5
25	7	5	7	6	6	4	1	3	4	1	3	4	2	3	3	2	3	4	5	5	5	7	4	2	0	2	98	24	7
26	7	5	7	6	6	4	2	3	4	1	3	3	2	1	2	2	3	4	5	5	5	7	4	2	2	0	95	22	7

SI = Shimble Index

AN = Associated Nos.

SI = 1860

AN = 136

Mean Associated Number = 5.23

Table 16.4 (b) : *Shortest Paths (Dispersion Matrix) 1981 Network*

		V1	V2	V3	V4	V5	V6	V7	V8	V9	V10	V11	V12	V13	V14	V15	V16	SI	AN
Pernem	V1		2	1	1	2	2	4	3	3	4	4	5	5	5	6	3	50	6
Bicholim	V2	2		2	1	2	2	3	1	1	2	2	3	3	3	4	2	33	4
Siolim	V3	1	2		1	1	2	4	3	3	4	4	5	5	5	6	3	49	6
Mapusa	V4	1	1	1		1	1	3	2	2	3	3	4	4	4	5	2	37	5
Calangute	V5	2	2	1	1		1	3	2	3	3	3	4	4	4	5	2	40	5
Panaji	V6	2	2	2	1	1		2	1	2	2	2	3	3	3	4	1	31	4
Vasco	V7	4	3	4	3	3	2		2	3	2	3	4	3	3	4	1	44	4
Phonda	V8	3	1	3	2	2	1	2		1	1	1	2	2	2	3	1	27	3
Valpoi	V9	3	1	3	2	3	2	3	1		2	1	2	2	3	3	2	33	3
Nargao	V10	4	2	4	3	3	2	2	1	2		2	3	1	1	2	1	33	4
Sanvordem	V11	4	2	4	3	3	2	3	1	1	2		1	1	2	2	2	33	4
Sanguem	V12	5	3	5	4	4	3	4	2	2	3	1		2	2	1	3	44	5
Quepem	V13	5	3	5	4	4	3	3	2	2	1	1	2		1	1	2	39	5
Cuncolim	V14	5	3	5	4	4	3	3	2	3	1	2	2	1		1	2	41	5
Chauri	V15	6	4	6	5	5	4	4	3	3	2	2	1	1	1		3	50	6
Cortalim	V16	3	2	3	2	2	1	1	1	2	1	2	3	2	2	3		30	3

SI = 614 AN = 72

SI= Shimbel Index

AN = Associated Number = 890

SI = Mean Shimbel Index = 38.37

AN = Mean Associated Number = 4.5

the existing centres and new towns.

Most of the towns located in the periphery, in the western plain and coastal parts are nearly clustered and have accounted for a less connectivity, for instance, it is simply not possible to reach either Benaulim (V25) or Candolim (V22) within seven steps. Obviously, at this stage one can note the role of nodel clustering and consequent increase in paths in changing the accessibility surface. It is sensitive to change in the number either of nodes or edges or direction of flow.

Impact of Transport on Industrial Development

Till 1961 the economy of Goa was mainly based on agriculture, trading and mining activities. The Industrial sector was limited to a less number of cottage and small scale units known for the production of soaps, tiles, processed cashewnuts, husked rice, coir, handloom, candle, coconut oil, etc. and service activities like tyre retreading, wood working, etc. The mining industry was the backbone of Goa's Economy. Portuguese Government during pre-liberation period did not pay much attention to provide sufficient industrial infrastructures for the basic and overall industrial growth. Due to some local resource problems such as non-availability of energy, water supply, lack of finance and limited local market, it was very difficult for the Goa state to take early stride in the development of industrial sector.

In 1965-66, there were 68 registered S.S.I. units in operation with 3823 estimated numbers of daily workers. However, Goa has made a great strides in manufacturing sector particularly after 1970. In quantitative terms the industrial sector has progressed from 46 (1961) service oriented and manufacturing units like workshop, soaps, cashewnuts, rice and flour mills to over 5,638 small scale units with an investment of Rs. 111.11 crores (plant and machinery) employing over 47,060 people by 1995. Goa also has 62 large and medium scale unit with an investment of over Rs. 415.56 crores, employing over 11,573 persons during 1993-94.

The total main workforce has increased from 2,45,221 in 1961 to 4,25,277 by 1995. Whereas, the industrial workforce has registered a spectacular growth that is from 5940 in 1971 to 47,060 by 1995. The growth in terms of percentage is 692.25 (1971-95) as against the 73.43 per cent (1961-965) of total main workforce. Substantial growth can be noticed especially in the decade 1971-81, that is 272 per cent (Table 16.1). Such rapid increase is due to remarkable jump in the number of industries from 621 (1971) to 2229 by 1981. It is very conspicuous that the industrial workforce is growing more rapidly than the general main workforce in the Goa state and thus, reflects the shifting of people from primary activities to the secondary activities and shift in the agro-based economy to industrial oriented economy. Hence, the per capita income is constantly increasing and kept the Goa state in the second position in Indian states as one of the states with highest per capita income.

The faster industrialization is mainly because of an excellent natural harbour and an all weather port, well developed transport system, pleasant and moderate climate, well developed infrastructural facilities such as, power, water, communication and urban amenities. Availability of developed industrial estates and locations in hassle-free, and literate, skilled disciplined and peaceful workforce along with ample financial sources. The analysis of talukawise distribution of small scale units highlights the influencing factors over a particular pattern of distribution through time.

Among eleven talukas of Goa, in 1981 the highest number of S.S.I. were concentrated in Salcete taluka that was 348 (24%) followed by Tiswadi 296 (20%), Bardez 263 (18%) and Phonda 174 (12%). About 74 per cent of the total small scale industries were established only in these four talukas. The spatial spread of the industrial units is highly controlled by the factors such as resources, physical setting, economic and socio-political factors. In 1984 some more industries were added to the existing list. Even during that period also the highest concentration again was in Salcete with 579 (21.6%) units, followed by Bardez 496 (18.5%). Tiswadi 483 (18%), and Ponda 330 (12.3%) units.

Further in the year 1993-94 the total number of industries have gone upto 5493, that is slightly more that three and half times the 1981. The position of Salcete taluka (1207 units) (22%) as the area of highet concentration of industries remained intact. This was followed by Bardez 1041 (19.4%), Tiswadi 897 units (16.8%) and Ponda 788 (13.7%) units.

There has been a tremendous growth in small scale industrial sector during the last 13 years of time. However, the growth seems to be lop-sided as it witnessed largely by four talukas, whereas rest of the talukas show a poor growth. The growth was in response to the removal of traditional constraints as lack of water and power, provision of better infrastructural facilities, well developed transport network and favourable Government policies and financial help.

Areal variation in the concentration of industries exists even at district level that is, out of 5,493 (1993-94) units (including small, medium and large units) 3,401 units (62.00%) are located in North Goa and remaining 2,092 (38.00%) units are distributed in South Goa Salecte in South Goa and Bardez (19.26%) in North Goa are the leading industrial talukas of Goa. The interesting feature of the industrial distribution is that out of 62 large and medium scale units, as many as 20 units (38.46%) are located only in Phonda taluka which is the centrally located taluka of the state. The pattern of factory workers present a clear picture of the degree of industrial develophent in different talukas of Goa. The Table 16.2 shows the total employment in large, medium and small scale units in 1994. Out of 42,766 (1994) persons employed in industries 7571 (17.70%) are in Salcete taluka followed by Phonda 7041 (16.46%), Marmugao 6667 (15.59%), Tiswadi 6604 (15.42%), Bardez 5516 (12.90%), Bicholim 3355 (7.8%) and Quepem 2672 (6.25%). Canacona accounts for the less concentration of industries and employees. It is quite conspicuous that the concentration, that is 78 per cent, of the total industrial workforce is highly confined only to five talukas viz; Salcete, Phonda, Marmugao, Tiswadi and Bardez, Rest of the talukas have less number of industries and employees.

Many economic geographers used different methods to study the degree of industrialization. For the present study one such method has been employed which express the industrialization in term of ratio of number of employees per 1000 inhabitants of an area. Here talukawise employees and total population have been taken to arrive with the result to classify the talukas as highly industrialised.

It is discernible from the Table 16.5 that the Salcete taluka (6.47) and Phonda taluka (6.02) have a high degree of industrialisation. It is depicted through the high values scored by each taluka. Marmugao (5.70), Tiswadi (5.64) and Bardez (4.71)

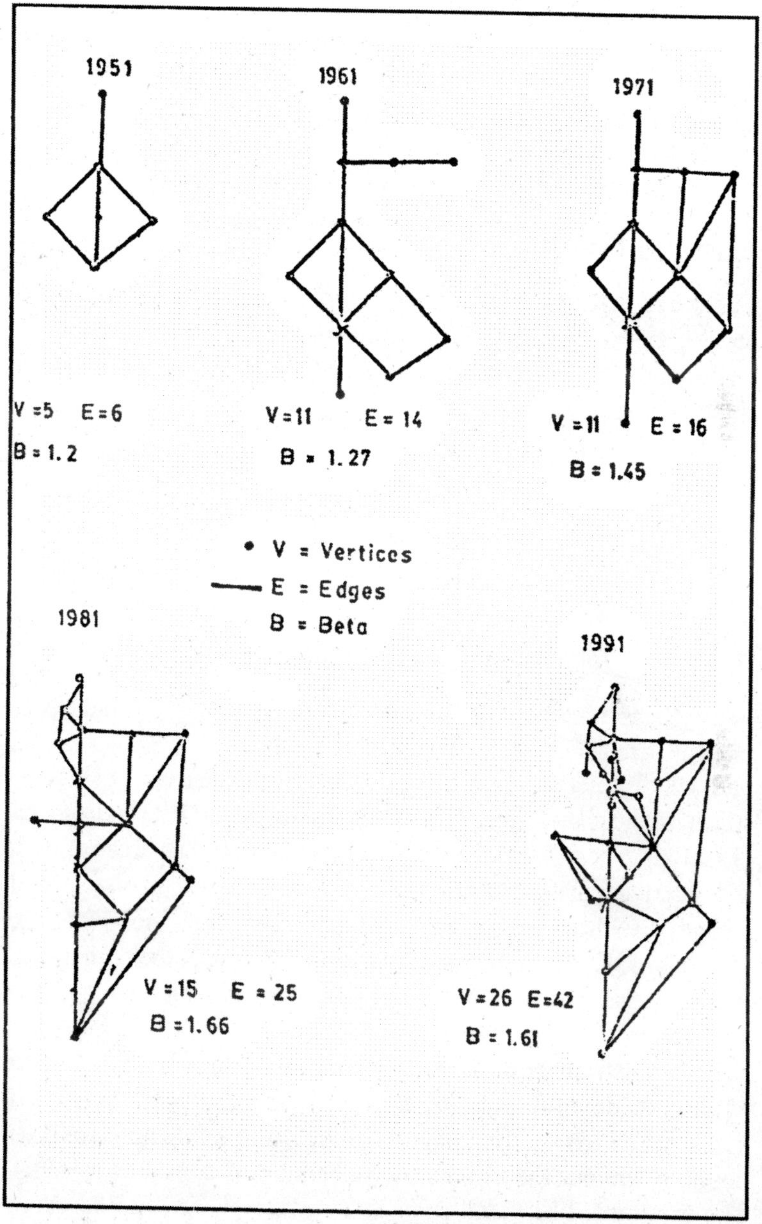

Fig. 16.2 : *Growth of Urban Centres and Transport Network*

talukas belong to second category of (moderate) industrialisation. The talukas viz, Bicholim (2.87), Quepem (2.88) and Sattari (1.29) have scored very low values and fallen in the category of less industrialised talukas. However, Pernem, Sanguem and Canacona have belong to the talukas of least industrialization. Such a situation might be because of topography, resource problems, sparse population and inaccessible locations.

Industrial Intensity

There are different ways of measuring the level of industrial developement. Mandal (1971) has calculated the industrial intensity on the basis of : (i) Number of registered factories, and (ii) Factory employment.

Industrial Intensity = x+y/2

where,

$$x = \frac{\text{No. of factories in a unit area}}{\text{Total No. of factories in the state}} \times 100$$

and

$$y = \frac{\text{No. of persons employed in the given unit area}}{\text{Total No. of industrial workforce of the states}} \times 100$$

Same method has been employed to assess the industrial intensity of the Goa state. The Table 16.4 shows that Salcete taluka with 19.80 per cent stands first in the industrial intensity, (Fig. 16.3) which followed by Bardez (16.08 %), Tiswadi (16.07 %) and Phonda (15.21%). These four talukas have been classified as areas of high industrial intensity. Marmugao is the only taluka in the second category with the score of 11.51 per cent. Bicholim (7.41%), (Quepem 5.49 %) belong to third group and rest of the talukas fall in the category of least industrial intensity with less than five per cent value.

Industrialization is considered as one of the indicators of economic development of a region. The intense degree of industrialization in different parts of the world would not have been possible without the efficient development of transport systems. The transport, like many other infrastructural facilities has been considered the cause and an effect of industrialization. The cheap,

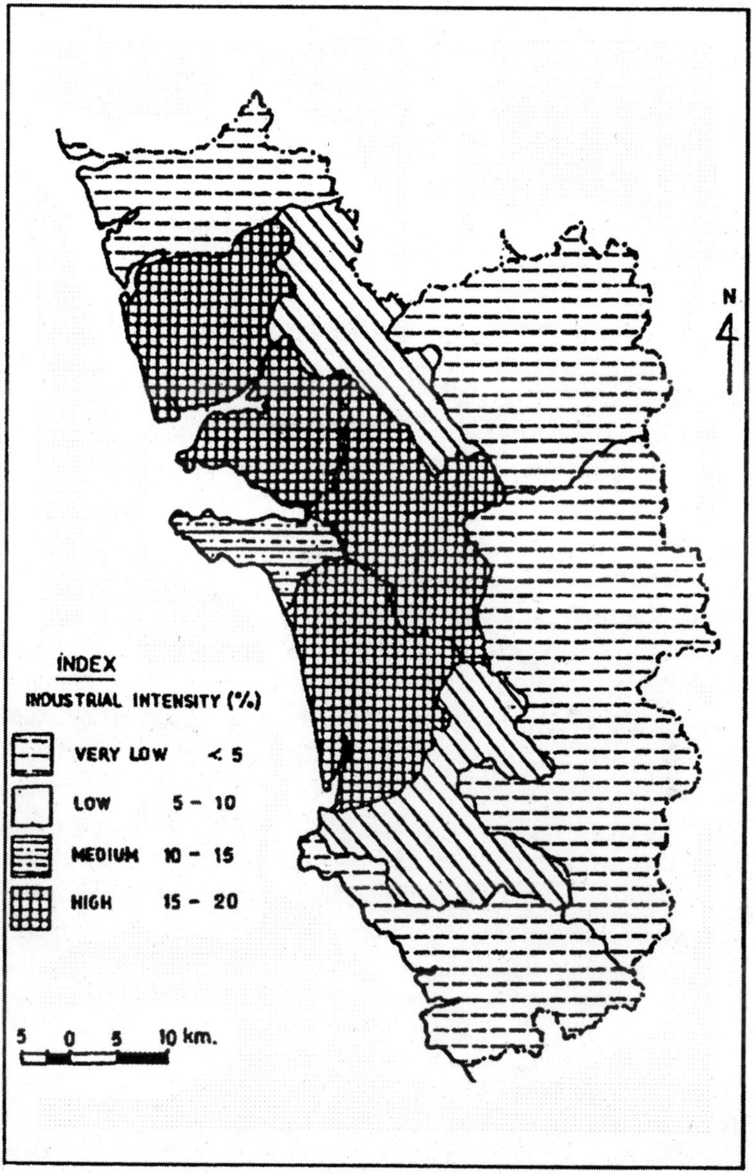

Fig. 16.3 : *Goa Industrial Intensity (1993)*

safe, quick and efficient means of transport system usually has a far reaching effect on overall regional economic development. The development of transport and industry are influencing one another and portraying a strong relationship between them. Such relationship can be expressed numerically with the help of coefficient of correlation technique.

The spatio temporal analysis has been done at two different stages: (i) state level, and (ii) taluka level, which yield an idea about the trend and intra-regional contrast in the development of transport and industrial factors. Network efficiency indices and indicators of industrialization have been composited (see Table 16.5 (a) and (b)) separately and arranged along the line nomograph (Fig 16.4). A visual comparison of the distance between composite indices of network efficiency (1951-1991) marked on the nomograph indicate that great achievement in the efficiency of transport was attained during 1971 and 1981 period. However, a substantial decline in the transport efficiency of the network of the year 1961 over the year 1951 was mainly due to inproportionate increase in the urban centres as compared to roads. 1961-71 and 1981-91 periods witness a remarkable growth in the transport development. The distance between the industrial indices marked on the monograph (Fig. 16.4) shows a considerable stride especially during 1971-81 and 1981-91. In these two decades (1971-91) the number of industries increased from 621 to 4763, that is by 667 per cent, whereas the industrial workforce has increased by 560 per cent during the same period.

It is conspicuous from the nomographs that there is a positive relationship between the transport and industrialization. Increase in one sector is reflected in the positive growth of the other sector. Especially after 1971 both sectors are growing hand (and intimately) for the overall regional development.

The taluka level analysis by comparing the components of transport and industrialization yields a greater spatial disparity in the inter-regional developments. However, most of the factors of both transport and industry, when copared, worked out a strong association or relationship. The coefficient of correlation (r) between accessibility (1991 network) derived by Simbel index and workforce is +0.61, whereas the coefficient of correlation

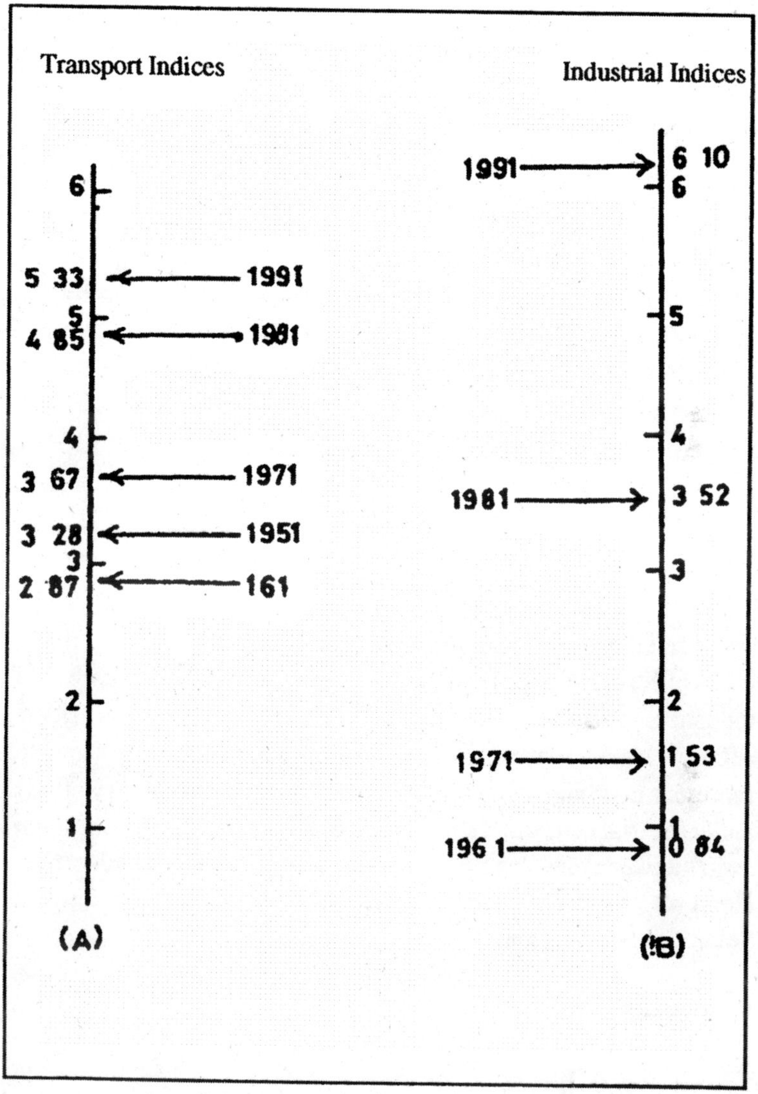

Fig. 16.4 : *Nomograph of Transport and Industry*

Table 16.5 (a) : *Degree of Industrialization*

S. No.	Taluka	Degree of Industrialization (No. of factory workers per 1000 (Index)	Categorization of Taluka based on index	
1.	Tiswadi	5.64	Salcote	I highly
2.	Bardez	4.71	Ponda	Industrialsed
3.	Pernem	0.66	Marmugao	II Medium
4.	Bicholim	2.17	Tiswadi	Industrialised
5.	Satari	1.29	Pardez	
6.	Phonda	6.01	Picholim	III low
7.	Sanguem	0.52	Quepem	Industrialised
8.	Canacona	2.39	Satari	
9.	Quepem	2.28	Pernem	IV very low
10.	Salcete	6.47	Sanguem	Industrialised
11.	Marmugao	5.70	Canacona	

between industrial workforce and accessibility is +0.68. Similarly, the coefficient of correlation between accessibility and number of industries is also +0.61 and accessibility and industrial intensity is +0.68. All these positive correlation values are very significant and represent a strong association between the accessibility, transport and industrial concentration. Similarly there is a strong association even between the density of roads and number of industries. However, weak but positive relationship is established between the length of roads and industrial factors.

Conclusion

Though the overall development seems to be much faster both in terms of transport and industrial development but it is spatially highly imbalanced. Well developed transport network with high density and accessibility is usually confined to the centrally located talukas viz. Salcete, Tiswadi, Marmugao, Bardez and Phonda. Similarly,

Table 16.5(b) : *Industrial Intensity of Goa (1993)*

Sr. No.	Taluka	Percentage of units to total (x)	Percentage of factory employment to total (y)	Industrial Intensity $\frac{x+y}{(2)}$	Talukas according to intensity
1.	Tiswadi	16.71	15.44	16.07	Salcete
2.	Bardez	19.26	12.90	16.08	Bardez
3.	Pernem	3.16	1.80	2.48	Tiswadi I high
4.	Dicholim	6.19	7.84	7.41	Ponda
5.	Satari	2.06	3.53	2.79	Marmugao II med.
6.	Ponda	13.96	16.47	15.21	Bicholim III
7.	Sanguem	1.81	1.42	1.61	Quepem
8.	Canacona	1.96	1.06	1.51	Satari
9.	Quepem	4.74	6.25	5.49	Pernam
10.	Salcete	21.91	17.70	19.80	Sanguem IV very
11.	Marmugao	7.44	15.59	11.51	Canacona Total

the same talukas together share 75 per cent of the state. The growth pattern is highly lop-sided as it is witnessed only by five talukas and rest of the talukas receiving less attention.

References

Almeida, J.C. (1990) Industrial Scenario of Goa : Then and Now, International Goan Youth Convention, Seminar, December 17-27.

Angle, S.P. (1983) *Goa : An Economic Review*, The Goa Hindu Association, Kala Vibhag, Panaji-Goa, India.

Bradnock, B.G. and Robinson H. (1978) *Geography of Transport*, 'Aspect' Geography Series, Mac Donald and Evans L.td., Landor.

Bradnock, R.N. (1976) Measurements of Accessibility and their relationship to development in Tamilnadu, *Indian Geographical Journal*, Golden Jublee Volume, Madras.

Census of India, District Census Handbook of Goa, Daman and Diu, 1971,1981, 1911.

Clerk, W.A.V. (1986) *Statistical Methods for Geographers*, John Wiley and Sons, Ner York.

Dalton, R. Garlick, J., Minshull, R. and Robinson, (1972). *Correlation Techniques in Geography*, George Philip and Sone I.T.D Joncon.

Gazetteer of India, Union Territory; Goa, Daman and Diu Part I (1979).

Industrial and Commercial Directory of Goa, 1980, 1990 and 1995, Goa Chamber of Commerce and Industry, Goa.

Indusrial Potential Survey of Goa State (1989), Small industries service Institute (SISI), Government Branch, Margao.

Mandal, B. (1971), Manufacturing Regions of North Bihar, *National Geographical Journal of India*, vol. XVII, India.

Regional Study of Goa : Transport and Communication, Department of Town and Country Planning, Govt. of Goa.

Rocket Ibrahim (1993) Transport Network Evolution and Economic Development in North-Western India (1865-1941) in *'Perspectives in Economic Geography : Transport Geography*, Ed. by Majid H., Anmol Publications, New Delhi, India, Vol.6.

Socio- Economic Indicators of Goa, Govt. of Goa, 1993.

Statistical Pocket Book of Goa, Directorate of Planning, Statistics and Evaluation, Govt. of Goa, 1987-88, 1990-91 and 1993-94.

Taaffe and Gauthier (1973) *Geography of Transportation*, Prentice Hall Inc. Engelwood Cliffs, New Jersey.

Yeates, M.H. (1968) *An Introduction to Quantitative Analysis in Economic Geopraphy*, McGraw Hill Book Co., London.

17

RAILWAY TRANSPORT IN VADODARA
A Historical Perspective

Salil Latey and *B.C. Vaidya*

Introduction

Transportation is regarded as an index of economic, social and commercial progress of a country (Chakraborty, 1998). Commerce and movement of people rests on well laid foundation of transport. No region or country can ever flourish if it is lacking adequate transport facilities. And, therefore, availability of means and modes of transport is very much needed for regional development. Better transport makes possible to Indian farmers to put their less accessible land to more productive use. Moveover, efficient transport also gives rise to large volume of interaction because the people can go easily to the nodes or market centres to avail all necessary facilities. Inadequate and inefficient transport facilities cannot initiate any sort of development in any region. Though transport system accelerates the economic activities, the rate of development depends upon the reliability, capability and frequency of the transport system. Bearing this significance of means of transport for local, regional and national all round development, author tries their mites with taking Vadodara railway as case study. In this chapter attempt has been made to examine the evolution of Vadodara railway junction since 1983 and assesses its impact on regional development in surrounding Vadodara.

Historical Background of Vadodara Junction

In 1860 the BBCI company (Bombay-Borada and Central India was established which was owned by Maharaja Ganapatrao Gaekwad, the ruler of princely state of Vadodara. One narrow and another metre guage railway line was constructed during this period. The narrow guage of 0.67 miles was constructed from Vishudmitri to Chota Udaipur, Jambusar-Kavi-Dahoj and some area of the present Amreli district in Saurashtra. Though the Mumbai-Surat-Vadodara Ratlam and Vadodara-Ahmedabad railway line was constructed in 1862, but actually was opened for traffic in 1922, while metre guage railway line connecting Ahmedabad-Delhi was introduced later on. After merging Vadodara state into Indian Union, BBCI became the part of Western Railway on 5th November, 1951. The railway staff college was started in 1952 for providing training for railway employee in our country.

Geographical Background

Vadodara is situated on the bank of Vishwamitri river extending from 22°17" North latitude and 73°15" East longitude. It is surrounded by the districts of Kheda, Panchamahal and Bharuch, Jawaharnagar, Karjan, Padra, Vaghodia and Dabhoi in the immediate vicinity. Physiographically, Vadodara city is situated between the Mahi-Narmada doab, a flood plain with the well known alluvial soil. Areas on both north and south side of the city are fertile with gently rolling plain. City enjoys semi-humid tropical monsoons characterized by seasonal rainfall by the south-west mansoon. Vadodara appears to be selected as a junction for Delhi bound trains for geographical reason. In those times, the British rulers had to travel by sea from England, around the Cape of Good Hope, to land at the first Indian sea port, Mumbai, which naturally became the Gateway of India. It was to be liked with Delhi which had become capital of India after 1857. There had been two possible options for route to this railway, either via Ahmedabad or via Surat/Bharuch. Physiography is playing a vital consideration here as toward Surat of Bharuch would be facing the obstacles of the mightiy Satpudas and Vindhyan ranges through which Narmada flows. The igneous and sedimentary type rock structure would have added to the cost of construction of tunnels, viaducts, bridges and embankments. Heavy rainfall and deep jungles would have made the things worse. Bifurcation from Ahmedabad meant passing through semiarid areas north of Ahmedabad which are lithologically not so strong to withstand the heavy traffic. The Aravalis acting as partial barriers, problems of wind erosion, harsh climate, water scarcity in the unproductive areas through Rajasthan went against this option. A small gap was found existing between the Vindhyan and Aravali ranges on the Malwa plateau, about 175 km. northeast of Vadodara through which the track has been routed with less hindrance. Thus, Vadodara become the junction for Delhi and Mumbai bound trains. The distance and stations from Vadodara area given in Table 17.1

Table 17.1 : *Distances from Vadodara and Number of Trains*

Stations	Distance in km	Years 1983	Years 1996
(A) Mumbai Side :			
1. Mumbai Central	392	12	18
2. Bandra Terminus	381	0	4
3. Vsai Road	344	0	5
4. Valsad	192	1	2
5. Surat	130	2	5
(B) Ahmedabad Side :			
1. Ahmedabad	100	9	13
2. Viramgam-Hapa-Gandhidham	400	2	3
3. Hapa	422	1	2
4. Rajkot	346	0	5
5. Porbandar	562	0	1
6. Okha	598	0	2
(C) Ratlam side :	NA	7	17
Total		34	75

Source : Railway Time Table.

Political Background

It is said that Vadodara had trade link with Roman ports from ancient Chandanawati – the city of sandalwood. The name Vadodara as presently known derived finally from Viravati-Vataptra-Vadapadraka-Badode-Bododa to be finally Vadodara today. It has been under Maratha rule of Gaekwads from 18th century. It was one of the first princely state of having its own railway. Maharaja

Sayajira Gaekwad III, who rules Vadodara for 64 years from 1875 to 1939, is considered as the architect of the modern Vadodara. He was at least 150 years ahead of his times. He encouraged talented people from all over India to take up services in Vadodara state. To name a few, they included Sri Aurobindo, Dr. B. R. Ambedkar, Shri Romesh Chandra Dutt. The realized that if Vadodara was to progress, its contact with the outside world should increase. Railway lines is his state exactly did this. The infrastructural facilities created at Vadodara junction are sufficient to meet even today's requirements. The State had its Planning Commission,. The greatest contribution of Sayajirao was the Baroda College, which became Maharaja Sayajirao University in 1949. The University imparts knowledge in each and every field of study.

Growth of Vadodara City as Railway Junction

It is interesting that city and the railway junction have promoted each other's growth. Initially, it was a centre for handling tobacco, cotton and other crops marketed to near and far off places. Later, it became industrial centre for handling textiles and agro-products. Vadodara's mineral wealth consists of calcity, limestone, quatite, feldspar, glass sand (largest deposits in India), quartz and graphite. Exploitation of these provides ample traffic to railways. Discovery of Natural Gas in Ankleshwar, Gandhar, North Gujarat and Mumbai high and setting of oil cargo handing port facility at Kandla brought out an industrial revolution along Vapi-Ahmedabad sector. With strategically located railway junction, Vadodara easily became the largest industrial centre from mid 60s due to handling of tobacco, cotton and other crops markets to following major ways.

1. Gujarat Refinery of Indian Oil Corporation, Jawaharnagar was started in 1964 is the country's largest. The products are motor spirit, superior kerosene, high speed diesel, light diesel, light diesel oil, aviation turbine fuel, benzene, toluene, naptha. It has its own railway siding at Karachiya.
2. Indian Petrochemical Corporation Limited (IPCL) was started in 1969 and produces aromatics olefins, synthetic rubber, low density polyethylene/polypropylene, ethylene glycol, detergent alkylates, acrylic fibre and, polyeter filament.

3. Gujarat State Fertilizer Corporation (GSFC) was started in 1962 and produces nitrogenous fertilizer, capro lactum argon, methanol.
4. Gujarat Alkalies and Chemical Limited (GACL)
5. Heavy Water Project was initiated to use as a moderator and heat transfer agent in natural urenium fuelled nuclear reactor.
6. Alembic Industries produce spirits, galancial preparations both Ayurvedic and pharmocopoeial.
7. Sarabhai Chemicals produce of antibiotics and life saving drugs.

The primary chemical industries have triggered growth of secondary chemical industries like petrochemical, paints and dyes, plastics and hundreds of ancillary industries in industrial estates around Vadodara. All this has been possible because of the railway as the lifeline and the centre for providing all infrastructural facilities which have developed hand in hand. With Mumbai-Pune-Nasik belt getting saturated Vadodara is now getting up the preforonces, with tremendous job and growth opportunities, migration is taking place from all the parts of India. Tourism and the hotel industry are the latest additions to the develop of Vadodara. The developments of Vadodara can be summed up by the list of following establishments in and around Vadadora, namely, Railway Divisional Office, Railway Workshop and Locoshed, Railway Staff College, Airport, M.S. University, ONGC, GEB Head Office, Vadodara Dairy, All India Radio, customs and Excise office, EME and Air Force bases, English and Gujarat newspapers. Tourist resorts and film city at Halol.

Development of Railways at Vadodara

The development of Vadodara has taken place right hand in hand with Vadodara city. The several fold increase in migration of people from all over India resulting changes have occurred since 1983. The major changes can be noted, namely. New trains to connect many cities in country for example, Muzaffarpur, Gorakhpur, Indor, Bhopal, Jaipur, Pune, Bangalore, Jagganathpur, Cochin, Trivandrum, Nagarcoil are now have direct trains. This has made a phenomenal increase in total trains number from 34 in 1983 to 75 in 1996, Increased in the frequency of the existing trains, for example,

Howarh and Chennai are now having a daily train. Extension of the destination for the six existing trains for example, Lucknow-Ratlam Awadh Express is now extended both ways to Gorakhpur and Bandra Terminus, Mumbai-Delhi Sarvodaya Express is extended upto Jummu Tawi. Moreover, Vadodara railways has introduced numberous facilities by providing computerised tickets to passengers. The platform number was increased one to four besides this one more platform is being constructed for viable and smooth running of railway. Additional foot overbridges provided to travellers, pictograms and parking facilities have been renovated. One to computerization creation, train tracking singalling system, the numbers have changed into four digits.The notable thing in case of this railway is that three superfast trains are running via Vadodara, namely, Rajdhani Express (Mumbai-New Delhi), August Kranti Express (Mumbai-Hazrat Nizamuddin) and Shatabdi Express running from Mumbai to Ahmedabad. Fig. 17.1 and Table 17.2.

Goods Transportation

The Vadodara railway is not only distinct for passenger transport but also significant to goods transport as well. There are five places, namely, Mumbai, Bhusawal, Surat, Ratlam and Ahmedabad for getting goods. The various goods are transported by goods trains. The major bulk comes from Mumbai mainly iron and steel reaching to Vadodara. The central Indian coal is transporting from Bhusawal side whereas fertilizers from Surat side. Moreover, local surrounding places are also sending raw as wall as finished materials to be disposed of and transported within India by Vadodara railway. The bulk traffic terminates here consisting of almost 1860 wagons/day of coal, iron and steel, cement, jaggery, foodgrains, dairy products and fertilizers. Table 17.3 shows the major commodities are being loaded at Vadodara.

Future Action Plan by Railways

It is interesting to note that certain plans have been putforth by transport authority to develop the Vadodara railway junction. The change screw pattern is being introduced. The third line is to be

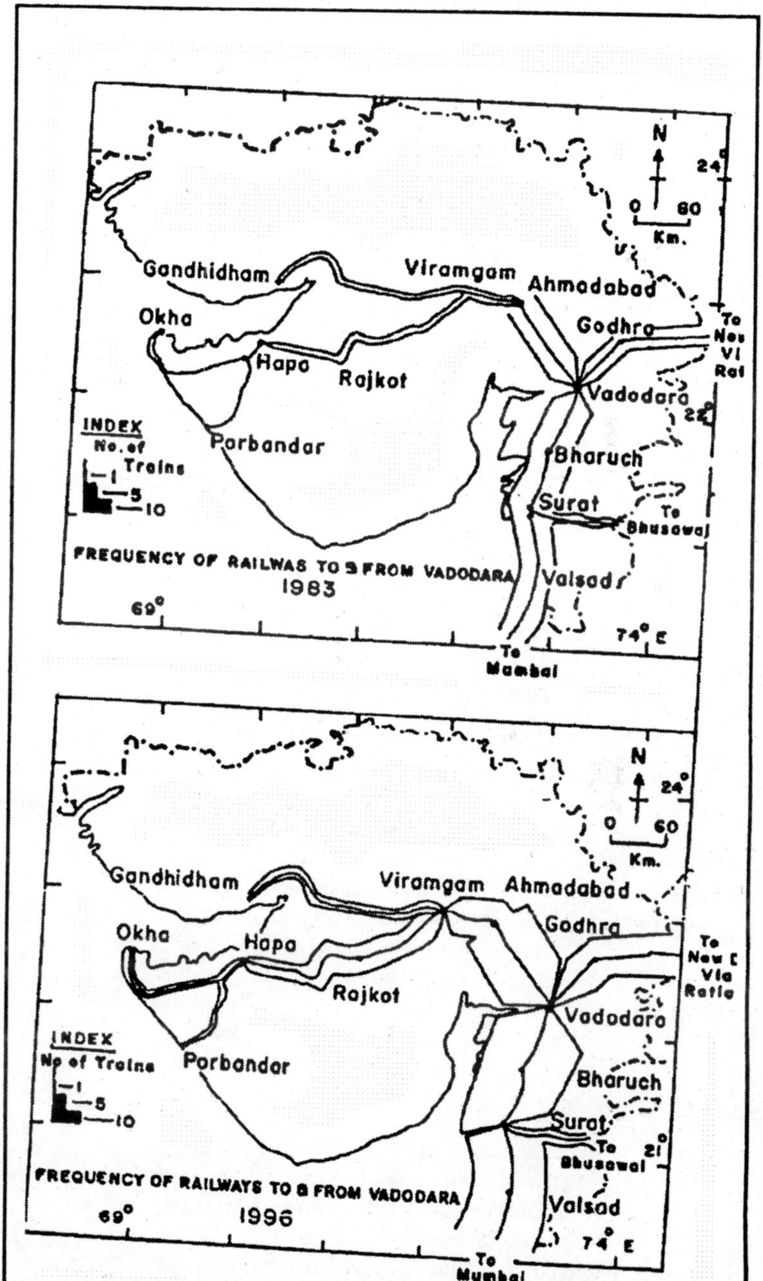

Fig.17.1: *Vadodara : Development of Railway Stations*

Table 17.2 : *Frequency of Trains, Centres, Resources, Industrial Connection with Vadodara*

City/Industrial Centre	Resources	Industry/Advantage/Problems	Number of Trains Years 1983	1996
1. Gandidham	Lignite, coal, gypsum llimestone, iron or bentonite, salt, oil natural gas.	Agricultural product, handicrafts, tourism, oil industry, Rcc. salt works, cement, organic, spinning, chemical, pipe making.	1	3
2. Okha	Bauxite, building mate	Cotton mills, mining, food processing, Chemicals, fertilizers, Machinery, Agro-based industries, oil mills, processing, of tea, salt, cement, Oil refineries, export of bauxite, chemicals, salt, cement import of minerals, gypsum, Tourism : Dwarka area.	Nil	Nil Oil
3. Porbandar	Porbandar stone, sand stone, calcite, stained gypseous clay Kmagnetite iron ore, salt, thermal power	Agro-based economy, tourism, marine fishing industries, cement, ceramics, cotton ginning textiles, oil mills, fertilizers, salt, shipyard, chemicals problem: Soil Erosion	Nil	Nil
4. Ahemdabad	Agricultural Products	Handicrafts, Manchester of east, cotton textiles, chemicals, machinery Tourism, most importance city in Gujrat.	9	13
5. Rajkot	Agricultural products in centrally located	Groundnut oil mills, machinery, textiles, ceramics, small industries, fast developing city.	1	5
6. Valsad	Agricultural Products, forest products	Highest density of industries, Agricultural and forest products, teak, timber chemical, pharmaceuticals problem : Heavy rainfall.	1	2
7. Surat	Building stone, lime-stone, mineral water, salt, forest, cotton, Agricultural Products Sugarcane, timber, rice power station: Ukai, Kakrapara	Agro based industries, textiles, fishers Diamond cutting, petrochemicals, steel hizira industrial complex and fertilizers	2	5
8. Ahmedabad	Agricultural Products,	Chemicals, textile & machinery,		

	Tobacco	Petrochemical products, heavy, machinery, milk products , Centres: Nadiad, Anad, Dakor Good transport & communication	Nill	Nill
9. Vadodara	Natural gas, thermal power, hydropower,	Refinery, chemicals, petrochemicals, pharmaceuticals, cables, heavy machinery, good transport network	Nill	Nill

Source : Statistical Abstract of Gujarat State

Table 17.2 : *Principal Commodities Loading at Vadodara*

Sr. No.	Products	Company	Station
1.	Petroleum Products	Gujarat Refinery	Karachiya yard
		IPCL	
		ONGC	Hazira
			Gandhidham
2.	Fertilizers	KRIBHCO	Hazira
		GNFC	Bharuch
		GSRC	Bajwa
3.	Imported Goods		Kandla
4.	Salt		Ahmedabad
			Gandhidham
5.	Dairy products	Amul, NDDB	Anand
		Panchmahal dairy	Godhra
6.	Military Traffic	EME, Air Force	Dhrangdhara
			Jamnagar

Source: Filed Work

constructed between Miyagam-Dabhoi-Samalkaya as a pass line by conversion of existing narrow buage line to broad guage line and electrification as well. Moreover, the railway authority has introduced new schemes for development of Vadodara railway and surrounding regions as below :

1. Full rake (a set of bogies generally, called train by general public) handling facility at Bharuch, Godhra, Nadiad, Ranoli and Bajwa.
2. Passenger utility like Microprocessor based on season Ticket Issuing Machine at Vadodara.
3. Additional line at Vashwamitri, Ankleshwar and one between Bajwa and Vadodara.

4. Computer reservation terminal extension.
5. Voice interactive system for reservation status enquiry.
6. Computer aided announcement for trains.
7. More main line Electric Multiple Unit MEMU rakes.
8. Satellite Reservation office at Alkapuri Vadodara.
9. Remodelling of Marshalling yard.
10. Latest high Horse Power High Speed WAG7 locomotives which can haul 4,700 tonnes of load and have a power of 5000 HP to gradually replace existing 132 locomotives of 7 categories, some of them are 25 to 30 years old.

Conclusion

The present study has been carried out to examine the Vadodara railway for the period from 1983-96. It is noticed that the significant changed have occurred to Vadodara railway and its surrounding. Most of facilities required to persons or travellers are available here. Even Pune going Ahimsa Express and Calcutta going Howarh Express are being announced in Marathi and Bengali language respectively. Thus, Vadodara has became truly a cosmopolitan in nature due to railway transport linkages with nearly every corner to India. And, therefore railway junction has been developed at Vadodra proves to be major in this regard.

References

Dixit, K.R. (1974) : *Geography of Gujarat*, National Book Trust of India, New Delhi.

Govt. of Gujarat (1971) : *District Gazetteers of Vadodara*, Surat, Kutch, Jamnagar, Junagarh.

Ministry of Railways (1983 and 1996) : Railway Time Table, Western Railway.

Ministry of Railways (1996-97) Train at a Glance.

Ministry of Railways (1993) : A Review of the Performance of Indian Railway (1951-1989), Railway Board, New Delhi.

Ministry of Railway (1995) : Indian Railway Year Book. New Delhi.

Ministry of Railway (1974) : Indian Railway, Century - Facts and Major Problems.

Ministry of Railways (1995) : Bharatiya Rail, Railway Board, New Delhi.

Sahai, J.N. : (1953) *Indian Railways 100 Years.*

18

TRANSPORT DEVELOPMENT IN THE NORTH EASTERN REGION OF INDIA

Anup Saikia

Introduction

Transport constitutes one of the most vital organs of the modern economic system. It allows to overcome the friction of space to move goods and people, albeit at a cost, whether in terms of energy, time or the surrogate in terms of monetary value. Is it necessary to move goods and people over space? Is there an economic rationale to this? Yes, without such a rationale no movement over space is necessary. The economic advantages of the movement must far exceed, or at least equate with the cost of friction. The advantages may be purely utilitarian, say the tourists, travel for aesthetic satisfaction or recreation, somewhat difficult to estimate in terms of benefit, but nevertheless discernable. The movement over space of people is far more complex due to the complexity of human affairs.

All movements over space require a medium, whether a railway track or a cart track or the waterway or the air which provides initial resistance in the opposite direction of the applied force[1]. This application of force involves energy, which involves financial considerations. Therefore, with the process of development and emergence of complex societies the expenditure of energy enormously increased. Often a direct correspondence is sought between the level of development of a society and the per capita consumption of energy. However, the consumption of energy is not only related to distance over which people and goods are to be transported, but also with the medium of transportation (the friction surface), the efficinecy of combustion technology (conversion of heat to mechanical energy), the efficiency of fuel and the efficiency of the transportation network.

In modern complex human society, urbanization can be considered as a special spatial feature, not that there were no urban centers in the ancient civilizations, but that the nature and magnitude of contemporary urbanization is cordinally different compared to its ancient counterpart. Often it is argued that the urbanization process itself is the result of energy economy in the society, where large scale consumption and production centers are located in close proximity to reduce the cost of transportation (and therefore energy). However, with the increase in the magnitude and complexity of the urban center and urban systems, further efficiency for smooth and low cost of movement became an important goal. The cost of time becomes a significant variables like India where investment on infrastructure is a major constraint, particularly in view of rapid urbanization and growing size of urban centers, intra-city movements of goods and people is gradually becoming a problem.

While in developing countries transport facilities are considered as an essential component of infrastructure in the developed realm it is seen as a basic and necessary social amenity. Whichever view is ascribed to the conclusion remains the same. Thus, transport services are intrinsic to contemporary civilization. The importance of transport can be understood in terms of its three capacities in fostering development (Maovenzadeh and Geltner, 1984). Firstly transport serves as a basic physical and economic infrastructure that enables raw materials, intermediate products and final products to be distributed to various locations, local, regional or international and also expends the geographical extent of markets fostering large scale production. Secondly, transport serves as a basic social and political infrastructure, facilitating the flow of ideas, people, information and goods, at the same time enhancing the political and social integration of territories, and Thirdly, transportation can serve as a force for industrialization through its requirements of infrastructure and equipment, which, if produced locally can boost industrial development.

Present day society which involves large scale production, regional and product specialization and international trade relies heavily on the transport sector so much so that a modern economy simply cannot exist without an efficient transport system. Transport and economic development are symbiotically linked and development of the former is a necessary precondition for the later (Rostow, 1964).

Study Area

The importance of transport as an infrastructure is all the more for developing regions and nothing epitomizes this more than India's northeastern region (NER) where the lack of a comprehensive transport network has been preceived as a hindrance and stumbling block to the development process (Deka, 1986). North east India is a hilly and mountainous tract of land 255,037 km^2 in area situated between 21° to 29° 30' N latitudes and 89° 46 to 97° 30' East longitudes and comprises of seven states, namely, Assam, Arunachal Pradesh, Manipur, Meghalaya, Mizoram, Nagaland and Tripura, accounting for 7.7 per cent of the country's area. It hosts about 35 million persons or about 3.7 per cent of the country's population. The region shares international boundaries with China, Myanmar, Bangladesh and Bhutan and is tenuously linked with the Indian mainland by a narrow strip of land in north Bengal called the "Siliguri neck", the width of which varies sfrom 21 to 33 km, a link not infrequently disrupted by heavy rains and floods. The difficult terrain consisting of tertiary mountains and Archean plateaus along with its location at India's north-eastern periphery and its position at the trijuncture of India, China and Myanmar all add up to its geostrategic significance (Saikia,1995). The population consists mostly of Mongoloid stock of the Austro-Asiatic and Tibeto-Chinese groups the later of both Tibeto-Burman and Siamese-Chinese branches (Taher, 1988) which are largely agricultural in terms of their economy, practicing as they do both settled and shifting forms of cultivation. The region presents a scenario in which both advanced capitalist techniques in the form of the tea plantations, and the most rudimentary form of agriculture in the guise of shifting cultivation, exist side by side. The region is poorly developed by any yardstick and has a lop sided industrial base which centers mainly around extractive activities that were developed to suit the colonial suction mechanism. Inspite of possessing substantial natural resources, the region remains a lagging behind enclave even in terms of the developing Indian economy, contributing a miniscule 2 per cent to the country's industrial output (Saikia, 1996).

Transport Scenario

It is often stated that lack of a well developed transport sector has been a hindrance to the region's development (Deka, 1986; Gopalakrishnan 1995). The following sections deal with two aspects: an assessment of transport development in the region and the gaps in this sector, between the region vis-a-vis the rest of the country; and of necessity, remains confined to an understanding of roads and railways as these two being the more important modes of transport.

Road Transport

Although roads are the Principal means of transport in the region, barring Assam, the other states did not far well in terms of the national aggregates (Table 18.1). It is evident that as far roads are concerned the region was poorly placed compared to the country

Table 18.1 : *Road Length in Nertheast Region (1st March, 1979)*

States	Total Length	% Surfaced to Total Length	Roads% per 100 Km²	Motorable Roads to Total Length	Motorable Roads per 100 km²
Arunachal Pradesh	11553	20.74	13.82	43.20	5.97
Assam	56983	14.73	72.19	64.09	6.52
Manipur	8842	17.19	39.47	72.30	28.54
Meghalaya	3690	39.97	16.40	98.56	16.16
Mizoram	2916	37.24	13.82	100.00	13.89
Nagaland	5785	24.06	35.06	96.30	33.76
Tripura	7836	15.57	74.63	68.87	51.40
Total NER	97605	17.91	38.28	67.03	25.66
All India	1604110	38.86	48.90	83.81	40.98

Source : Basic Statistics of North Eastern Region, 1980, NEC

as a whole. There can be doubt that the nature of the terrain can and does act as an impediment to road development, yet in a scenario where neither railway nor airways are very developed, the difficulties of terrain can hardly be a tenable explanation. Thus, an area 7.7 per cent of the country's total possessed only 6.08 per cent of the country's total road length, only 4.86 per cent of its motorable length and only 2.8 per cent of its surfaced roads. The only respect in which the region fared better than the national average was in terms of road length per unit population.

However, this also is not true indicator of road development since the all Inida population densities are much higher than these of the NER, which is the only reason why small population size states like Arunachal Pradesh and Nagaland had several times higher kilometrage per unit population than the national norm (Table 18.2). Yet even by such a yardstick, the relative advantage of the NER appeared to be narrowing down by the onset of the early nineties.

Table 18.2 : *Road Length in per unit Population*

States	Road Length per '000 Population	
	Year 1979	1992
Arunachal Pradesh	21.43	8.74
Assam	3.19	3.09
Manipur	7.15	3.70
Meghalaya	3.17	4.45
Mizoram	7.48	5.37
Nagaland	10.11	7.21
Tripura	4.37	4.74
Total NER	4.14	3.73
All India	2.54	2.44

Source : Basic Statistics of North Eastern Region, 1980, & 1995 NEC

Although, the position of the northeastern region as a whole was poor in terms of road length compared to the rest of the country, as regards individual states, Assam was relatively well placed possessing 3.5 per cent of the country's roads (as against Tripura and Manipur were also fairly placed. However, general roads were inadequate, and the view has been held by a senior army officer that : "these states suffer from woefully inadequate communications, as only limited surface communications have been built mainly to meet the army's logistic requirements" (Sood, 1992). In fact, : except for roads built for defence requirements after the Sino-Indian War of 1962, the internal network of roads, and civil aviation facilities in the North East or civil air links are marginal and generally short of the requirement of the people" (Dixit, 1994).

Not many changes have resulted since the 1970s, and even today the position of roads, which continue to be the most important mode of transport, leaving ample scope for improvement. Thus, although in absolute terms roads increased from 97,600 kms to 116,500 kms in 1992 (Table 18.3) in real terms vis-a-vis the rest of India the position continued to be poor.

Table 18.3 : *Road Length in NER (1979)*

States	Total Length (Km)	Surfaced Roads	Percent Surfaced	Roads Per 100 sq km
Arunachal Pradesh	7520	2496	33.19	8.74
Assam	68913	10841	15.73	3.09
Manipur	6765	2741	40.52	3.70
Meghalaya	7832	2931	37.42	4.45
Mizoram	3708	1268	34.19	5.37
Nagaland	8805	6842	77.70	7.21
Tripura	13008	4448	34.19	4.74
Total NER	116551	31567	27.08	3.73
All India	2065209	1043365	50.52	2.44

Source : Basic Statistics of North Eastern Region, 1995, NEC

From the foregoing it would appear that in addition to terrain, population is an important catalyst to road development, and Assam and Tripura with higher densities do posses greater density of roads. Other than this centrality of location seems to help Assam and act as a sort of detriment to the peripheral states. Arunachal's condition in this context is particularly bad and till today there exist district headquarters that have no road communications, as is the case with Anini the district capital of Dibang Valley district.

Railway Transport

Railway made their entry into the region when British interest in the tea industry necessitated the initiation of changes in the inefficient transport modes and the inland waterways were replaced as the most important mode of transport. British colonial profit motives apart, there can be little doubt that the introduction of railways caused a veritable revolution in the region's hitherto laggard transport system and paved the path for economic development. British interests in developing tea industry as a profitable venture necessitated improving the existing transport network. Railways were introduced in the region, yet naturally the layout and spread of the lines were planned in such a manner as to be of primary benefit to the plantation areas, even if this meant ignoring erstwhile centres of commerce and population. Instances where the planters interests directed colonial Government policy in transport development were legion.

Colonial policy was geared towards extracting tea (and utilising related items as coal and oil for the plantations) and what was required were transport networks which linked the plantations to the port of Chittagong or Kolkata for further transfer to the external markets (Barpujari, 1992). This was in conformity to the railway policy laid down by Dalhousie whose objective was to link up interior part of each Presidency with its chief port to bring the produce of the interior into the ports. At least one result of this policy was that the roads of the region, until then in extremely poor state of maintenance were improved. More significantly the development of railways took place and it was this medium of transport that was instrumental in achieving the colonial objective.

The need for a faster and more efficient means of transport than that provided by roads and riverways to link the export-oriented tea enterprise with sea ports led to the construction of the two trunk routes – the Assams Bengal Railway and the North East Frontier Railway (Dasgupta, 1988). Railways served not to extract tea but also to import labours and foodgrains to the estates and at times for the transfer of the upper Assam coals and oil that estates used as fuel (Dasgupta, 1988).

Railways even aided the extraction and exploitation of forest produce in the region. Whereas in the pre-railway times extraction of wood from the forest was carried out using elephants and thereafter via the riverways, with the introduction of railways in upper Assam saw mills extracted the logs through railways (Deka, 1979) which were introduced in 1882. The two main rail lines the

Assam Bengal Railway and the Eastern Bengal Railway ran through

Table 18.4 : *Road Length in NER (1979)*

State	Broad Gauge	Meter Gauge	Total
Arunachal Pradesh	—	1	1
Assam	266.41	2200.42	2466.83
Manipur	—	1	1
Meghalaya	—	—	—
Mizoram	—	2	2
Nagaland	—	9	9
Tripura	—	45	45
Total NER	266.41	2258.42	2524.83
All India	36504.00	21997.00	62486.00*

* Including 3985 kms narrow gauge

Source : Basic Statistics of North Eastern Region, 1995, NEC

740 miles of the province and linked the valley with Kolkata and Chittagong ports respectively by the 1920s : the feeder lines serving the purposes of connecting the tea estates with the steamer ghats, stations on the main railway lines and the trunk roads (Dasgupta, 1988).

The trend of development of railways in the region under colonial tea interests were similar to those in the rest of the country, where rail links connecting two interior areas with the metropolitan centres existed while none between the interior/rural areas themselves existed, the network being a reflection of the imperial suction mechanism (Raza and Aggarwal, 1986). Whatever, the motive the fact undeniably remained that colonial policy saw a tea change in the regional economy with the superimposition or a modern communication system particularly in the valley. In fact, "The development of a modern system of transport opening up the remote tea area and linking the province with other parts of India, began as a direct response to the requirements of the tea industry. If the pressure exerted by the tea interests and the Govenment's concern for their welfare initiated the process, its economic viability and subsequent extension were largely sustained by the growth of the industry (Sarma, 1964).

Improvements in road and railways also resulted and since the tea industry was the prime factor behind improvements in the transport sector, the impulses of modernisation remained mostly confined to the Brahmaputra Valley and to the Cachar district. The peripheral hills area remained in Subbarao's terminology, areas of isolation. Commercial interests of the colonial masters did not penetrate as far as the hills, although administrative and security factors did send impulses to such peripheries yet the magnitude of these vis-a-vis the Brahmaputra Valley were almost insignificant. Herein lay the seeds of the pattern of development that was to be found well into the second half of the twentieth century.

Even today most of the rail lines are concentrated in Assam reflecting the importance of the colonial legacy to transport network development in the region. The fact that the basic structure has not changed significantly since reflects the tardiness of post independence railway development in the region. Only Assam and Tripura had railways of certain length within their borders and the other states either had only access to a railhead or did not have even such a facility. Out of the total track length of 2524 kms only 10.5 per cent or 266 km was broad guage. The region accounted for less than eight per cent, about 10 per cent and 4 per cent of the country's broad guage, metre guage and total track length respectively (Table 18.4). Although there are plans to upgrade the rail network, the fact that precious little had been done in recent years is unmistakable. In fact apart from extension of the broad gauge section from New Bongaigaon to Lumding, significant improvements have hardly occurred since 1966 (Gopalkrishnan, 1995).

Concluding Remarks

The region has a transport system that needs considerable development and upgrading if it is to aid in bringing about a semblence of economic and regional development. Although recently several new projects have been announced under Rs. 6,100 crore economic package for the region (Times, 1996: Sentinel, 1996), notable among these being completion of the third bridge over the

Brahmaputra, namely, the road-cum-rail Naranarayan Setu Bridge at Jogighopa which will link the south bank of the Brahmaputra with Garo hills of Meghalaya; construction of a fourth bridge at Bogibeel near Dibrugarh at a cost of Rs. 200 crores, along with attempts to link all state capital cities of the region's seven states, the current status of transport in the north-eastern region as an infrastructure and a catalyst to development are not enviable. The sooner the improvements to this important sector's prospects for all-round development and the region's economy can be in a better position to further growth impulses whether from foreign direct investments or developing the tourism industry or tapping the ample water resources that the region is blessed with. However, for the present as the 17th Report on Industrialisation of Backward Areas of the Parliamentary Standing Committee on Industry has noted : "Not even an apology of a railway network exists over large parts, and roads are equally sparse" (*Assam Tribune,* 1996).

Clearly for this integral sector of the region's economy has to be substantially developed if any dent on the region's stunted pace of development is to be made and the ideals of balanced regional development espoused in the Five-Year Plans are to be realised. An efficient transportation sector would not boost commercial and industrial activities and allied activities such as the tourism sector in the region but would enhance the free flow of goods, people and ideas to and from the region. Present day problems such as transhipment of goods and people from metre guage to broad gauge are hindrances that not only cause delays and inhibit carrying capacity and throughout but frequently contribute to cost escalation as well. The conversion to broad gauge of the rail track existing that is being undertaken in a phased manner is a step in rectifying present impediments to smooth traffic flow in the region, although this again is a move long over due.

References

Haggett, P. and Chorley (1969) : *Network Analysis in Human Geography,* Edward Arnold, London.

Hoyle, B. S. (1973) : *Transport Development in Under Development Countries,* Mac-Millan London.

Medhi, S. B. (1975) : *Transport System and Economic Development Strategy for Assam*, Publication Board, Guwhati.

Raza, M. and Aggarwal, Y. (1986): *Transport Geography of India*, Concept Publishing, New Delhi.

Singh, V. K. (1983) : *Railway Freight Rates in India*, Verman and Company, Saharanpur.

Sexena, H. M. (1980): Road Transport Connectivity : Pattern and Economic Development in Rajastan, *Geographical Review of India*, Vol. 402, No. 2, Kolkata.

Thakur, R. (1983) : *Profile of a Development Strategy for Indian's* North East, Span Publication, Guhati.

Wernet, C. (1985) : *Special Transportation Modelling*, Sage Publication, London.

R. I. T. E. S. (1985) : *Regional Transport Survey of North Eastern Region*, Railway Board.

19

KONKAN RAILWAY TRANSPORT
A Study in Regional Development

B.C. Vaidya

Introduction

The role of transport in the process of regional development is of utmost significant as it accelerates the development of the region. Historically, transport network can play a key role in spatial diffusion. On the other hand, areas which have witnessed a slow pace of transport development have retarded its progress and remained relatively isolated permitting stagnation of for a longer time. The process of development have consequently, operated on a low key performing as major constraint on the processes of overall transformation of the region. It is, however, not very clear as to which comes first. It is quite possible that transport development brings its faster socio-economic changes in the region. Secondly, it is equally probable that the overall processes of regional development compel a situation demanding faster transport development. And thirdly, possibility may exist wherein both these development processes may continue simultaneously sustaining and strengthening each other, resulted to awake faster growth of region. Thus, transport intensification and development keeps relationship positively at global, national and regional level.

The existing situation is even more complex in case of Konkan region of four states. The nature of physiographical condition prevailing in the region has in the past acted as a serious constraint to develop transportation as comparable to the other parts of the country. In regards to perception of transport is not viewed as an unmixed blessing as part of infrastructural development though transport facilities are needed urgently for a quicker pace of socio-economic development of the Konkan region. The economists, planners and large section of the region feel that more than fulfilling regional aspirations of development the particular role of transport development caters the needs of an overall national growth by depriving the region of its vital resource base through effective modes of transport. The present treatise has attempted to assess the significance of Konkan Railway and examine the potential of Konkan Railway for future development of the region. Moreover, modest attempt is made to suggest feasible measures for better transport network in study region.

The Region

The region consists of part of four states, namely, Maharashtra, Goa, Karnataka and Kerala. The Konkan Railway from Roha to Mangalore crossing these districts in Raigarh, Ratnagiri and Sindhudurg districts in Maharashtra, Goa State, North Kannada, Shimaga and South Kannada districts from Karnataka which is extending of an area of about 5,400 sq. km and has population of about two crores. This railway route lies in the strip of Western Ghat and western coast. The entire area is highly undulating and of rugged topography with steep slopes with ridges, spurs and saddles on one hand and broken coastline with many creeks and caves on the other hand. Moreover, this area has hundred of streams, rapids and river and valleys lying perpendicular to the coast. This region receives rainfall from southwest monsoons and the whole area get flooded with water but becomes dry immediately in summer and therefore, this region facing acute shortage of water every year. The Konkan region could not make economic and social development due to lack of or absence of good and efficient transport system. Transport infrastructure along coast is totally inadequate. Only National Highway No. 17 (Mumbai-Goa) passing from this region. There is a less link of State Highway, Zilla Parishad and Village roads. The entire Konkan region is endowed with rich minerals, namely, iron ore, bauxite, chromite, silica, sand and many variety of fruits. In order to have social and economic development. Konkan Railway seems to be key in this direction. And therefore, present chapter is attempted to provide background for the development of Konkan region.

Existing Transport Situation

It is seen that transport infrastructure in Konkan region is quite inadequate. In order to make speedy development of region requires all modes of transport like air, water, rail and roads are complementary to each other and may convert to region poor backward strip in a boom belt. At present, there is one National Highway No 17 runs parallel to the coastline which is fully saturated and long stretches are damaged due to heavy rain. The Arabian coast of Konkan has a good ports but many of them are filled with sand hence cannot fully developed for water transport. Even then Govt. of Maharashtra has tried to overcome and has set up Maritime Board. The coastal water transport may develop in near future. Water transport is a bulk and cheapest transportation. There are many ports along coast which could be connected by regular ferry system. Some of them can be noted as Revers, Alibag, Murud, Harne, Reddy, Dhabol, Jaigad, Deogad, Ratnagiri, Vengurla, Port with Mumbai at north and Panji and Mangalore in south. Due to lack of sufficient road and other modes konkan region could not make fast and constant development neither in pre-independence period nor in the period of planning. The entire konkan has now been receiving much attention toward development of infrastructure like transport by which agriculture and industry may promote. The Konkan Railway Corporation Limited was set up on 19th January, 1990 under the Indian Company Act during the regime of Prime Minister V.P. Singh (1989) when Prof. Madhu Dandwate was holding Finance Minister, George Fernandes as Railway Minister and Shri Ramkrishna Hegde as Deputy Chairman of Planning Commission of India.

Organization of Konkan Railway Corporation

The Konkan Railway was a gigantic project. The corporation is headed by Chairman and Managing Director who is assisted by three full Directors with its headquarters at CBD Belapur, New Mumbai. There are in all 14 Directors on the Management Board of Konkan Railway Corporation including Chairman and Managing Directors and three full time Directors (Fig.19.1 and Fig 19.2). During the construction phase, the entire line of 760 km was divided into seven sectors for construction, each of about 100 to 120 km in

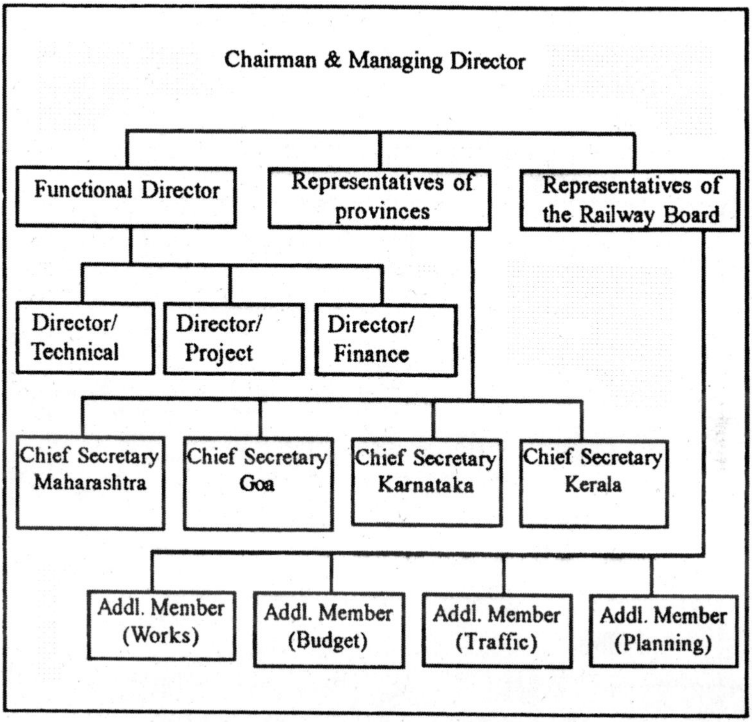

Fig. 19.1 : *Konkan Railway Board of Directors*

lengths. with headquarters at Mahad, Ratnagiri (North), Ratnagiri (South), Kudal, Panji, Karwar and Udupi. Each sector was under the control of a Chief Engineer. Moreover, for normal operations and running of service of the entire length of 760 km had divided into two regional offices at Ratnagiri and Karwar. Both regions are headed by Regional Railway Managers and are presently working. The project has provided by direct employment to 2,000 officers and staff and direct and indirect employment through contractors of about 25,000 persons The project was a challenge to young engineers who had to work in an unequal terrain to complete the project as per schedule with the technology to keep the cost low. The schedule was delayed to some extent mainly due to shortage of finance at the time of initial stage, opposition from so called environmentalists and interested parties from Goa and some tunnels and bridges took more time to complete as the terrain is very complex.

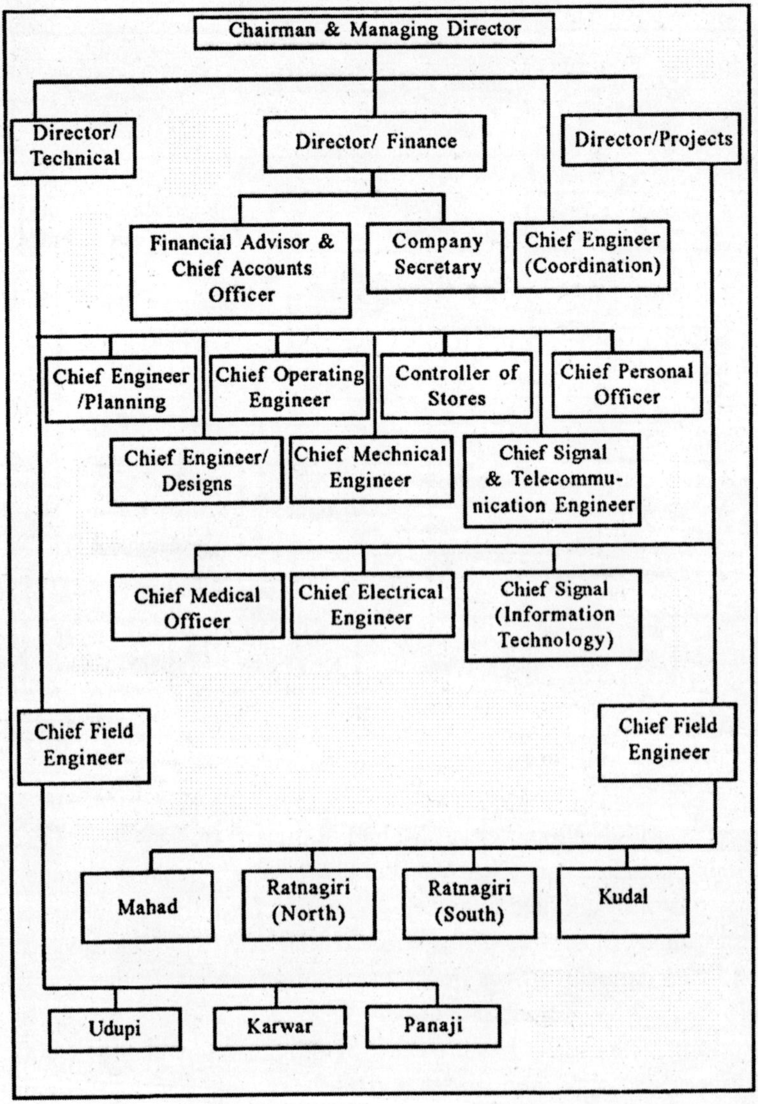

Fig. 19.2 : *Konkan Railway Organisation (Construction Phase)*

There are five components shared in formation of expenditure. One is Central Govt. and other four states, namely, Govt. of Maharashtra, Goa, Karnataka and Kerala. Table 19.1 reveals states, length, districts and talukas on Konkan Railway. Out of total 760 km, lies in Maharashtra has 50.04 per cent followed by Karnataka (35.07%) and Goa (13.09%)

Kerala has not a railway route of Konkan Railway, though state is sharing expenditure for better link with coastal region and speedy transport both for goods and people movement. Moreover, Mangalore is well-connected with Southern Railway. The entire Konkan Railway covers six districts of Maharashtra and Karnataka and nine talukas from Goa. The three districts from Maharashtra are, namely, Raigarh, Ratnagiri and Sindhudurg district Uttar Kannada, Shimago and Dakshina Kannada districts from Karnataka. Nine talukas cover from Goa are, namely, Pernem, Bardez, Tiswadi, Marmugao, Salecete, Quepem and Canacona. The Konkan Railway has many bridges both major as well as minor. There are altogether 171 major bridges of which 84 lie in Maharashtra, 51 in Karnataka and 36 in Goa. (Table 19.2).

Table 19.1 : *States, Length and Talukas/Districts on Konkan Railway*

Sr. No.	States	Length in kms	No. of Districts	Districts/Tahsil
1.	Maharashtra	382	3	Raigarh District Ratnagiri District Sindhudurgh District
2.	Goa	105	9	Pernem Tahsil Bardez Tahsil Tiswadi Tahsil Marmugao Tahsil Salecete Tahsil Quepem Tahsil Canacona Tahsil
3.	Karnataka	273	3	Uttar Kannada District Shimago District Dakshina Kannada District
4.	Kerala	Nil	Nil	Nil
	Total	760 km		

Source : Konkan Railway Corporation Limited, Belapur.

The Table 19.2 shows that more than 50 per cent expenditure on construction of Konkan railway was mainly shared by Govt. of India (51%) followed by Maharashtra (22%) and Kerala (15%). Two states, namely, Goa and Karnataka shared 6 per cent each on Konkan Railway. On entire Konkan Railway has 63 railway stations to provide facility for goods and people movement.

Table 19.2 : *Tunnels, Bridges and Stations on Konkan Railway*

Units	Major Bridges	No of Stations	Tunnels	Expenditure in per cent
Govt. of India (Ministry of Railway)	Nil	Nil	Nil	51
Govt. of Maharashtra	84	30	69	22
Govt. of Goa	36	10	10	06
Govt. of Karnataka	51	23	13	06
Govt. of Kerla	Nil	Nil	Nil	15
Total	171	63	92	100

Source : Ministry of Railway, Govt. of India, New Delhi.

The starting station on this route is Roha lying in Raigarh district and Satarda is the last station in Maharashtra. The Konkan Railway route enters in Goa through Pernem station and ends with Canacona railway station. Later on, this route goes to further Karnataka through Uttar Kannada district where Asnoh(H) is first and Mangalore is last end for Karnataka and Konkan Railway too. Out of 63 railway stations, 30 lie in Maharashtra followed by Karnataka (23 railway stations) and Goa (10 railway stations). Table 19.3 shows statewise railway stations and length from Mumbai CST.

Major and Minor Bridges

On Konkan Railway, many major and minor bridges have been constructed. More than 150 major and nearly 1800 minor bridges appear on Konkan route. While constructing major and minor bridges the area under study was surveyed and studied by either consultancy group or by own authority. It was decided not to disrupt the natural flow of water while constructing the brigdes. Of course, Konkan Railway Corporation faced many difficulties like absence of rock in river course which caused stumbling block in construction of bridges on Konkan Railway. Minor bridges can be defined as bridges with a single span of less than 12 meters or multiple span of less than 18 meters. While major bridges have a single span of 12 meters or more and multiple span of 18 meters or more.

There are 171 major bridges in number on Konkan route, approximately in every 4.2 km. Sharavati bridge is longest on Konkan

Table 19.3 : *Railway Stations and Distancen on Konkan Railway*

Sr. No.	Name of State	No. of Railway Stations	Name of Railway Stations and Distance
1.	Maharashtra	30	Rohal(83), Kolad(96), Indapur (h)(130) Mangaon (113), Goregaon (121) weer (130), Karnajadi (146), Diwan Kawathe(164), Khed (181), Anjani(205), Chiplun(211) Kamathe (221, Sarvade (229),Aravali Road(239), Sangameshwar Road (253), Ukashi (267), Bhoke (279), Ratnagiri (287),Nivsar (302), Talavade (318), Vilavade (334), Rajapur Road (350), Vaibhavwadi Road (376), Nandgaon Road (383), Kankavali (398), Sindudurgh (416), Kudal (426), Sawantwadi Road (447), Madure (H) (454), Satarda (460).
2.	Goa	10	Pernam (469), Thivim(479), Karmali(497), Neura-O-Grande(503), Verna (513), Majorda (518), Saravali (H)(520), Madgaon (525), Bali (542), Canacona (558).
3.	Karnataka	23	Asnoh (H) (576), Karwar (584), Harwada(598), Ankola(612), Gokurna Road (H),(620), Kumta (639),Honnavar (653),Manki(H), (671)Murdeshwar (679),Bhatla(694) Shiroor(702), Byndoor (H), (709), Bijoor (715),Senapura (729), Kundapura (743), Barkur (759), Udupi (775), Padubidri (791), Mulki (808), Surathkal (817), Thakur (821), Kankandi, (838),Manglore(843).
4.	Kerala	Nil	Nil

Soure: Konkan Railway Corporation Limited, Belapur
Note: Figures in brackets indicate distance from CST Mumbai

railway (2065 meters) on Sharavati river in Uttar Kannada district in Karnataka state and this was constructed within 34 months in June 1994 followed by Kali Nadi Bridges (1.4 km.) within 34 months in June 1994. Both rivers Sharvati and Kali Nadi lie in tidal zones and being near to sea, there is no floods as such and less likelihood of deep scours. Measurements of scour at Kali Nadi bridge site indicated a depression of 200 mm in river bed at pier locations. The maximum depth of standing water in Sharavati river was about 7 metres and 6.5 metres in Kali nadi river. A 288 piles in Sharavati and 180 piles in Kali nadi had been used and were socketed in hard rock with varied depth of 15 m to 40 m. Among the works at Halladi-I and -II, Seetha and Madisala had appeared a special

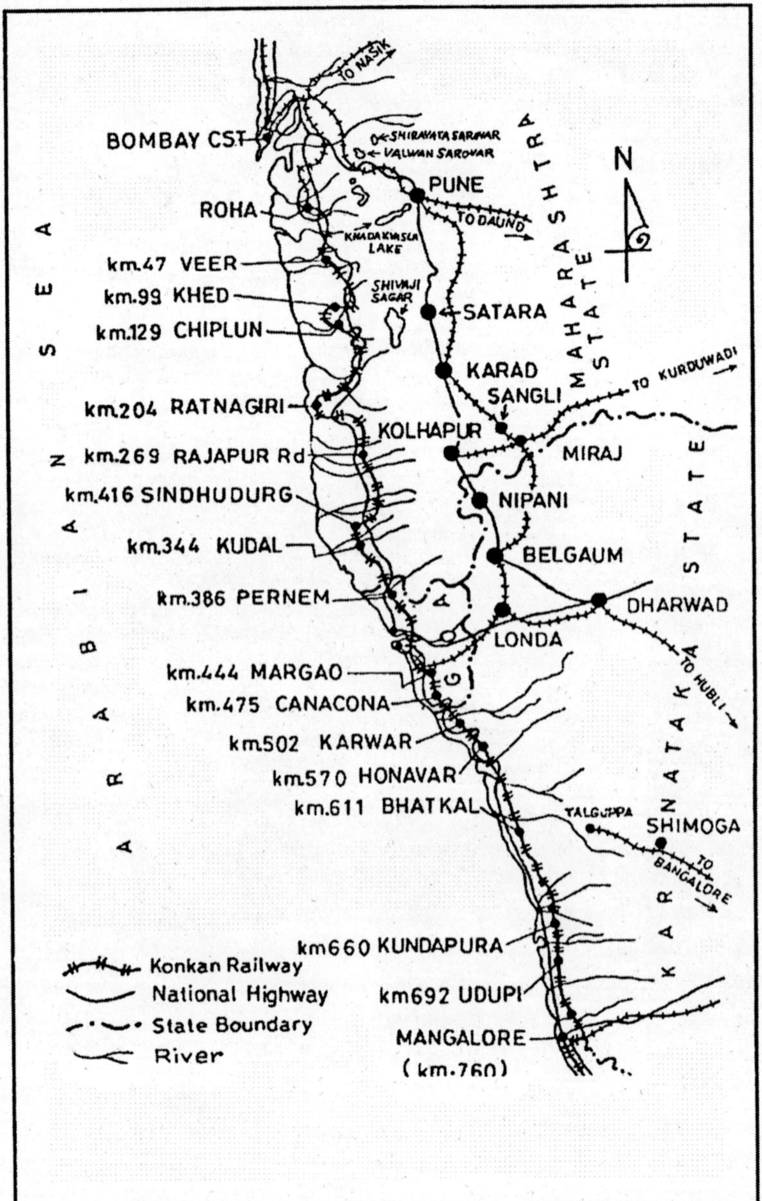

Fig. 19.3 : *Konkan Railway*

launching scheme. The railway alignment crossed the river Halladi almost parallel to NH-17 crossing. However due to presence of island the bridge was split into two sections as Halladi-I with 21 spans of 22.8 m. and Halladi-II with 10 spans of 22.8 m. Post tensional girders were used on these bridges and construction was completed in less than a year. Shastri bridge consist of 12 spans of 30.5 m box girder at CH : 147.717 to 142.113 between Sangmeshwar and Ukshi stations. Ayni bridge in Ratnagiri north section has transition curve and foundation with one metre diameter, bored cast *in situ* piles, four under each pier and six under abutments socketed in hard rock. Khedshi bridge consisted of six spans of 20 m PSC girders supported on tall, RCC hallow rectangular tapered piers upto a maximum of 4 m. height. The major bridges are shown in Table 19.4.

Table 19.4 : *Important Major Bridges on Konkan Railway*

Sr. No.	Name of Bridges	Name of Rivers	States/Districts
1.	Kundalika bridge	R. Kundalika	Maharashtra Raigarh District
2.	Savitri bridge	R. Savitri	Maharashtra Ratnagiri District
3.	Vasishtri bridge	R. Vasishtri	Maharashtra Ratnagiri District
4.	Shashtri bridge	R. Shashtri	Maharashtra Ratnagiri District
5.	Kajli bridge	R. Kajli	Maharashtra Ratnagiri District
6.	Mandovi bridge	R. Mandovi	Goa -------
7.	Zuari bridge	R. Zuari	Goa -------
8	Agnoshni bridge	R. Agnoshni	Karnataka Uttar Kannada Dist.
9.	Sharavati bridge	R. Sharavati	Karnataka Uttar Kannada Dist.
10.	Kumbar Hole bridge	R. Kumbor Hole	Karnataka Uttar Kannada Dist.
11.	Udaywara bridge	R. Udaywara	Karnataka Dakshin Kannada Dist.
12	Madisal bridge	R. Madisal	Karnataka Dakshin Karnataka Dist
13.	Sachikatta bridge	R. Sachikatta	Karnataka Dekshin Kannada Dist.
14.	Machkundi bridge	R.Machkundi	Maharashtra Ratnagiri District.
15.	Karli bridge	R Karli	Maharashtra Sindudurgh Dist.
16.	Gad bridge	R. Dad	Maharashtra Sindudurgh Dist.
17	Kali Nadi bridge	R. Kali	Maharashtra Dust., Raigarh Dist.

Source : Konkan Railway Corporation Limited, Belapur.

Panvel Nadi bridge was constructed on a deep and wide valley of an almost dry nala near Ratnagiri, was typical of the rugged landscape in and around the plateau which is famous for the orchard of Alphonso mangoes. This bridge is an arch bridge with central span of 300 m and rise of 60 m in reinforced concrete or steel. In this bridge a continuous span of PSC box girder having a uniform outer profile and span arrangement of 1×30 m. 9×40 m. 1×30m. within an overall length of 423.25 m was supported on PTFE bearings at abutment A1 and on all the piers. This concept was successfully applied in French Railway. Mandovi and Zuari are the two most important rivers in Goa state. Two bridges acrossing these were important for foundations presented problems due to absence of rock strata at a reasonable depth and adequate navigational clearance had to be provided under these bridges. Mandovi was crossed on two channels on either side of Diver Island, while Zuari was crossed just up stream of existing road bridgc opened for traffic in 1993. Though these rivers shared less catchment area, but flow is flat land with wide, shallow, estuaries, creeks with the tidal zones. Two spans of 124.2 m though steel girder was adopted for the Zuari river and vertical clearance increased to 13.7 m while vertical. For both bridges have launched 49.5 m box type girders weighting about 800 tonnes, was done by a special floating crane "Sahyadri". For launching of girders in Kali Nadi, Gad Nadi and Shastri bridges steel launching girders were used. This type of girder was used generally by contractors for shorter bridges. The tallest viaduct bridge across Panval Nadi (64 height) bridge is of special in nature for its construction. There are other bridges can be notable are, namely, Swarna, Sita, Agnashini, Shastri, Gad, Vasisthi and Savitri.

Besides this major bridges on Konkan Railway more than 1800 minor bridges have been constructed with linear waterways (Table 19.5). The minor bridges on Konkan railway can be defined as bridges with a single span of less than 12 metres or multiple span of less thn 18 metres. All bridges on Konkan Railway are designed to M.G.B. loading standard of 1987, bridges substructure and foundation code. Generally, 1.2 meters diameter RCC pipe, NP class IV has been adopted for a barrel length upto 20 meters but higher banks or barrels longer than 20 meters, 1.8 meters has been adopted. The foundation of minor bridges are generally in firm soil or rock. In few cases, RCC pipes were laid on raft foundation.

Table 19.5 : *Minor Bridges of Kokan Railway*

Sr. No.	Railway Sections	No. of minor bridges
1.	Roha - Khed Section	283
2.	Khed - Ratnagiri Section	177
3.	Ratnagiri - Sawantwadi	222
4.	Sawantwadi - Canacona Section	261
5.	Canacona- Mangalore Section	456
6.	Kundapura - Mangalore Section	200
7.	Other minor bridges (No. minor of road crossing bridge	247

Source : S.V. Selelkar, 1999.

Tunnelling work on Konkan Railway

On Konkan Railway, there are altogether 92 tunnels with total length of 84.8 km. This accounts for 11 per cent of the total length as tunnels. The construction of tunnels on this route was the most critical as construction of tunnels was not straightforward, involving many uncertainties and unforeseen problems. And, therefore survery of this region was made by Central Railway for 58 tunnels of about total length of 28.31 km, while, Southern Railway had projected a length of 37.01 km of tunnelling of 73 tunnels in 1986. After taken task by Konkan Railway Corporation, suggested 92 tunnels after comprehensive survey of this region. The notable fact is that Konkan Railway route has longest tunnel in India, namely, Karbude (6506 m) near Ratnagiri. This closely followed by the Nathuwadi tunnel (4389 m.) Tikhe tunnel (4077 m.) and Berdewadi tunnel (3976 m.). Most of long tunnels have been constructed on hard rock by good tunnelling material although joints and columnar formation in rock did give rise to unforeseen problems such as falls of loose rock or carving of the top. Out of 84.8 km. of total tunnelling 54.9 km belongs to Ratnagiri district alone and ouit of 92 tunnels 63 tunnels lie in Maharashtra, 16 tunnels in Goa and 13 tunnels belong to Karnataka. Table 19.6 reveals the important tunnels length and State. Most tunnels in Maharashtra, South Goa and north Karnataka were in basalt where there was difficulties for tunnelling. But for seven tunnels four in Goa and three in Karnataka had problems due to predominance of white and whitish yellow lithomargic clay or kaolinite, with isolated boulders known as schistose.

Karbude tunnelling was started in July, 1991 and completed in 74 months. This tunnel alignment is straight and formation is on rising grade of 1 in 150. The rock structure for this tunnel was Deccan trap through out except for the last 240, at the Ratnagiri end where it was partly in leterite and lithomagic.

The atmost care was taken for ventilation while passing trains through the tunnels on Konkan Railway. The longest tunnel on the Indian Railway was 2.2 km long tunnel on Central Railway near Khandala. The tunnel has a rising grade of 1 in 37 and sharp curve of 5 degrees with the result that when diesel tankers pass through tunnel, the fumes cause suffocation to workmen inside the tunnel and thus, ventilation problem arises when two diesel engines haul the load.

Table 19.6 : *Important Tunnels on Konkan Railway*

Sr.No.	Name of Tunnels	Length in metres	State/District
1.	Nathuwadi	4389	Maharashtra
2.	Karbude	6506	Maharashtra
3.	Parchure	2628	Maharashtra
4.	Chiplun	2033	Maharashtra
5.	Sarwade	3427	Maharashtra
6.	Honnavare	1500	Karnataka
7.	Bhatkal	855	Karnataka
8.	Byndoor	1876	Karnataka
9.	Padi	1860	Goa
10.	Pernem	1561	Goa
11.	Barcem	3343	Goa
12.	verna	700	Goa
13.	Sarzova	620	Goa
14.	Loliem	820	Goa
15.	Gopalwadi	1208	Maharashtra
16.	Berdewadi	3976	Maharashtra

Source: A Treatise on Konkan Railway, Belapur.

It was therefore, necessary to study the ventilation in long tunnels of more than two km long when used by diesel engines, specially on up gradients hauling maximum load. A consultancy was given to UNITES to study this problem and recommends solutions. Based on these study Railway Board nominated a committee to suggest a final solution for ventilation. These solution were implemented on Konkan Railway. And thus, many methods were used for excavation of tunnelling, of course, depends on type of the rocks on spot. These methods consist of preparation of the face, drilling holes, charging with explosives, blasting desassing, scraping or trimming for loose rock and removing the muck. All these operations are called a cycle.

In hand rock the work consists of repetitive cycles of drill, blast degas, trim and re-muck on a full face of the tunnel, on the other hand, in soft soil the tunnel excavation is done in two stages heading and benching. In soft soil, tunnelling had many hazards like roof collapsing or water mixed flow of water. In case of Honnavar tunnel, shield tunnelling was employed. On Konkan Railway different types of shields are used for various types of soil in tunnelling and one adopted for Honnavar is called the blade shield.

Konkan Railway has nine tunnels more than 12.2 km and longest is Karbude (6.5 km.). This tunnel was designed to ensure a minimum of 3 m per sec. air velocity in any part of 6.5 km. Motor operated mechanical dempers of six m diameter at two ends are provided in the air handling room to change the direction of airflow inside the tunnel and to prevent inflow of air when the exhaust system was being operated. This enabled purging out of entrapped pollutants from extreme ends of the tunnel. Karbude tunnel alignment is straight and formation is on a rising grade of 1 in 150. The rock structure for this tunnel was on Deccan trap throughout except for the last 240 m at the Ratnagiri end where it was partly in laterite and lithomergic. Nathuwadi tunnel (4389 m) is under Koshedi mountain range rising upto 450 m above sea level. Other important tunnels are Sawarde (3,429 m) Parchure (2,628 m), Pernem (1560m.), Old Goa tunnel (500 m.), Varna (850 m.), Padi (1917 m.), Mudgere (835 m), Honnavar (1500 m) , Bhatkal (864 m) and Byndoor (1962 m) tunnel.

Time and Distance Saving Relevance

There is a substantial reduction of time and distance from Mangalore, Cochin, Madgaon and many other places located on Konkan Railway. Earlier people have to be travelled via Pune as there was a single railway route via Pune linking Mumbai to Bangalore. To note a example of two places, namely, Cochin to Mumbai earlier distance was 1849 km consuming timing of 36 hrs, now due to introduction of Konkan Railway this distance was reduced to 1336 km and save 513 km in distance and thus, train from Cochin to Mumbai reaches in 24 hrs by Konkan route. Another significant reduction in case of Mumbai-Mangalore distance by 1127 km and saves 15 hrs. Earlier distance between Mumbai to Madgaon was 629 km has reduced to 444 km and travelling time to 10 hrs. Thus this railway route does not only saving time and distance but also substantial saving transport cost of passengers and transport cost of goods as well.

Recently load trucks service transport by Konkan railway was introduced which gained significant. This is first kind of transport in India. In the beginnging of this service truck owners were reluctant to give positive response for transporting loaded trucks by Konkan Railway. After realizing of considerable reduction of transport cost and other benefits started truck transportation by Konkan Railway. The loaded trucks are loaded on railway train at Kolad near Roha and then finally after transportation, the trucks are unloaded at Surathkat near Mangalore. This truck service transport by train saves both travelling time and cost. The earlier truck transportation, road was time consuming of 72 hrs besides many difficulties, like tier puncturing, accident, petrol or diesel consumption. But of truck transportation by Konkan Railway overcame all these difficulties of an accident, no tier puncturing and most saving of petrol or diesel. The saving of timing by trucks transport is much more significant, for example, from Kolad near Roha to Surathakal takes only 24 hrs where earlier was taking 72 hrs and, thus saves 48 hrs.

Table 19.7 : *Major Reduction in Distance and Time*

Stations Sr. No.	Earlier Distance in km	New Distance in km	Saving Distance in km	Earlier Hours Riquired	New Hours	Saving Hours
1. Manglore-Mumbai CST	2041	991	1050	41	15	26
2. Manglore- Ahmahabad	2653	1435	1218	58	32	26
3. Manglore-Delhi	3033	2326	679	65	39	26
4. Cochin- Mumbai CST	1849	1413	436	36	24	12
5. CST Mumbai- Madgaon	629	444	185	20	10	10

Source : Press Handout, Konkan Railway Corporation, Limited. Belapur

Thus, because of considerable saving of time and petrol or diesel, many long distance trains have started on this route. Table 19.8 reveals eight long distance trains are presently running on Konkan Railway. The train from Kurla to Erunakulam reaches within 27 hours by this route, whereas Rajdhani Express from Trivendrum to Hazrat Nizamuddin takes only 30 hours. Matsygandha Express has recently started and reaching in 19 hours. Earlier route via Pune was a long distance route and time consuming too . But reducing both in distance and time now resulted to gear up the movement of people and commodities of Konkan region .

Table 19.8 : *Long Distance Trains and Distance*

Sr. No.	Name of Trains	Stations From--To	Time Required Earlier (Hrs)	Time Required New (Hrs)	Saving Hrs.
1.	Netravati Express	Kurla to Erunakulam	36	27	06
2.	Rajdhani Express	Trivendrum to H. Nizamuddin	44	30	14
3.	Mangla-Akashdeep Exp.	H. Nizamuddin to Ernakulam	54	26	28
4.	Gandhi Dham Express	Gandhidham to Nagarcoil	48	32	16
5.	Rajkot-Cochin Express	Rajkot to Cochin	51	41	10
6.	Rajkot-Trivendrum Exp.	Rajkot to Trivendrum	55	47	08
7.	Matsygandha Express	Kurla to Mangalore	29	19	10
8.	Konkan Kanya Express	Mumbai CST-Manglore	30	20	10

Source : Konkan Railway Corporation Limited, Belapur.

Besides these, long distance trains of more speed, many other trains for short distance either passengers or express have started on Konkan route. Table 19.9 lists 19 short distance trains are presently existing for movement of people and goods. Earlier there was absence of railway route hence, people had to depend on National Hinghway-17 for transport being no other alternatives to this region. Due to introduction of short distance trains on Konkan route have fulfilled the dream of local people who are interested to travel for short distance for daily purpose.

The Konkan Railway is a gigantic project in India, which was constructed by applying many innovative methods and techniques that is certainly a outstanding in nature. The basic principle of BOT (Built-Operate-Transfer) was used in Konkan Railway. This Konkan route has more than 12 lakhs prestressed concrete monoblocks sleepers. This is first kind of route where

171 major and 1919 minor bridges were constructed on this route on 760 km. This is the speciality of this Konkan Railway that 84.8 km. tunnelling work spreaded over 92 locations, applying approximately tunnel in every 4.2 km. Karbude is the longest (6.5 km.) in Indian sub-continent lies on the Konkan route. Moreover longest river bridge namely, Sharavati bridge lies on this Konkan route Table 19.9.

Table 19.9 : *Short Distanct Trains and Distance*

Sr. No.	Name of Trains	Distance in Km.
1.	Diva-Chiplun Passenger	171
2.	Dadar Ratnagiri Passenger	279
3.	Kurla-Madgaon Express	509
4.	Sawatwadi-Ratnagiri Passenger	160
5.	Madgaon-Manglore Passenger	318
6.	Diva-Sawantwadi Passenger	404
7.	Pernem-Sawantwadi Passenger	22
8	Udapi-Manglore Passenger	676
9.	Roha-Veer Passenger	48
10.	Udapi-Kundapura Passenger	33
11.	Veer-Khed Passenger	52
12.	Chiplun-Ratnagiri Passenger	77
13.	Khed-Chiplun Passenger	30
14.	Rantagiri-Sawantwadi Passenger	161
15.	Kundapura-Canacona Passenger	181
16.	Canacona-Madgaon Passenger	32
17.	Pernem-Madgaon Passenger	57
18.	Manglore-Udupi Passenger	67
19.	Ratnagiri-Madgaon Passenger	238

Source: Konkan Railway Corporation Limited, Belapur.

Material and Matter of Proud

The Konkan Railway was constructed by applying innovative methods and materials of more than 610 thousand tonnes of cement, 115 thousand tonnes of steel, 14 thousand tonnes structural steel, 87 thousand tonnes rail. No single wooden sleepers have used in entire work. RCC sleepers are the major consideration on Konkan route. 88 million cum tonnes earthwork evolved and this entire materials out of tunnels are so huge wall of 12 meters width and one meter height can ce constructed on the coast line of Maharashtra for 720 kms.

Table 19.10 : *Salient Features of Konkan Railway (Roha-Mangalore)*

Sr.No.	Salient Features	Description
1.	Guage	Broad Guage (1676 mm.)
2.	Length	760 km
3.	Ruiling Gradient	1:150 (compensated)
4.	Major Stations	63
5.	Major Bridges	171 (With a total linear waterway of 21.50 km)
6.	Minor Bridges	1819 (Liner waterway of 5.73 km)
7.	Number of Tunnels	92 Total length of 84.8 km.)
8.	Longest Tunnel	Karbude Tunnel (6.5 km.)
9.	Talest Viaduct	Panval Viaduct (64 metres height)
10.	Longest Bridges	Sharavati River Bridge (2065 meters)
11.	Track structure	52 kg long welded ralls
12.	Number of Sleepers	12,87,000 prestressed concrete Monoblocks

Source : Konkan Railway Corporation Limited, Belapur.

There are more than 13 lakhs pressed concrete monoblock are used all over 760 km length. In track construction, modern Japanese pressure welding technique was applied. Konkan Railway route is without any joint, no fish plates, and hence causing less chances of removing fish plates. Moreover, no noise would arise due to absence of joint. One notable novelty is that the track ballasts are not used anywhere on the track. The entire track of rai!way was built with speed potential of 160 kmph and utmost care has been taken to maintain visibility and to keep less number of curves with a radious curves more than 1250 metres and gradient of 1:150.

Konkan Railway route was constructed to face earthquake at 6+ Ritcher scale or any tremer alignment. R.C.C. hollow pillars have been used for construction. The minimum number of road crossing have been kept along the NH-17. There are more than 247 crossing both road bridges and under bridges on entire Konkan route. In order to larger smoke discharge by engines in tunnels, specific ventilation of huge shafts at the top of tunnels have been constructed centrifugal fans of 150 kw have been fixed to exhaust the inside air of Karbude tunnel (6.5 km length) and Jet fans are provided for artificial ventilation in Nathuwadi, Parchure, Boserem and Karwar tunnels on Konkan Railway route.

The speed of train passing through tunnels is allowed to 50

kmph as against normally to 15 kmph. Moreover, four aspects high speed signals are provided at stations to maintain required high speed of trains. To handle vice communication data and T.V. signals, optic fibre cables have been used for telecommunication purpose. Besides this, Konkan Railway was well-connected with both Southern Railway with Mangalore and Central Railway with Mumbai. All performing activities of trains like scheduling, tracking, monitoring or rolling stock and assets management are provided by telecommunication. Utmost care has been taken considering future prospect of industrialization and preservation of landscape of Konkan region. New tunnels, bridges and stations buildings are additional cultural landscape to natural scenario and all effort have made for less plain cutting no discruption to volume of water stream been direction and slope. A massive tree plantation scheme on strips of land along route of Konkan Railway has been started to enrich natural beauty of Konkan region.

Future Prospect

The introduction of Konkan Railway in west coast has opened up new era to extract available human resources and will exert considerable inpact to make substantial increase in economy of this region. Resource based industry, agriculture, tourism would flourish to large extent besides living standard of people. The industrial growth in this region is expected to increase both employment potential and new industries (medium and large). This would open the markets both north and south India for the industrial products. The entire Konkan area is vast storehouse of minerals like iron ore, bauxite, chromite, silica, sand and the train service will give a fill lip in to ferrous and non-ferrous industries. Forest and mineral products, the luscious mangoes, coconuts and cashew have potential in this area. This rich treasure house had not been fully exploited previously due to the absence of an inadequate practical and economical means of transport. Among the major industries there are petroleum refinery planned plant at Mangalore, steel plant near Mangalore (being planned of 2 million tonnes, Kaiga nuclear plant near Karwar. The proposed 1000 MW Thermal Station at Dabhol petroleum refinery near Chiplun, Gas Cracker at Nagothane near Roha. A pig iron plant with capacity of one mfa has been set up at Redi near Banda (Maharashtra). A steel plant is also carry up at Satarda (Maharashtra). M.I.D.C.of Taloda, Roha, Patalganga, Mahad, Usar, Natothana, Pen, Panel, Rasayani, Alibad, Chiplun, Phophali, Khandpodi, Lote, Parshuram, Ratnagiri, Jaigad, Deorukh, Rajapur, Dapoli, Kharachi, Mirjole, Kudal, Kankawali, Majagaon, would strengthen many ancillary units, SSI units and medium types of industries. Konkan Railway may boost for tourism industry. In Goa, it is easier to visit to beautiful beaches formers temples, ancient Churches and unique cuisine of the area besides its "sussegad" life style. In Maharashtra, forts, hill stations and beaches of Arabian coast can be visited by the Konkan railway.

Concluding Remarks

It is safely said that Konkan Railway as passing through backward and less transport infrastructure region along coast may develop fastly in future. The forest and mineral resources can be tapped on large scale of this region and may start local process of manufacturing units or transport these resources to other place of industry. Local human power can be supplemented to this manufacturing process. Though parallel National Highway- 17 appears along coast but it remained badly damaged in rainy season due to heavy rainfall in this region. This can be overcome by this railway route and provides transportation facility for both people and goods throughout the year and hence industrial growth based on local raw material is expected to increase and will boost the employment potential to Konkan region. Thus, whole area would get industrial web besides other training centres for manufacturing. Moreover, local credit societies, co-operative banks would open in this region for loan purpose for small scale industry and cooperative processing and thus marketing units may flourish on a large scale. The starting of Konkan Railway in this region may ensure the immense potential of the coast in the fullest extent. In addition to this would bring prosperity to this region and may welcome changes in the life style of the inhabitants. Thus, this railway opened up new era of socio-economic development of this backward region. The Konkan Railway was dedicated to our nation as National Railway on 1st May 1998 by Honourable Prime Minister Atal Bihari Vajpayee. In conclusion, this railway is of immense beneficial to people of India for the movement of goods and people as well.

Refenceces

Bhatnagar, N.L. (1956) : *Transport in India and Abroad*, Jain Prakashan, Nath & Company, Meerut.

Dastane, S. & Hardikar (1992) :*"Maharashtra"*, 1992

Gatade, D.G. (19): *Industrial Geography of India*, Printwell Publication, Jaipur.

Govt. of India (1997):" Indian Railway : A Glorious Half Century " *Yojana*, Nov.1997.

K.R.C.L.(199): "Konkan Railway Corporation Limited" *Press Hand out*, 24th August, 1999, Belapur.

Pathak, S.P. (1998): "A Dream Comes True", *Indian Railway* , July - August, 1998, Ministry of Railway, New Delhi

Rajaram, B. (1998): "The Konkan Railway", *Journal of Indian Railways*"April, Ministry of Railways, New Delhi.

Salelkar, S.V.(1999) : *"A Treatise on Konkan Railway"*, Konkan Railway Corporation Limited, Belapur, 1999.

Sreedharan, E. (1994) :"Konkan Railway - A Trend Setter, *Journal of Indian Railways*, April, 1994. Ministry of Railway, New Delhi.

Sreedharan, E. (1997) :"Success Story of Konkan Railway, *Journal of Indian Railways*, April. 1997, Ministry of Railway New Delhi.

Sreedharan, E. (1993) :"Konkan Railway Project" *Indian Railways Journal*, April, 1993 : Ministry of Railways, New Delhi.

URBAN TRANSPORT IN MAHARASHTRA

Vidyadhar Deshpande

Introduction

The Urban scene in India has noticed phenomenal change in last four decades. The total population has grown from 3.5 times from 1901 to 1991. Out of this, urban population has increased nine times in the same period. Urban areas having 26 per cent of the total population of the country, are important as these contribute about 55 per cent of the country's G.D.P. and this proportion is expected to grow to about 80 per cent in next two decades.

Trade, Commerce and other economic activities thrive in the areas where accessibility is high. Urban Development takes place around such activity generators. Transportation is, thus, crucial for the economic growth of the cities. Urban mobility is a basic requirement of urban economy. As a result of this, urban transport is gaining more and more importance with the passage of time. The process of urbanization in developing countries resulted in the sudden rise in urban population in last two to three decades. This is quite different from the process of urbanization in developed countries which was gradual and accelerated a bit in twentieth century. The rapid urbanization in developing countries like India has posed serious problems before the city managers and administrators in the field of Housing, Infrastructure, Social amenities and Transport.

Urban Transportation has a very important role to play in the overall urban development. This necessary and prerequisite service provides life line to the urban centre and helps to sustain the city by assisting socio-economic and cultural activities.

An Overview of Urban Transport in Maharashtra

In Maharashtra Mumbai is served by the suburban railway systems.

Pune and Pimpri Chinchwad are also served by suburban railway system to some extent but its impact is not noticeable. In Mumbai the suburban railway system was developed gradually keeping pace with expansion of the city to meet the demand of the commuters over a period of time. Railway commuters in Mumbai have increased from 7.83 lakhs in 1950-51 lakhs to 50 lakhs in 1998-99. Though the number of trains per day have been increased from time to time to meet the evergrowing commuter demand the constraines on line capacity, physical space and paucity of resources have made it quite impossible to improve and expand the system adequately. As a result of this, the suburban trains carry over load and the commuters have to travel in inhuman conditions The comprehensive traffic and transportation plan prepared for Pune reveals that out of total 24,04,000 daily personal trips, 2,22,800 trips i.e. only 9.55 per cent of the total trips were performed by rail. This is very small percentage as compared to 50 per cent in case of Mumbai.

Road Based Public Transport

Road basedn public transport due to ready infrastructure plays a key role in providing mobility to the people. This system needs low initial capital outlay and has addition or incremental quality. The rail based system, though effective cannot be developed due to its capital intensiveness particularly on the background of inadequate financial resources. It is, therefore, needless to say that the best available alternative is the bus system. Such system if efficiently operated and made reliable may divert personalized vehicle trips reducing congestion on the streets, accidents and increasing overall productivity.

In Maharashtra Mumbai , Thane, Pune, Pimpri, Chinchwad, Kolhapur, Solapur are the urban areas which are served by the respective Municipal Transport services. In general the bus services managed by these Municipal Transport Authorities barring the exception of Mumbai are highly unreliable and, therefore, people choose to have personalized mode of transport even at the risk of their lives on the congested roads of the cities. All the cities have witnessed phenomenal rise of two wheeler vehicles. Had the bus transport was reliable this rise perhaps could have been delayed if not altogether avoided.

One of the major allegations against these transport undertaking is that they are loss making propositions, and used to be highly subsidised either by Government or by the local bodies. However, this can be avoided by efficient management and by introducing the concept of privatization so as to make the system more reliable and efficient.

Personalized Transport

It will be no exaggeration to say that there is auto two wheeler explosion in Maharashtra and this trend will continue unless efficient and reliable mass transport system in urban areas are introduced and operated. Heavy growth of this mode can be judged by the fact that in Maharashtra state two wheelers have grown from 12,532 in 1961 to 32,87,581 in 1998 indicating phenomenal rise during the period of less than four decades (Table 20.1). In 1998 auto two whelers were 65 per cent of the total vehicles on the road. In Pimpri-Chinchwad the same percentage in 78.77 per cent. Whereas in Kolhapur it is 72.26 per cent and in Pune it is 72.85 per cent as per statistics of 1997. This explosion of auto two wheelers is bringing heavy pressure on energy consumption and also creating lot of environmental problems.

Intermediate Public Transport

In addition to bus transport in certain cities, three wheelers have contined to grow for the simple reason that they provide flexible (time and space) service at the fixed tariff schedule. In Maharashtra auto rickshaws have grown from 379 in 1961 to 3,24,624 in 1998 indicating tremendous rise over less than four decades. In Pune and Pimpri-Chinchwad, they form 6.83 per cent and 2.76 per cent of the total vehicles on the road respectively.

Importance of Public Transport

The overall growth of private modes of Transport and increasing unabated trend shows clearly that the urban streets will be cluttered with many more vehicles especially during the peak hour rush period. This will definitely affect the urban mobility and will increase the travel time unless it is catered through mass transport system. The

Table 20.1 : *Yearwise Motor Vehicles on Road in Maharashtra*

Sr. No.	Category	1961	1971	1973	1980	1985	1990	1998
1	Motor cycle/Scooter	12532	86749	157092	314571	745193	1588663	3287581
2	Cars	50589	118808	171826	216640	290386	411559	694125
3	Taxi Cabs	6086	17970	21777	30701	39196	42489	73210
4	Auto Rickshaws	379	3888	8049	24802	69048	119186	324624
5	Stage C- Contract Carriages	4135	7561	9629	13955	18252	21953	37473
6	Public Carriers	17177	37128	50139	70991	109689	148237	329654
7	Private Carriers	5943	16275	22875	30927	35456	39778	
8	School Buses	269	478	534	581	772	993	1470
9	Ambulances	119	423	613	852	1446	2181	3549
10	Private Services Vehicles	177	953	1368	2095	2943	4384	8328
11	Trailors	1554	7376	10089	20549	38109	56833	135783
12	Tractors	899	8335	11130	21684	38269	57157	135004
13	Others	285	1086	1149	1187	2882	4807	9191
	Total	100144	307030	466270	749535	1391641	2498220	5039992

Source : Motor Transport Statistics of Maharashtra 1998, Transport Co., Mumbai

car has a capacity of four passengers, scooter/motor cycle has capacity of two passengers whereas the bus carries about 50 passengers.

The bus transport has certain merits, fixed schedule, fixed tariff and fixed route provide a certain level of service which satisfies the demand including the peak hour demand, on large scale. In addition, high capacity makes it possible to reduce vehicle density on the streets. The bus transport can serve growing cities to a great extent till a situation warrants rail based rapid mass transport system. Like in Mumbai they even serve the role of equal partner in dealing with the traffic demand of the commuters. Buses are, therefore, looked upon as an important tool for solving problems of urban transportation. Moreover, it is quite easy to systematise the bus operations. But system, therefore, can be considered as an important means of intracity movement of the people. It is, therefore, needless to say that the emphasis should be laid on this mode of transport and the bus systems are optimised and managed efficiently.

It is common experience that the bus transport always depends on the aid in the form of grants and subsidies from the Government and parental organisations. It is hence all the more necessary that the efforts are made to attract the people from their personalized transport to public transport. Efficient bus system management is one part of operation of this mode of transport, the other important part being efficient traffic management of the city which will help speedier movement of the bus fleet. In urban traffic planning exercises, buses will have to be given some kind of priority over other vehicles. Such efforts were made in Mumbai to have separate lanes for only BEST buses. The national transportation policy has also recognised the importance of this mode of transportation so as to divert traffic from the personalized modes to public transport.

Performance of the Municiplal Transport in Maharashtra

There are six urban centres in Maharashtra who have Municipal Bus Transport. A few urban centres are catered through Maharashtra State Road Transport Corporation buses. The six urban centres are Mumbai (BEST), Pune (PMT), Pimpi Chinchwad (PCMT), Solapur (SMT), Kolhapur (KMT) and Thane (TMT).

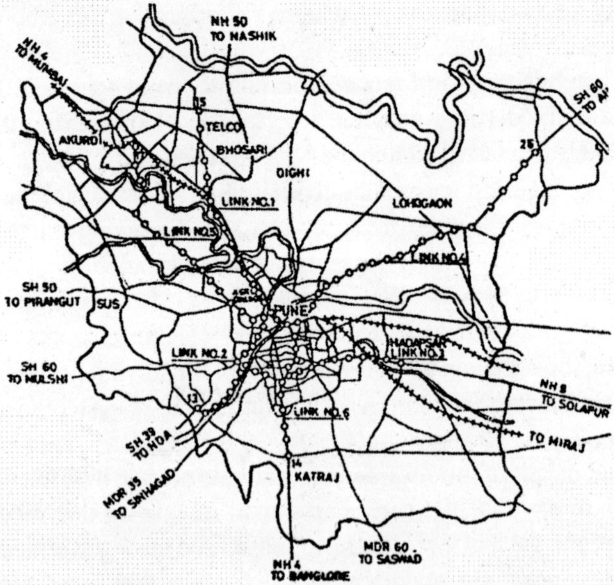

Fig. 20.1 : *MRTS for Pune Metropolitan Area*

As in March 1997 BEST has a fleet strength of 3,719 whereas Pune Municipal Transport had 810 buses, Pimpr-Chinchwad had 238 buses, Solapur Municipal Transport has 156 buses, Kolhapur Municipal Transport has 160 buses and Thane Municipal Transport had 227 buses. Financial performance of these undertakings is given in Table 20.2 indicate that all these transport undertakings suffer a loss except T.M.T. As regards the utilisation of fleet in 1997 i.e. bus utilisation per km. for buses on road, it is maximum in case of P.C.M.T. 279.2 followed by P.M.T. 218.5, S.M.T. 241.4, K.M.T. 235.4, BEST 203.6 and T.M.T 193.2 Table 20.2 indicates vital statistics of these six Urban Municipal Transport undertakings which throw light on their overall performance.

It can be seen from the performance statistics of the above

six urban transport undertakings that they are highly energy and cost inefficient. There is lot of scope for improvement. The problems of road congestion, pollution and increasing financial and energy costs are likely to reach intolerable levels unless corrective measures are taken immediately. It is needless to say that for obvious reasons of avoiding wasteful personalized trips, public transport in cities needs some special attention and favours from Government. The fare policy and subsidy option provides a valuable tool to achieve this objective.

Problems of Urban Transport in Maharashtra

Transportation Problems

Problems of urban transportation can be broadly categorised. They are almost the same everywhere. They can be stated as inadequate infrastructure, lack of strong mass transport system, growing personalized vehicle ownership specially two wheelers, resulting congestion on the already inadequate street network, reducing journey speeds, increasing accidents and frustration etc.

Traffic Problems

These are a product of multiple factors ranging from non controls to unplanned growths. Some of them seen on roads are mentioned below :

(i) The urban traffic is extremely heterogeneous. Different modes, fast as well as slow are intermixed. Within fast modes also speeds vary considerably. The extreme heterogeneity and fast and slow modes sharing the same way reduce overall journey speed, added to it the inadequate carriageway and lack of discipline create frequent traffic jams during peak hours when, in fact, the transport system is required to be the most efficent.

(ii) Sufficient spaces for on-street and off-street parking of vehicles are almost never provided with the result number of vehicles are parked on the carriageway itself, reducing carriageway capacity. This results in overcrowding and congestion leading to jams.

(iii) Lop sided and directional imbalanced flows create paucity of capacity in one direction and under-utilisation of capacity

Table 20.2 : *Urban Transport Performance at a Glance*

Municipal under-taking	Fleet Stre-ngth	Fleet Utilis-sation %	Bus utilization per bus on road per day (kms)	Bus utilization per bus held per day	Rate of break break 10000 effective (kms)	Rate of accidents per lakh effective (kms)	Earnings (Palsa/ km)	Occupa-tion Ratio %	Total Cost (Paisa/ km)	Profit/ Loss (Paisa/ km)	Fuel Consum-ption kms per Litre
Mumbai (Best)	3356	94.4	203.6	192.2	1.98	0.57	2100.0	82.73	2514.2	-413.6	3.00
Pune	775	80.8	247.6	200.3	1.30	1.80	1281.1	96.61	1427.8	-146.7	3.44
Thane	217	88.9	187.8	167.0	3.83	2.76	1678.9	101.08	1670.0	9.0	3..31
Kolhapur	155	85.2	243.1	207.0	32.91	1.65	1080.0	62.73	1170.4	-90.4	3..54
Solapur	146	66.4	235.0	156.1	N.A.	N.A.	1064.4	54.68	937.1	-127.3	3.62
Pimpri Chinchwad	220	74.1	270.7	200.5	3.38	0.70	1092.6	76.66	1145.8	-53.1	3.52

Source : Profile and Performance, Facts 96-97 - CRT, Pune

in the other. Such tidal flow situation are never attended to.

(iv) Heavy peaking producing a strong skew in the peak and non peak flows results either in provision of higher capacities to meet the peak demand only resulting in uneconomic losses.

(v) Overcrowding resulting out of inadequate infrastucture with a failure to augment the same from time to time.

(vi) Failure to update or provide suitable facilities by way of standard geometric designs and segregation-directional and modal to optimise level of service and reduce conflicts.

(vii) Lack of traffic operational controls to suit the changing needs of the dynamics of traffic.

(viii) Presence of hazardous elements endangering the safety of travelling public (stray animals).

(ix) Pedestrians the important component of traffic are always neglected. Many times footpaths are not provided and when provided are of inadequate width. Further, there are almost always heavy encroachments on footpaths like hawkers, vendors etc. As a result they become inadequate to cope up with high volume of pedestrian flow forcing pedestrians to walk on the road carriageway causing further congestion and ceating hazardous situations.

(x) Environmental deterioration is an impact of all the above factors. Noise and smoke pollution is everyday experience in big cities.

(xi) Lack of co-ordination between the agencies responsible for Traffic Management and those with other civil duties is too well known. Approaches and dstyle of functioning of various authorities as far as traffic is concerned is at best collous and unscientific to hilt.

(xii) Lack of Financial Resources : Traffic and transportation planning and implementation both have greatly suffered due to inadequte resources. With the result the traffic and transportation infrastructure in cities has almost remained static whereas the demand on it is incessantly increasing. It is no wonder traffic snorts and bottlenecks is the result and mobility of average urbanite is the victim.

Trent of Urbanization in Maharashtra

Urbanization coupled with industrialisation has resulted in the growth of urban centres. Maharashtra in highly urbanized state with 38 per cent of the population living in urban areas, and it is expected by 2001 this per centage will go upto 45 to 48 per cent.

There are 229 cities and towns in Maharashtra. Out of these 15 are cities with Municipal Corporations, 18 are 'A' class Municipal Towns, 47 are 'B' class Municipal towns and remaining 163 are 'C; class Municipal Towns. Distribution of urban population in various classes of cities as per 1991 census show the following pattern :

Table 20.3 : *Classification of Towns according to Population*

No.	Classification of Towns	Population (in lakhs)
1	Corporation cities	184
2	'A' Class Municipal Town	44
3	'B' Class Municipal Town	27
4	'C' Class Municipal Town	34
	Total	289

Remaining urban population lives in the fringe areas, cantoments etc. in urban agglomerations. Bulk of the urban population about 70 per cent stays in corporation towns and 'A' class Municipal towns. There are 33 cities/urban agglomerations each of which have a population over one lack. These 33 cities are the places where transportation problems can be encountered in causing 3 to 4 decades and they deserve immediate attention.

Traffic and Transportation Planning

Transportation planning has to be well co-ordinated with the landuse planning and development. Urban Transportation System as an integral part of the urban systems, makes maximum contribution to sound urban growth and meets transportation needs equitably, economically and efficiently. However, for neglect of transportation planning in a scientific manner for long time, large cities and towns are facing traffc and transportation problems. Proper transportation planning can effect substantial economy in the traffic operations by creation of requisite facilities for smooth and safe movement of traffic amply supported by parking lots, truck terminal facilities etc.

Transport Planning Efforts in the State

The Urban Transportation Planning is done for two phases, namely, long-term and short-term. Comprehensive Transportation Planning with emphasis on demand projection and devise supply therefore is long-term exercise whereas traffic system management planning for traffic operations plans with emphasis on optimising use of existing facilities is short-term exercise. The main aim of this planning exercise is finding solutions to immediate problems with short tems objectives.

Out of 33 one lakh plus cities mentioned earlier the major urban transportation problem would be mostly confined to the million plus cities and also those cities which would reach this population threshold in recent future. The biggest city in Maharashtra is Mumbai for which elaborate traffic and transportation studies have been carried out during last 3 to 4 decades. The transportation wing of Mumbai Metropolitan Regional Development Authority is looking after the traffic and transportation problems of Mumbai. However, similar attention, was not unfortunately paid to the remaining metropolitan areas in the state. This lopsided approach to the urban transport problem in the state was realized by Directorate to Town Planning and had decided to concentrate its attention on the transportaion studies and eventually evolution of solutions for these cities. With this in view Directorate set up a Traffic and Transporation Cell with two objective, namely, (i) to undertake transportation planning studies for metropolitan cities in the first instance and (ii) to study traffic and transportation implications on the land use planning through the development plan making process and suggest transport related proposals in the land use plan. In Maharashtra Development Plans of Municipal Towns are prepared with due attention to existing status of transportation system and proposals are framed for good road network and reservations for parking lots and transport ternimal bus including terminals and truck terminals etc.

This cell was entrusted with the work of preparation of comprehensive Traffic and Transportation Planning for Metropolitan cities. Accordingly C.T.T.Ps were prepared for Pune and Nagpur Cities. After this, some trafic and transportation studies were made for Kolhapur, Solapur, Akola, Jalgaon, Bhusaval, Pimpri-Chinchwad, Aurangabad etc. and junctions improvement schemes for many junctions in important cities in Maharashtra.

Comprehensive Traffic and Transportation Plan for Pune

This Survey was done long back in 1980. This survey was based on 25 per cent sample survey of households. This study was augmented by carrying out following studies :

(i) Outer cordon (ii) Goods Vehicle (iii) Rail Commuter (iv) Screen line survey.

This data was further augmented by travel demand studis and travel characteristic studies. Based on these studies proposals of the transportation plan were framed. The salient features of this plan were as under :

(i) Comprehensive cycle network estimated to cost Rs. 1400 crores. at 1981 prices. (ii) A High Capacity Mass Transportation Route (HCMTR) together with terminal facilities costing around Rs. 100 crores at 1981 prices and (iii) Proposal for sites for parking, bus terminals, bus depots, interchanges truck terminals etc. The cost of implementation of T and T plan was estimated to be Rs. 193 crores at 1981 prices. The cycle network proposal of this plan was taken up for implementation by Pune Municipal Corporation with the assistance of HUDCO and the State Government. Some work on the HCMTR was also done by the Pimpri-Chinchwad Municipal Corporation. Part of the land required for HCMTR was also acquired by the P.C.M.C. Inspite of the medical prescription through the C.T.T.P on Traffic and Transportation ailments of this area nothing much could be achieved due to lack of serious and sincere efforts to implement the proposals of the C.T.T.P. Now the situation is also changing rapidly. The cycle network which was most relevant in 1980 is no longer important as the cycles are slowly diminishing in number on the advent of the two and three wheeler auto rikshaw. They are outnumbering the cycles and forming major i.e. more than 70 per cent traffic load on major roads of extended Pune Municipal Corporation limit. This is also the state of affairs in other cities. These proposals will have to be examined in the changed circumstances. The development plan for the extended Municipal limits of both Pune and Pimpri-Chinchwad Municipal Corporation is under preperation. Various traffic and volume count surveys are under preperation. However, the results are yet not available.

The Municipal Administration of both P.M.C. and P.M.C.C. has rightly understood the importance of rail based Public Transport System and they have carried the feasibility report for the development of L.R.T. (Light Railway Transport) through RITES. This study was done for the area admeasuring 809 sq.km. earmarked as Pune Metropolitan Area (PMA) in the sanctioned Regional plan of Pune of 1971. This study was done for Pune Municipal Area to serve the forecast travel demand by the year 2026. During this study, it is observed that highest vehicular volumes recorded were 7420 P.C.U. during Morning Peak Hour. The highest volumes of passangers in passanger vehicle was 16675 during MPH and 23070 during Evening Peak Hours. The share of auto two wheelers in observed vehicular volumes ranged predominantly between 20 per cent to 80 per cent while that of bicycle was 11 per cent to 40 per cent and of bus 0.2 per cent to 6 per cent, The road capacity urilizatoin estimated at 60 locations across P.M.A., 48 were functioning at volume to service volume ratio more than 1.00 indicating congestion.

Out of major intercity roads radiating from Pune, NH-9 to Sholapur carried the heaviest passenger volume at 61,138 trips per day. Journey speeds during peak periods on typical working days on major corridors ranged between 5 kmph to 60 kmph. The surveys further revealed that average number of vehicle owned per household was 0.74, for bicycle 0.51, for auto two wheelers and 0.03 for car. 51 lakhs trips were performed daily in 1996. Residents of P.M.A. made 41 lakhs trips with trip origin and destination in P.M.A., Per capita trip rate for all intra P.M.A. trips was 1.14, whereas corresponding inter zonal P.C.T.R. was 0.79. Inter city trips constutued 14 per cent of total trips. Bus and rail trips constituted 22 per cent of all intra P.M.A. trips while by 2 whelers, car and auto rickshaws were 41 per cent and walk trips 37 per cent out of intra P.M.A trips, work trips constituted 57 per cent trips for education 35 per cent and others 14 per cent

Based on the above survey and potential trip demand. RITES have suggested following six lines :

Table 20.4 : *Potential trip demand*

Line	Description	Length (km)	Year of Commissioning
1.	Agriculture college to Telco	12.543	2002
2.	Agriculture college to Varje	10.103	2006
3.	Karve-Road Swargate-Hadapsur	11.560	2011
4.	Agriculture college to Lonikand	22.840	2016
5.	Agriculture college to Chinchwad	16.110	2021
6.	Agriculture college to Katraj	12.760	2021
	Total	85.916	

Based on projected traffic of 36000 PHPDT by the year 2026 Rail based Mass Rapid Transit system (MRTS) is being proposed for Pune City. It is proposed that initially there will be four train formation with six vehicle train with frequency of three minutes. Each vehicle of 30 m length and 2.26 m vide will be able to carry 375 passengers with standees at 8 person/sq.m Total Capital cost of all these six lines will be Rs. 4326 crores and it is estimated that cost of operation and maintenance would be Rs. 35.00 crores at 1998 prices.

At present within P.M.A. the cost of travelling per trip (over 7.6 km) is Rs. 5 and average cost of traval by personalized two wheelers per trip works out to Rs. 16. It is expected that Rs.5 passenger trip of 7.6 km i.e. 0.66 paise per km which is comparable to present P.M.T. fares would be proper fare for Mass Rail Transit system. If the above proposal as suggested by RITES is implemented in true spirit it will help reliving the traffic pressure in Pune Metropolitan Area.

Comprehensive Traffic and Transportation Plan for Nagpur

This work was undertaken in 1985 based on the interview of 5 per cent of the household in Nagpur city. Like Pune also this data was augmented by carrying out other suverys. As part of this plan 65 important road intersections in Nagpur city were taken up for improvement after carrying out traffic studies. More than 50 per cent of these designs are implemented on sites.

Other Proposals

Considering the importance of rail based transit system the proposals of forming L.R.T. systems for Nagpur and Thana are under consideration with the aid form the international Funding agencies and German Govt. Similarly to ease out the traffic and transportation probems of Mumbai, Mumbai Urban Transport Project-II is formulated. For this project, costing around 6000 crores rupees assistance from World Bank is expected. The preliminary work of project formulation, formulation of Rehabilitation policy for the rehabilitation of the people likely to be dishoused due to various components of this project etc. are already completed and project can take off any time. Some prerequisites of the World Bank before actual funding such as formulation of Mumbai Rail Development Corporation and feasibility studies etc. have also been completed. It is expected that the project will be completed in two phases and will need about 7 to 8 years from its start. Once completed, this project will help in solving the major traffic problems within the M.M.R. area.

In addition to MUTP-II project, the proposal to form Mumbai Metro Corporations is also under the consideration of Govt. of Maharashtra. Traffic problems of Mumbai are so accute that if we do not form Mumbai Metro Corporation. Now, one of the studies suggested that, we will have to pay dearly for this lapse. By conservative estimate it is estimated that delay in formulation of such corporation will cost us 1 crore rupees per day. Couple of proposals in this regard i.e. composition of the corporation, its partners, proposed alignment of Metro system, its cost, equity participation by the partners, fare structure and the innovative measures for the required revenue generation etc. are under scrutiny of the state government and it is expected that very soon decision on this issue will be taken by the government.

Transportation System Management

The National Commission, on urbanization has recommended Transportation System Management Techniques which offer low cost solutions to many urban Transport problems through optimal utilization of the available road space and other urban infrastructure in our cities. The T.S.M. measures operate mainly through

(i) Improving vehicular flow, interalia, by segregation of fast and slow moving traffic and adoption of traffic engineering techniques such as channelization,
signalization bus-stop locations, one way and reversible streets etc.

(ii) According preferencial treatment to public transport modes on the road.

(iii) Reducing peak travel demands.

(iv) Regulating parking of vehicles.

(v) Promoting high occupancy of motor cars.

(vi) Raising the efficiency of transit systems to achieve increase in ridership.

The basic object of T.S.M. techniques is to maximise the movement of people on the roads rather than merely maximising the movement of vehicles. Some of the effective T.S.M. techniques tried in Maharashtra are given below :

(1) Widening of rod by compulsory acquisition on payment of compensation or by giving Transferable Development rights (TDR) in lieu of land surrendered for road, free of cost.

(2) Flexible laning.

(3) Traffic signals Management.

(4) Staggering of working hours to avoid peak hour congestion etc.

Honestly at present T.S.M. are not attempted seriously and systematically barring few exceptions. So far generally the improvements are done by the Police Department with the help of the local authorities on cost sharing basis. These efferts are piece meal and do not make real impact on the problems of traffic congestion. By undrtaking T.S.M. technigues chronic traffic probems can be taclelled effectively. For that purpose the local civic administration should be alive to the traffic problems and must equip themselves with manpower to handle these problems.

Road Based Public Urban Transport System

The best way to reduce the traffic density on roads is the bus based public transport system no. of vehicles required for catering to one million passenger kms varies widely between buses, intermediate public transport modes (i.e.auto rickshaw and taxis) and personalized modes of the transport (i.e. auto rickshaw and taxis) and personalized modes of the transport (i.e. cars and auto two wheelers). While the number of buses required for moving one million passenger kms is 111 or 333 pcus the corresponding number of cars and two wheelers for same passenger kms is 130 times the requirement of buses in terms of passenger car units. Introduction of bus system where not already existing and strengthening it where existing will cause considerable reduction in the number of personalized vehicles required to cater to the projected passengers transport demand in the urban areas, as large number of passengers will switch over to public transport. This will bring down pcu/km. In addition to this there will be sharp reduction in energy requirement there by releasing the pressure on foreign exchange requirement and also reduction in the level of environmental pollution.

After careful and detailed study the C.I.R.T. has suggested following norms of bus fleet for different categories of cities.

Table 20.5 : *Population and bus fleet*

City Population	Norms / One lakh population
Above 30 lakhs	50 with railway system
	75 without railway system
10 to 30 lakhs	50 buses
5 to 10 lakhs	30 buses
3 to 5 lakhs	20 buses
1 to 3 lakhs	10 buses

CIRT has observed in cities with one million plus population that with above norms the total saving in fuel for vehicles running on the road will be to the tune of 3.72 million litres. It would mean tremendous saving in foreign exchange requirement and improvement in quality of environment. It is, therefore, needless to say that the efforts should be made at all levels to augment the bus fleet in 6 urban centres having transport undertakings and bus sysyems should be introduced in more and more urban centres. Ofcourse the augmentation of bus fieet and capital expenditure for introducing the bus system in more and more urban centres will involve an outlay of huge amount. From where to bring these outlays is the measure problem. However, resources can be mobilized through following sources :

(1) Loans from multinational agencies like World Bank.

(2) Issue of debentures.

(3) Through privatiztion.

(4) Through fiscal support and subsidy by Government.

(5) Road pricing and user restraints.

(6) Parking restraints.

(7) User taxes.

(8) Vehicle ownership restraint.

Reccommendations

On the basis of above disscussion a few reccommendations are made as under.

(i) As envisaged in the Development Plans, the proposals inrespect of road widenings and proposed roads should be implemented scrupalously, so as to augument transport infrastructure.

(ii) Sites for off street parking, bus depots, terminals etc. should be acquired and developed as expeditiously as possible.

(iii) Circulation improvement schemes, such as foot paths, central verges, medians or dividers, right turning lanes exclusive bus bays etc. should be undertaken wherever identified.

(iv) National and state high ways should, as far as possible be diverted to avoid congested areas and they should be provided with service roads of adequate width.

(v) The demand mangement can be done by taking following measures :

(a) Area licensing is proved to be effective system. According to this system low occupancy vehicles pay charge for entering a congested part during rush hours. This has been tried successfully in Singapore. This system encourages use of Public Transport and sharing of private cars and taxis and discourages avoidable system is capable of absorbing the demand so diverted, with the same comfort and reliability.

(b) Intoducing car free zones where only public carriers like buses, taxis, ambulances, fire lighting vehicles etc. would be permitted to operate.

(vi) *Urban Data Base*

To deal with the urban transportation probems, a comprehensive data base is required which is not available readily at present. It is, therefore, necessary to built a sound data base in the urban transport sector. It is also necessary that urban transportation plans should be prepared after carrying out survey with detailed costing of various projects. So that definate efforts can be made to raise the resources to implement the proposals of the plan.

(vii) *Unified Transport Authority*

At present the traffic plans are prepared, barring few exceptions by the local authorities with the help of local traffic police. In order to co-ordinate the planning and implementation more effectively, in transport sector there is a need to establish unified Transport Authority for the state having manpower from various fields such as Transport Planners, Engineers, Administrators and Police. As per the conservative estimates the tentative cost of establishment of such Authority is one of the recommendations of the National Commission on Urbanization.

References

Census of India - 1991

Comprechensive Traffic and Transportation Survey, Pune. Traffic and Transportation Cell, 1978.

Information gathered from M.M.R.D.A. officials 1998.

Motor Transport statistics of Maharashtra 1997, compiled by Transport Commissioner.

Survey done for the Development plan of extended Municipal limit of Pune Municipal Corporation (38 villages), unpublished 1999.

Survery findings of Feasibility study done by RITES for developing LRT for Pune and Pimpri Chinchwad area 1998.

Traffic and Transportation policies and strategies in Urban India. Ministry of Urban affairs and Employment Govt. of India 1998.

21

ROAD NETWORK OF DROUGHT PRONE AREA IN ANANTAPUR DISTRICT, ANDHRA PRADESH

Y. V. Ramanaiah, K.Krishnaiah and *B.S. Reddy*

Introduction

No wonder that the transport system may be called as the sinews and lifeline in terms of economy and integration of a country. Transportation is the blood stream for any developed economic system (Hoyle, 1973 and Morrill, 1974). Among the various means of transport, it has been well said that the roads and railways are the main arteries for any nation, and the traffic flows along them is the blood stream that gives it life.

The economic structure of the modern century has been built upon cheap transportation. Road network provides the most fundamental and cheap mode of transport in the rural dominated settlement structure of India. In this settlement structure, roads play a vital role not only in aiding economic and social development, but also in creating new links between towns and villages and thereby bringing villages into the mainstream of economic life. Hence, the economic advancement and culture of the nation can be easily assessed with a mere glance of its roads. The development of transport network, practically, road transport could be made possible ever large clumps of people and economic activities, and, therefore contributed to a great increase in the number of larger settlements with well built up area and productivity.

In the backward areas like Anantapur district, Andhra Pradesh where agriculture is the predominant source of economy, the transport network preferably the road transport could do more in improving the socio-economic conditions of the people. In these areas where population is thin with dispered distribution of rural settlements utmostly called for well connected road network to break the traditional outlooks of the people and develop technological insights in exchanging the functional services. It is in this direction an attempt is made to examine the structural characteristics and development of road network in the drought prone area of Anantapur district, Andhra Pradesh. This attempt is likely to provide greater scope to formulate better strategy for the development of road transport system which is the main infrastructure organ for an integrated area development of the region.

Study Region

Anantapur district is one of the most backward regions of India located in the southwestern part of Andhra Pradesh state. In terms of geographical area, it is the largest district in Andhra Pradesh with a low density of population of 133 persons per square kilometre dwelling in 934 rural settlements and 12 urban settlements.

Anantapur district has an undulating terrain dotted with ridges and clusters of rocky hills which are characteristics of Deccan Plateau. Since it is located in the rainshadow belt of Western Ghats, the rainfall is very low (544 mm) and erratic. The soils are predominantly red and poor in fertility. Agriculture is the important occupation of the people but it is precarious. The frequent occurrence of droughts, prolonged dry spells and crop failures and scanty irrigation facilities have had a debilitating effect on agricultural economy of the district. Groundnut is the predominant crop in the cropping pattern of the district. The district is fairly well known for some of its mineral deposits, but the area lacks in mineral based industries. In view of prevailing precarious nature of agriculture, low irrigation development, lack of industrial base and poor socio-economic conditions as well as dispersed distribution of rural settlements, there is a need for the development of infrastructural facilities in general, and transport network in particular. It is an established fact that the road transport network necessarily plays a key role in improving and modernising the important economic activities like agriculture and industry as well as in establishing growth pole centres to serve various functional services. For providing a comprehensive plan for the development of road transport network in Anantapur district, it is imperative to gain insights into various structural characteristics of road network and the changes therein. It is also necessary to understand the growth of road network and its main problems and constraints which hinder the form and function as well as efficiency of road network. A study on accessibility and the pattern of traffic flow help to pinpoint the problems of regional road transport. The network analysis in the study is made for different points of time, namely, 1961,1971,1981 and 1988.

Evolution of Road Network

The evolution of road transport network dates back to the 17th century revealed that there were only few trade routes and high ways in the district. The important highway passed through central part of the district was Kasi-Rameswaram highway. In the olden times due to strategic location of this district in the Deccan plateau and mighty Kingdom of Vijaya Nagar emperors, many historical, political and pilgrim places like Hemavathi, Penukonda, Gooty, Hindupur, Rayadurg, Penna Ahobilam, Tadipatri, Lepakshi and Kadiri have been connected by highways. Most of the roads developed during Vijaya Nagar rulers were used for the movement of troops, military and arsenals.

It is pertinent to note that there was not a single road fit for wheeled traffic upto 1853. During British rule, a few attempts have been made to widen the old passes for the movement of military wheeled carriages. After the creation of Trunk Road Department in 1845, new roads have been constructed, for example, the road from Bellary to Cuddapah passing through the taluks of Gooty and Tadipatri. After the formation of Anantapur district in 1882, the Local Fund Board of the district took up some measures in metalling the roads. By 1881-82 the Local Fund Board was maintaining 1320 kms. of road network in the district. During the First and Second Five-Year Plans only the important roads were widened and black topped.

At present, the total road network in Anantapur district consists of 8020 kms. of which 860 kms. highways and 7160 kms. other types of roads. This comes to around 42 kms. Per 100 sq. kms. of geographical area and 3.2 kms for every 1000 population in the district. The district is traversed by one National Highway

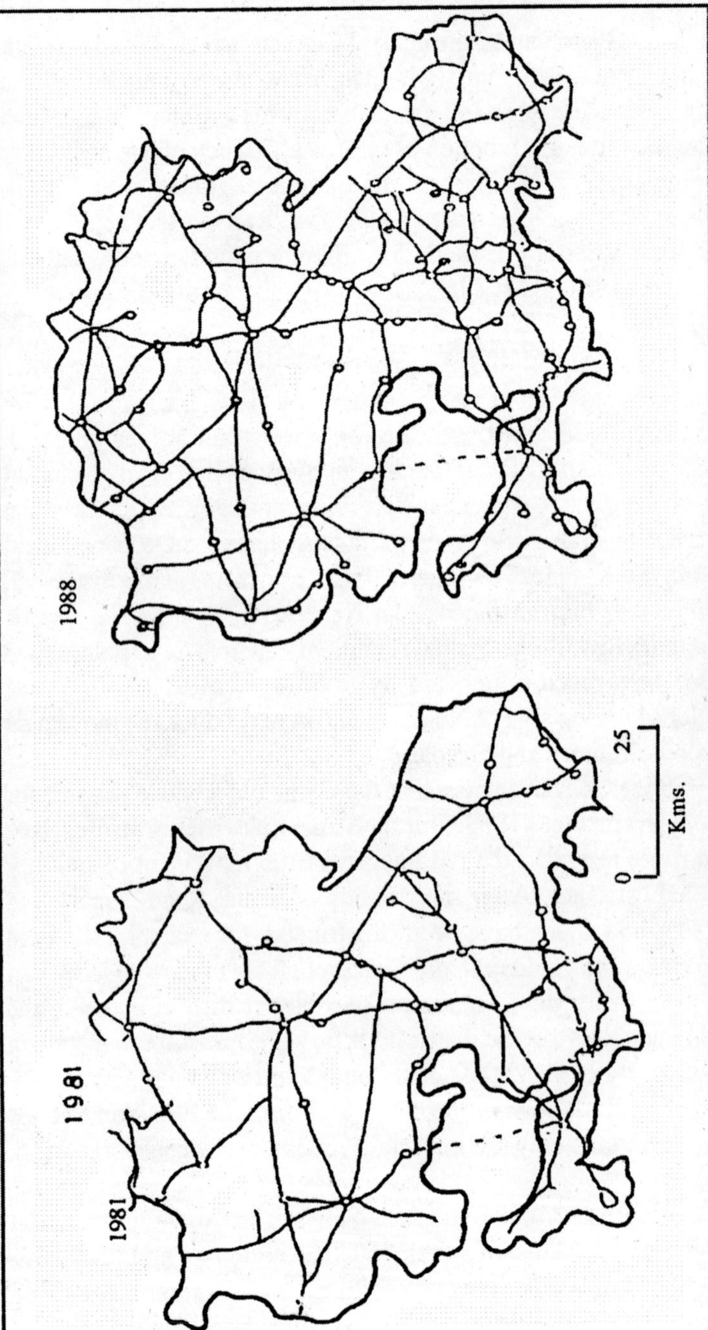

Fig.: 21.1 : *Anantapur District-Road Transport Network*

No. 7 which is running north-south direction. It is called Banaras Cape Comorin National Highway connects Anantapur with many important nodes like Kurnool, Hyderabad, Nagpur and Bangalore in outside the region. Between 1976 and 1986 there was a significant increase of road length in different types of roads in the district as it is evident from the fact that the state highways increased by 1004 kms. And the Panchayat Samithi roads increased by 1004 kms. of road length, Zilla Parishad roads increased by 593 kms.

Density of Road Network

The density of road network indicates the length of road per unit of geographical area which descerns the degree of connectivity and accessibility of road network system. The average density of road network in Anantapur district accounted for 0.4 kms/sq.km. The highest density of road network is noticed in Amarapuram while the lowest in Kudair mandal. There are 28 mandals showed the road density of more than the district's average while 35 mandals have below the average. The mandals located in northern and southern parts of the district are found with low density of road network of less than 0.3 kms/sq.km. Here, the dispersed distribution of settlements and hilly terrain conditions are caused for low density.

The spatial structure of road network is analysis in terms of topological and geometric components (graph-theoretic measures). In the present study all the urban settlements, administrative centres and the settlements having more than 5,000 population have been taken as vertices which come to 83 nodes. The national and state highways, major districts roads and other district roads are considered as edges. The graph theoretic measures such as Alpha, Beta, Gamma, Eta, Theta, Cyclomatic number and Pie indices as well as Binary, Shortest path Matrices, Konig Number, Shimbel Index and Detour indices are measured.

The study has revealed that there has been an increasing degree of road connectivity from poor to complex pattern. The Alpha and Gamma values represented a significant change in the structural pattern of road network from Grid to Delta pattern. The lower values of Eta and theta showed an improvement in the degree of connectivity. As increasing the values of Pie index from 5.8 in 1961 to 11.6 in 1988 indicating considerable increase in the compactness of the road network structure in the district. The node and link ratio assessment has been measured by employing Binary, Konig, Shinibel and Detour measures. In the year 1988, the Binary Matric revealed that Tadipatri urban node is identified as the most connective settlement with eight direct linkages followed by Kalyandurg, Uravakonda, Anantapur, Rayadurg, Kadiri and Penukonda nodes. It is inferred that the urban settlements showed more nodal relationship than the rural nodes in terms of direct road linkage.

The degree of centrally which is measured with the help of Konig number has revealed that N. Gate, Ananatapur, Pamidi are identified as the most central nodes since they have the low value of seven Konig numbers. Virtually, all these nodes with high centrality are located in the central part of the district. In contrast, Pedda Vadaguru, Kokanti and Valsa are identified as the least central nodes due to high values of Konig number. It is found that the nodes which are located on the National highway in the central part of the district have consisted more centrality irrespective of their rural of urban characteristics. The big urban settlement like Anantapur, Dharvamavaram, and Penukonda have also consisted more centrality and connectivity because of their urban characteristics and higher-order service centres.

From the Detour index analysis, it is found that Kadiri node has the lowest Detour value of 118 indicating the high nodal efficiency in saving journey time. Higher Detour values are found with Putlur and Yellanur nodes. The nodes located in northeastern and western parts of the district have shown high Detour values indicating low nodal efficiency. Here, it is stated that the transport road network in terms of directness is much influenced by physiography and spacing of settlements in the district.

The application of Shimbel index with the help of shortest path matrix has revealed that Ananatapur and Bathapalli nodes are identified as the most connective nodes in the district. In contrast, Valsa node with a maximum value of 592 has been identified as the least connected node because it is located in the foremost corner of the district. From the analysis of Shimbel index, it is found that the urban settlements possessing lower values indicating higher connectivity than the rural nodes. The towns located at the centre

Table 21.1 : *Changing Spatial Structural Characteristics of Road Network in Anantapur District*

Years	Vertices	Edges	Road length kms.	No. of Circuits	Cyclo Matic. Number	Beta	Alpha	Gamma	Eta	Theta	Pie	Diemeter	Density
1961	29	44	1250	15	16	1.5	0.3	0.54	28.4	43.1	5.8	215	0.06
1971	37	56	1502	18	20	1.5	0.3	0.54	26.7	40.5	6.9	216	0.08
1981	54	76	2018	26	23	1.4	0.22	0.49	26.5	37.7	9.2	220	0.10
1988	83	216	2620	54	134	2.6	0.83	0.89	12.1	31.6	11.6	226	0.14

Source : Author

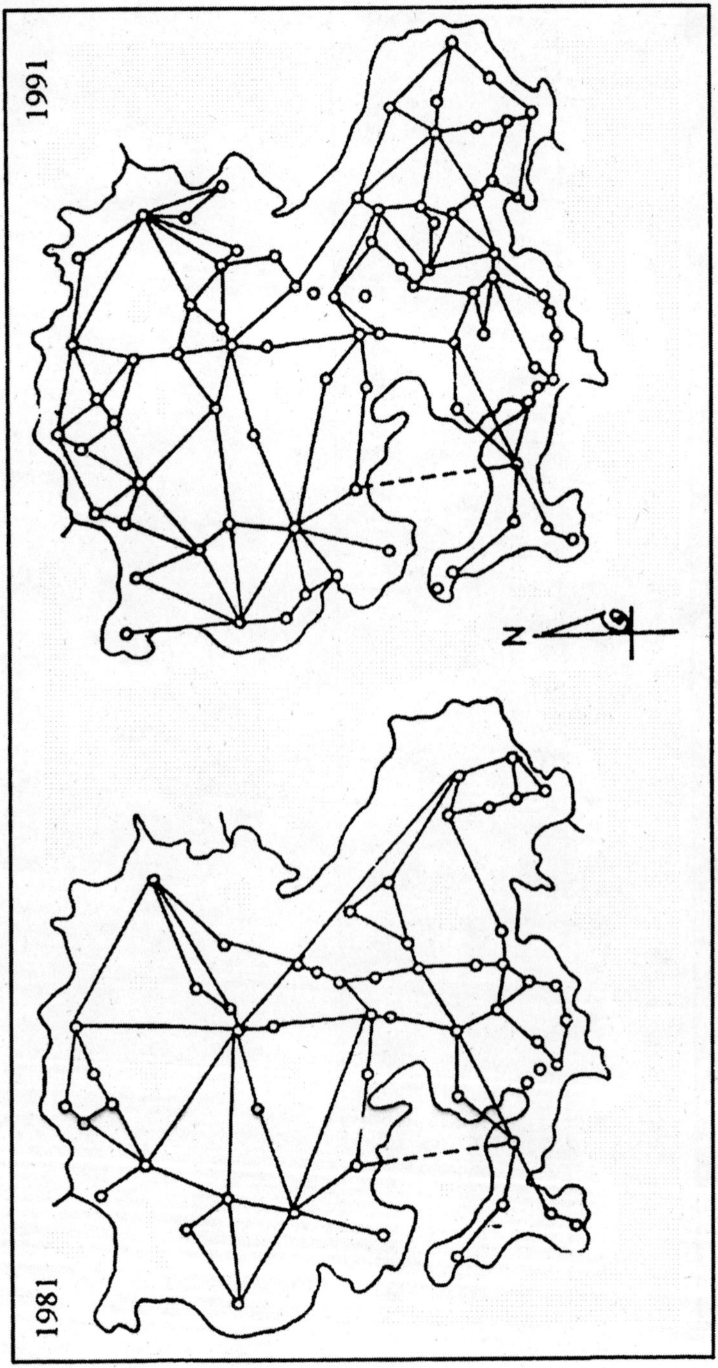

Fig. 21.2 : The Topological Graph Theory Form of Road Network

have been witnessed with greater connectivity than the towns at the peripheral areas. For example, Hindupur urban node located at the peripheral part has witnessed with a high value of Shimbel index indicating the low connectivity. Anantapur town which is identified as the most central node with greater connectivity which is not only due to its central location but also due to its higher-order functional service system.

The efficiency of road network also depends on direction and shape of the route or an edge. The "Straight" line is the most efficient route that any age of the network can make . In the present study, the route shape analysis is made for 11 urban nodes with an aim to measure the regularity and shape efficiency of road in terms of direct connection with the other neighbouring nodes. From the analysis, it is noticed that Hindupur node with a low other nodes are regular and high efficient in their route shapes followed by Uravakonds, Penukonds, and Anantapur nodes with low mean values. In contrast, Dharmavaram, Guntakal, Tadipatri and Gooty with high mean values indicating the low efficiency are irregularity of route shapes in the district.

Accessibility and Traffic Flow

In the present study, the accessibility of road network is also measured with the help of Isodromes at 2 kms., 5 kms, and more than 5 kms., distances from the edges. About 42 per cent of the total number of settlements in the district are located within 2 kms. isodromes showing the most accessible parts from the point of road facility, 24 per cent of the settlements are located in between 2 and 5 kms. isodromes in dicating moderate accessibility of the settlements interms of road facility. About 34 per cent of total settlements are located beyond 5 kms. isodromes indicating the most inaccessible in terms of road facility. These settlements located in the north, northeastern and southeastern parts of the district are found as the most inaccessible areas in terms of road facilities. Owing to thin distribution of population dispersed and isolated distributed of rural settlements, lack of sufficient and required cause ways and bridges across the rivers and streams and poor traffic flow have had a bearing influence on arrangement of road facilities to each and every corner of the place in the district. In

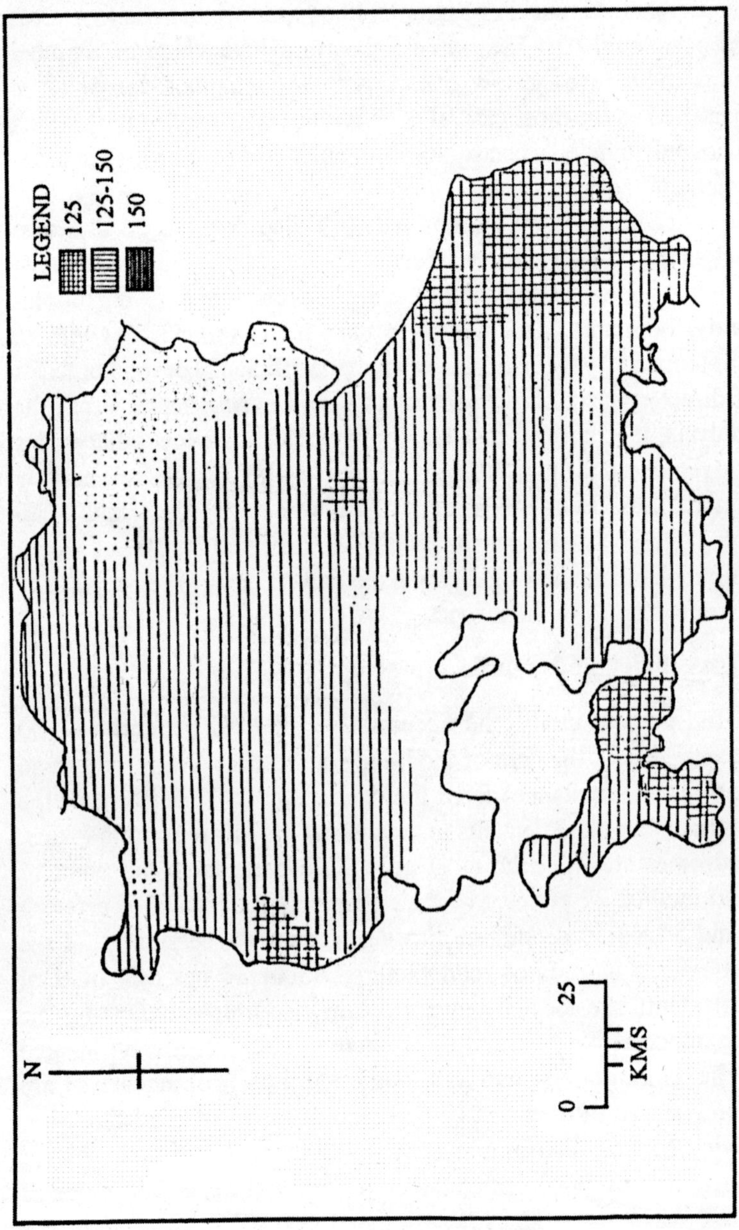

Fig. 21.3 : *Anantapur District Detour Index*

future, these inaccessible areas are to be well connected with the other areas in the district for overall development of the region.

Traffic flow means simply the volume of movement of goods, people and message. It is the core element of circulation system which reflects the complex nature of economic interdependence. In the present study, the flow analysis is made on the basis of passenger traffic flow on road network. The data is based on the number of trips flying on different routes which cover the public transport systems only. It is observed that high volume of bus traffic flow noticed on the National Highway connecting important urban centres in the district. The highest passenger flow is observed on Anantapur to Bathalapalli route. The low volume of bus traffic flow is noticed in northwestern and southwestern parts of the district because these areas are at the marginal ends not only for Ananatapur district but also for Andhra Pradesh state. The movement of people may be less from such marginal ends towards the district head quarter and other important urban settlements in the interior parts of the district for daily services because many of the daily services would be served by the important settlements in the neighbouring states. Another important reason that a low traffic flow in the western border of the district occurrences of drought and famiae conditions.

Conclusion

The socio-economics structure for the 21st century is to be built upon cheap mode of transport, for which, road is one of the most fundamental prospects of mankind. The major problems of road transport development are the construction and widening the transportation lines in an undulating terrain, bridging the streams, surfacing of roads with black top to cope up with the heavy traffic. Out of 934 villages, 826 villages have been connected with road system and bus seniees, whilethe remain 108 villages are yet to be connected by road transport. The old cause ways and culverts which were constructed years before have not been in good condition to meet the present traffic flow.

To overcome some of the shortages in road network, an effective plan has to be made for effective functioning of road transport system in the district. Construction of new roads in the inaccessible

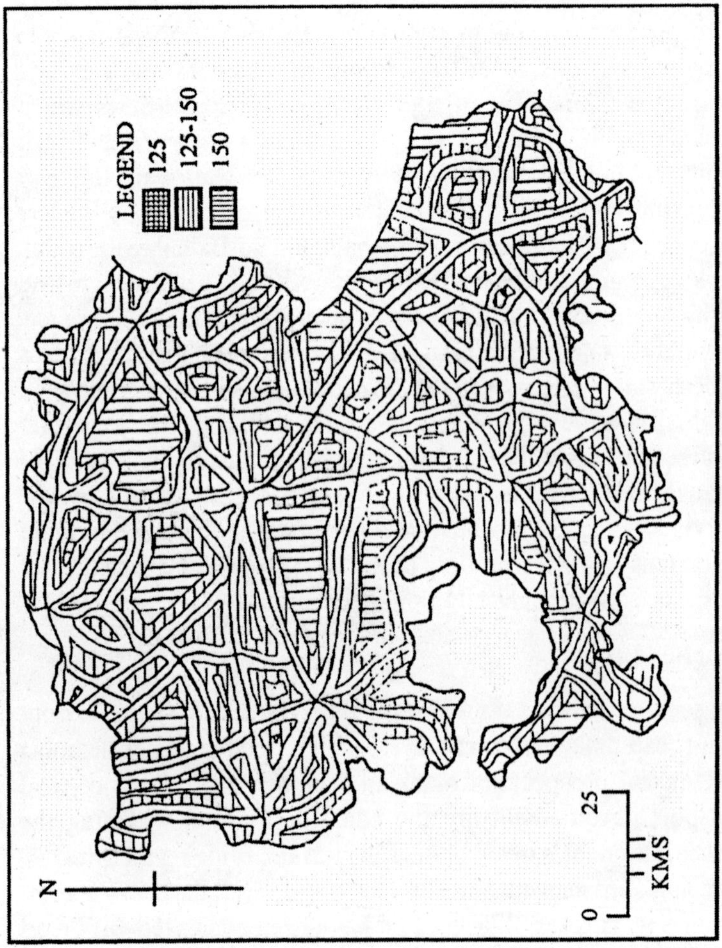

Fig. 21.4 : *Anantapur District : Accessibility Roads*

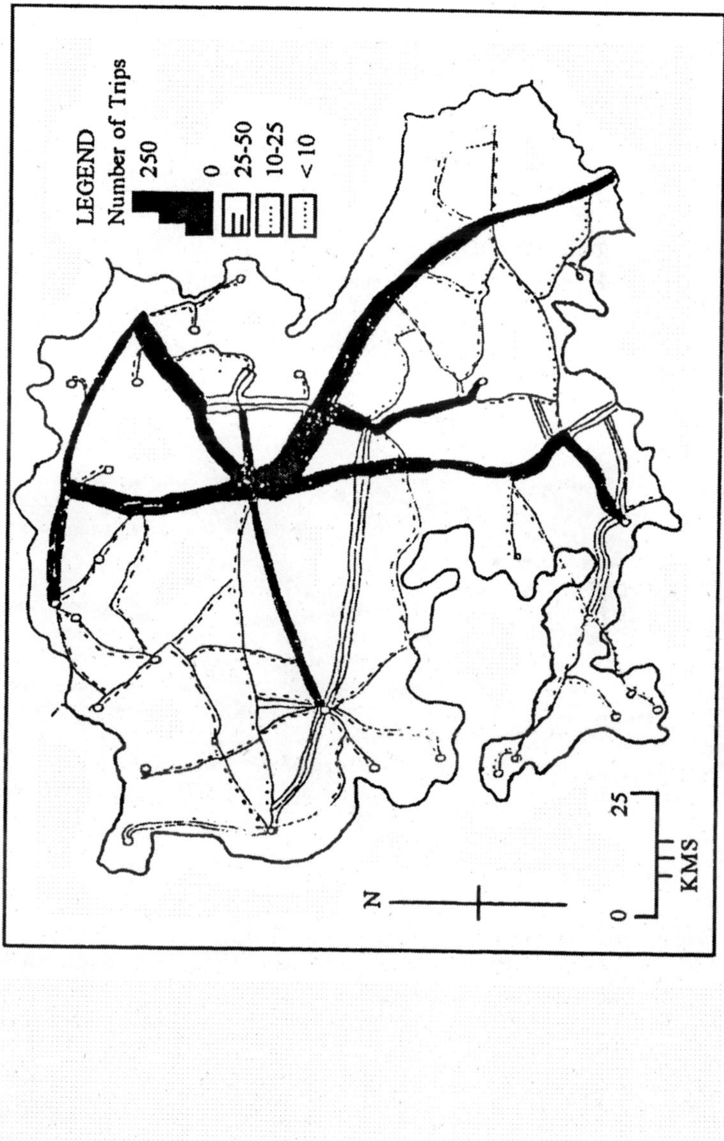

Fig. 21.5 : *Anantapur District Bus : Passenger Traffic Flow*

area, widening of roads in heavy traffic zones, repairing of old bridges, cause ways at important junctions, reducing the steep gradient on side edges of the roads, development of the plantation in all along the road sides, construction of bus stations with minimum facilities at important nodal centres, metalling and black topping of panchayat and zilla parishad roads and proper maintenance of road signal system are necessarily be taken up for the development of road transport in the district.

A well distribution of road network in the drought prone Anantapur district enable to serve. The needs of agriculture, industry, commerce and trade as well as to create new growth pole centres for well functioning of various services to the people of this region. Hence, it is needless to emphasize that the roads are the only means linking all the settlement, people and services and bringing into the mainstream of economic life of this backward district.

References

Haggett, P. and Chorley (1969). *Network Analysis in Human Geography*, Edward Arnold, London.

Hoyle, B.S., (1973). *Transport Developments in Under Developed Countries*, Macmillan, London

Morrill, R. L. (1974). *The Spatial Organisation of Society Duxburg*, Masschusetts.

Ramaswami, G. (1976) Road Network in Tiruchi, Tanjavur Planning Region : An Analysis', published in Golden Jubilee volume of the *Indian Geographical Society*, Dept. of Geography, University of Madras, Chennai.

Saxena H. M., (1980) Road Transport Connectivity Pattern and Economic Development in Rajasthan, *Geographical Review of India*, Vol .42 No. 2 Kolkatta.

Taffee, E. J. and Gautheir H. L., (1973) *Geography of Transportation*, Prentice Hall, London.

IMPACT OF TRANSPORTATION ON TOURISM DEVELOPMENT OF GULMARG AND PAHALGAM IN KASHMIR

T.A. Kanth and *M. Y. Bhat*

Introduction

The economic development of any region depends largely on efficiency of transport system if pursued optimally without creating ecological imbalances. Transport and regional development are inherently interwoven and interdependent due to the complementary of economic sectors. Economic development is a consequential product of transport efficiency bolstered by the system of backward and forward linkages. Transport being the derived demand forms an indispensable factor of production tied to all developmental processes in the economy. The functional significance of transport was admirably summarized by Marshall as most important activity in every stage of advanced civilization (Milne and Laight, 1965). Transport gives rise to division of labour, area specialization, extension of markets, utilities of space and optimization of production units (White & Senior, 1983). No industry is as vital to society as this can be expected to function in a void (Hibbs, 1970). In Socio- economic ecosystem, transport is regarded as its one of the subsystems (Wagner, 1960).

As economic ventures of a region expand and diversify, they continue to raise the demand levels for transport. Tourism being one of the economic activities involving travel cannot survive without transport sector. Travel and transport are inter-changeable resulting from push and pull factors present at the trip ends. Transport as an essential element of tourism is regarded as the principal service being rendered to tourists for the geographical spread of tourism.

The valley of Kashmir contains enormous potential for tourists and is connected by a National Highway (NH-1A) with rest of the country. Its vast tourist resources are still unexplored and unutilized due to the inefficient road network and transport frequency. The valley has further scope for expending its tourist industry if transport system is strengthened and reorganised uniformly in all potential regions. Since the pattern of tourist nodes, namely, Gulmarg and Pahalgam have registered maximum arrivals as compared to other nodes. Such flow density has created unprecedented problems of congestion, aesthetic disequilibrium and environmental degradation. It is, therefore, necessary to examine the reasons for such a situation and to suggest some policy measures in which a more positive and symbiotic relationship could be established benefiting both the environment and tourism in the context of appropriate development. Such environmentally sound tourism development could be achieved only through the assessment of carrying capacity in a particular region/node. The present study is, therefore, contextually related to the study area and attempts to evaluate the movement pattern of tourists while reaching the central node (Srinagar). Trip generation and flow density has also been examined. The impact of flow concentration on the environmental setting has been examined by way of detecting the land-use changes and working out the recreational carrying capacities by computing the user-density levels at sample nodes.

Recreational Carrying Capacity : A Conceptual Review

A particular kind of disamenity measurement on which several studies have been conducted is the psychological tolerance of recreationists to crowding which started with the work of Lucas, 1964. According to him different categories of recreational users have different spatial as well as contextual perceptions of wilderness and tend to avoid contact with other users (Lucas, 1973). The techniques for measuring disinterest or enthusiasm was developed by Hendee, 1973. In order to derive a reasonably accurate predictive index of congestion (Stankey & Lime, 1973; Cicehetti & Smith, 1973; and Manchik, 1973) have suggested different mathematical models. The carrying capacity studies demonstrate that recreational capacity is a function of management policy, whereby annoyance can be avoided by zoning access, restricting numbers and directing users to diverse areas (Hendee & Lucas, 1973; Stankey, 1974).

Nevertheless, much work still need to be done to improve our understanding of capacity levels. Peterson (1974) has made a valuable contribution by asking users to relate the desirability or undesirability of large number of recreational experiences to their perceived probability of occurrence. It is in this backdrop that Gulmarg and Pahalgam were selected to study for examining recreational carrying capacity in relation to their tourist flow and environment.

Data Base and Methodology

In order to achieve the above objectives, relevant data have been acquired from both primary and secondary sources which include road map of Kashmir and Digest of Tourism statistics. The data regarding the volume of tourist traffic visiting various nodes within the valley has been generated by recording the total number of tourist buses, cars and other flow regimes. In order to examine the nodewise network efficiency, graph theory indices were evaluated to determine the degree of accessibility 'Shortest Path Matrix', 'Associated Number' and 'Shimbel Index' have been examined. In addition, Gravity Model has been applied to measure the volume of potential trip generation and the magnitude of interaction between the central node 'origin' and the sample nodes 'destination'. To determine the trip interactions Olsson (1965), Taylor (1975) and Tocalis (1978) also used this model which is defined as :

$$T_{ij} = G\, O_i\, D_j\, d_{ij} \text{ (where, } \beta = 2 \qquad \text{................ (i)}$$

$$\text{or} \quad \frac{T_{ij}}{O_i D_j} = G d_{ij}^{\beta} \qquad \text{............. (ii)}$$

where, T_{ij}-refers to trip interaction between '*i*' & '*j*'

O_i — Tourist flow

D_j — Value of composite potential of destination node

G — Constant (i. e. 1000)

D_{ij} — Distance between *i*th and *i*th node

Applying logrithms on both sides of eq. (ii), we get;

$$\log \left[\frac{T_{ij}}{O_i D_j}\right] = \log G - \beta \log d_{ij} \qquad \text{................ (iii)}$$

or $$\log (T_{ij}) - \log (O_i D_j) = \log G - B \log d_{ij} \qquad \text{.........(iv)}$$

Finally eq. (iv) has been employed to compute the values for Tij, whileas, the composite scores for the derivation of 'Dj' were worked out using the Smith's index given as:

$$\sum_{i=1}^{n} Zi$$

where, I_i = composite index of 'I' level of indication for '*i*th node'

Z_i = Standard scores of i to number of graph theory indications for node 'I'

The valley of Kashmir has been one of the largest tourist receiving areas in the country. Tourist flow to the valley has increased from 10579 in 1951 to 557977 in 1989 at an average growth rate of 10.70 per cent per annum. Both Gulmarg and Pahalgam being potentially rich in tourist resources have also assumed exponential growth in tourist flow with the annual growth rates of 8.78 and 11.73 per cent respectively. However, the tourist flow during 1988-89 has shown a negative trend due to the political instability in the state (Table 22.1). The hierarchy of tourist nodes in the valley was worked out on the basis of six broad indices derived from 72 sub-indicators. The broad indices were viz. natural beauty, accessibility, boarding and lodging, marketing facility, administrative care and social amenities, adventurous sports and other entertainment. After determine the hierarchy of tourist nodes, Gulmarg and Pahalgam have emerged as second order tourist nodes (Table 22.2).

Table 22.1 : *Distribution of Tourist Flow*

Tourist Node	Number of Tourists Visited				
	1985	1986	1987	1988	1989
Gulmarg	123184	139244	166516	176610	166481
%Change	—	13.04	--19.58	6.06	-5.73
Pahalgam	83298	98569	129461	139624	122389
%Change	—	18.33	31.34	7.85	-12.34
Kashmir Valley	503614	589716	721654	722035	557977
%Change	—	17.10	22.37	0.05	-22.72

Sources : Srinagar Road Transport Corporation (SRTC) and Private Travel Agencies. This exclude those tourists who have visited individually by other non-vehicular means

Table 22.2 : *Tourist Flow and Indices of Tourist Potential*

S. No.	Index	Composite Score	
		Gulmarg	Pahalgam
1.	Tourist flow as %age of total valley arrivals	24.46	15.54
2	Natural Beauty	1.81	1.87
3.	Accessibility	7.94	8.49
4.	Boarding and Lodging	6.78	6.78
5.	Marketing Facility	4.96	4.96
6.	Administrative & Other Infrastructure	5.55	5.08
7.	Adventurous sports and other entertainments	6.05	4.18
8.	Composite Index	32.21	30.40
	Ranking Order in the Valley	II	II

Note : The Composite scores of the above indicators were derived from the weights assigned with the help of a scale (0-4) (Carvajal & Patri, 1979) based mainly on the magnitude of presence or absence of a particular component in case of natural beauty whereas in case of other indicators the centrality of a function was taken into account.

Movement Pattern and Flow Density

Trip generation between the origin and destination primarily being the function of distance and nodal characteristics, the sample nodes have depicted variation in the magnitude of trips. The study reveals that the inter-nodal potential for trip generation in case of Srinagar-Gulmarg has been 6.461 which is comparatively higher than 5.483 computed in favour of Srinagar-Pahalgam. The spatial pattern of tourist traffic has been highly varied among the nodes whereby Gulmarg and Pahalgam registered 24.5 and 16.5 percent of total tourist flow respectively in 1989 (Table 22.3) This could be attributed to distance factor and accessibility index. Besides, trip generating potential in case of Gulmarg has been more strong recording the composite value of 32.2 leaving Pahalgam behind with 30.4.

The intensity of trips emanating from Srinagar to generater nodes does correspond to the magnitude of their resource potential. Since recreational sites and hill resorts fall under the category of high density land-uses. Gulmarg and Pahalgam have recorded the higher levels of flow density triggered by the tourists. This evidence is depicted by the desire lines plotted in (Fig 22.1)

Table 22.3 : *Tourist Flow, Accessibility and Trip Generation*

Tourist Node	Tourist (1989)	% flow to total	Distance from Srinagar (kms)	Accessi-bility Index	Composite Index	Volume of trip Generation (log Tij)
Gulmarg	136481	24.5	39	7.94	32.1	6.461
Pahalgam	92289	16.5	96	8.49	30.40	5.483

Physiographic Characteristics

Both the sample sites consist of shales, limestones, sandstones, quartzites, mudstones, slates, agglomerate and pyroplastic pıoducts, phyllites, schists and Bafliaz volcanics (Shah, 1978) which are mostly unfossiliferous that of Gulmarg (Wadia, 1934). All these formations are covered with glacial and lacustrine deposits of pleisticine. moraine materials are the major surface features which render the area lithologically a weak zone.

The study area is divided into five slope regions. Comparing the slope, soils, natural vegetation and landuse characteristics, the real differentiation in characteristics based on slope vegetation cover. The soils of the region vary in origin from alluvial to lacustrine and glacial, and can be divided into two types on the basis of physiographic divisions (Kaul, 1965) soils of valley basin and side valley (upto 1850 m) and highland soils (1850-3500 m) which is the region of immature mountain soils where rigorous soil erosion takes place. Here it is worthwhile to mention that both the tourist places Gulmarg and Pahalgam are falling under the same altitudes range. Hence, the encrochment of new land through construction of roads and buildings at these resorts make the surrounding soil more weak and susceptible to erosion.

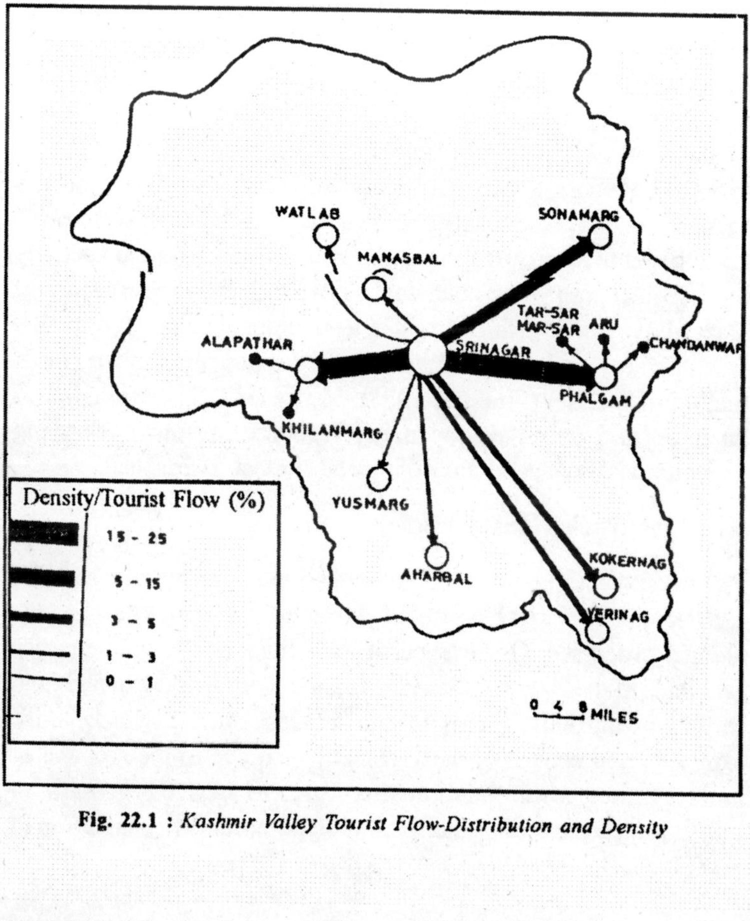

Fig. 22.1 : *Kashmir Valley Tourist Flow-Distribution and Density*

Tourist Infrastructure and Environmental Carrying Capacity

There has been a constant increase in tourist flow to Kashmir valley in general and some developed hilly resorts in particular. Such influx has enforced the Government agencies as well as the local people to develop a large scale infrastructure base in terms of roads, huts, hotels, guest houses, clubs, parks and other installations of tourist use. This was done with the aim that more and more tourists would visits these places and enhance their duration of stay. Consequently, the infrastructures development has no doubt boosted the tourist flow but simultaneously threatened the environment of these hilly resorts, which are highly fragile and ecologically less accommodating in nature.

Mountain environments are generally understood to have uneven topography, high altitude, inadequate soil cover, uninformity of precipitation and durinal variation in temperature, that render the complex mountain environments more fragile. Moreover, factor of low carrying capacity, high fragility, and vigorous mass movement of material also determine the extent of interactive systems in highlands (Weber, 1979). The main factor in attracting the tourists to mountainous resorts has been the utilization of natural resources. Land being the foremost concern of man for any economic activity has become more vulnerable, depicts itself in its two functions : (a) Hazard avoidance (which implies to minimise human and capital loss), and (b) guidance on suitability (for segregation of land of its major kind of use) (Pirazizy, 1988). Hence landscape evaluation is an important factor in improving resource inventory, carrying capacity decisions, and environmental impact assessment.

Road Network Construction

Provision of infrastructure at hill sites especially the accessibility and transport facilities have ramified their influence upon the dynamic natural processes. Occurrence of landslides is one of the major problems in the Himalayas in general and roads leading to highland resorts in particular. This is due to slope failure which cause extensive and consist of weak rocks which due to the seepage of underground water become instable. Moreover, growth in transportation network enhances the slope instability in addition to the clearance of forest cover on large scale. There are about 196 tourist buildings and 20 kms of road at Gulmarg and 284 tourist buildings and 23 kilometres of road at Pahalgam. Their construction has involved the mass clearance of trees from the local forests and also stones being extracted from the surrounding landscape. Such activities have far reaching environment impact especially on lithology and soil resources. Soft rocks are excavated manually while harder portions are blasted which weaken the geological structure of the construction sites and give rise to various landslides.

The important factor in causing landslides is the instability of slopes and the altitude of the area through which the roads pass. The main road system in both the sample sites have their origin and destinations at different elevations. Hence, the range of height through which they extend is different. About 72 per cent of landslides have occurred in comparatively lower elevation zones which mainly consists of weak sediments of valley floor type. A large number of landslides were experienced in the (1967) period, (which coincide with the tenure of Fourth, Fifth and Sixth Five-year Plans) than in the pre 1967 period (Mukerji, 1986) due to rapid expansion of road network, particularly in higher reaches where slopes are usually tree bound, vegetation cover is thin, mechanical weathering is intense and rocks are weak. The lower elevation zones of these areas experience a large number of landslides as a response of developmental activities (Pirazizy, 1989).

The network of roads in Gulmarg and Pahalgam tehsils is about 238.39 kms, and in tourist towns it is 43 kms. Mismanagement and cruel execution of their construction has led to destablilization of hill sides and production of immense debris. It is estimated that construction of just one kilometre long road requires removal of about, 40,000 to 80,000 m^3 of debris (Singh, 1985). Moreover, during the rains, landslides and rock-falls damage the roads and create more debris. The Border Road Organization (BRO) alone has been clearing debris from the roads under its charge at the rate of 724 m^3/km. per year in the Kashmir Himalayas (Patnaik, 1981). If on an average 550 m^3/km. per year of debris is generated on the already constructed roads, the 43 kilometres long road network of the area under study must be proving about 25,80,000 m^3 of debris every year in addition to tonnes of debris at the cost of new road construction.

The other aspect of environmental impact has been on the picnic grounds, parks and wilderness areas (Wall & Wright, 1977). Many of these have been concerned with the trampling effects on solids and vegetation by various activities such as Skiing (Baiderin, 1978) trail-bike riding (Crozier, *et al.* 1978), walking (Liddle,1975), forest and meadows (Dale and Weaver, 1974). These effects could be well understood by examining the intensity of users of the sample sites or working out their carrying capacity/equilibrium density.

User Density and Carrying Capacity

Before suitable region or localities are developed they must be checked for their capacity to absorb tourists, new facilities and activities are saturated (physical capacity), the environment is degraded (environmental capacity) or visitors' enjoyment is diminished (perceptual or psychological capacity) (Piperglou, 1967). These concepts are now generally accepted but the difficulties in measurement and quantifying the threshold has restricted its application (Barkham, 1973). The recommended densities for determine the carrying capacity are worked out as 1000 persons

Table 22.4 : *Tourist Flow Seasonal Population*

Sample site	Average Tourist per day	Seasonal Population	Local Resident Population (1981)	Total Population	Area at disposal (Total area-forest area) (hectares)
Gulmarg	2300	650	511	3461	544
Pahalgam	2700	875	2626	6234	1350

Source 1. Directorate of Tourism, Jammu and Kashmir Govt. Srinagar.
2. Town Directory, Census of India, 1981.
3. Personal survey.

per hectare or 10 m^2/ person (an Foras Forbatha, 1973), 600 holiday makers/hectare or 15m^2/persons Greece (Pipoglou, 1967), 25 m^2/ user (Acau, 1976), 1450 users/hectare i.e. 7 m^2/ user (Park and Recreation Information System, California 5m^2/ user in Yugoslavia (Andric, 1962). In the same manner user density during summer months for Gulmarg and Pahalgam was worked out by applying the following formula.

$$(UDd)\,i = \frac{Pi + PRP + Ps}{AD}, \quad \text{where } AD = (Ta - Fa)$$

(UDd) *i*	=	User density/hectare/day for ihe more
P*i*	=	User population of ith character (tourists)
PRP	=	Permanent resident population (locals).
Ps	=	Seasonal population under tourist service (include shopkeepers, hotal workers, pony owners, helpers and other tourist employees).
T*a*	=	Total area (hectares)
F*a*	=	Forest area (hectores)
AD	=	Area at disposal for users (hectores)

Hence, user density for Gulmarg was estimated about 6 persons/hectare i.e. 1667. m^2 per user/day while as for Pahalgam it was 5 persons/hectare or 2000 m^2 per user/day. Therefore, the results have clear that Gulmarg and Pahalgam have not yet reached the stage of equilibrium density or saturation level while comparing with already recommended densities. It also depicts that these tourist resorts have further carrying capacity for tourists to absorb, but at the same time if new areas from the same landscape (area at disposal) are encroached for building construction and other developmental activities. The tourist threshold (carrying capacity) would get ultimately minimized/ shrinked and would finally result in visits's enjoyment minimization and environmental degradation. The above discussion revealed that the multiple growth of tourist infrastructure as well as the flow have created the problems of great ecological concern. The sprawl of tourist towns in an oil-drop fashion has reduced the environment carrying capacity and simultaneously threatened the geo-ecological milieu of these very nodes. Unplanned and cummulative expansion of tourist facilities together with the imbalanced tourist flow are the basic eco-degradation multiliers that have been active during past few years. The study indicated that both Gulmarg and Pahalgam have undergone a considerable environmental change through the construction of tourist buildings, gardens and parks, roads and foot paths, catering establishments and the like.

In order to conserve the environmental resources and to maintain the ecological balance of tourist nodes, the following policies should be implemented. These include the regulation of tourist flow uniformly for both the seasons of summer and winter could be achieved by the development of winter sports especially at Pahalgam where tourist flow has been more confined to summer months. Degradation of landscape and forest cover be stopped by curbing further construction activities specifically at the hill resorts. However, tourist supporting facilities be provided only after the evaluation of carrying capacity of the nodes. Also Cost Benefit Analysis (CBA) of the tourist projects be made so that environments costs would not exceed the economic gains. This could be achieved by adopting the principle of environmentally sustainable tourism development.

References

ACAU, Atelier, Cooperrelif d' Architecture et d'. Urbanisme (1976), "Contribution a l' Etude des programmes de Nouvelles Stations de Vacances. Les Cahiers du Tourisme", C-7, Chet, Aix-en-Provence, transport in D. Peace's *Tourist Development, Longman* Publication, New York.

Andric, N., *et al* (1962), "Aspects Regionaux de la Planification Touristique", *Tourist Review*, 17(4), pp, 230-36, Translated in *Pearce's Tourist Development*, Longman Publication, New York.

An Foras Forbartha (1973), "Birthas Bay : a Planning and Conservation Study", An Foras Forbartha, Dublin.

Baiderin, V. V. (1978), "Effect of Winter Recreation on the Soil and Vegetation of slopes in the vicinity of Kazan", *Soviet, Journal of Ecology*, 9(1).

Barkham, J. P. (1973), "Recreational Carrying Capacity : A Problem of Parception", *Area*, No. 5, pp. 218-22

Carvajal, B. and Patri, J. (1979), 'Principios Basicos Para la Obtencion du un Indice de Jerarquizaction Turistica, Aplicado a La Provincia, Antarctica Chilena", *Inform, Geog. Chile*, No. 26, pp. 65-80, in D. Pearce's *Tourist Development.*

Cicchetti, C. J. and Smith V. K. (1978), "Congestion, Quality Deterioration And Optional Users: Wilderness Recreation in the Spanish Peaks Primitived Area", *Social Science Research*, No, 2.

Crozier, M. J. (1978), "Off Road Vehicle Recreation : the Impact of Off-road Motorcycles on Soil and Vegetation Conditions, *Proceeding 9th N. Z. Geog.*

Conf. N.Z. Geog. Soc Dunedin.

Dale, D. and Weaver, T. (1974), "Trampling Effects, on Vegetation of the Trail Corridors of North Rocky mountain forests", *Journal of Applied Ecology*, No. 11, p.767-72

Hendee, J. C.& Lucas, R.C. (1973): "Mandatory Wilderness Permit : A Necessary Management Tool", *Journal of Forestry*, No. 71.

Hibbs, J. (1970), *Transport Studies An Introduction*, John Baker, London.

Kaul, M.L (1956), Study of Soils at Different Altitudes in Kashmir State, Unpublished theses, SSAC, Indian Agricultural Research Institute, New Delhi.

Liddle, M.J. (1975): "A Selective Review of the Ecological Effects of Human Trampling on Natural Ecosystems". *Biological Conservation*, No. 7,

Lucas, R.C. (1964), "Wilderness Perception and Use: The Example of the Boundary Waters Cance Area". *Natural Resources Journal*; No. 3. p. 394

Lucas, R.C. (1973), "Wildness, Management Framework". *Journal. of Soil and Water Conservation*, No. 28 p. 150-54.

Menchik, M.D. (1973): "Optional Allocation of Outdoor Recreational Use in the Presence of Ecological Carrying Capacity Limitations and Congestion Effects". *Papers of the Regional Science Association*, No. 30, p.77-96.

Milne, A.M. and Laight, J.C. (1963), *The Economics of Inland Transport*, Pitman, Londan.

Mukerji, A.B. (1986) Landslides and Man, Development and Destruction in Western Himalayas, in Shafi and Raza's (eds) *Spectrum of Modern Geography*, Concept Publishing Co., New Delhi.

Olssion, G. (1965), *Distance and Human Interaction : A Review and Bibiography*, Regional Science Research Institute, Philadelphia.

Paterson, G.L. (1974b), "Evaluating the Quality of the Wilderness Environment : Congruence between perception and Aspiration". *Environment and Behaviour*, No. 6, p.169-93.

Patnail, N. (1981), Role of Soil Conservation and Afforestation for Flood Moderation in international Conference on Flood Disaster, December 1981, Ind. net Science Academy, New Delhi (Mimeograph).

Piperoglou, J. (1967), "Identification and Definition of Regions in Greek Tourist Planning", Papers, Regional Science Association.

Pirazizy, A.A. (1988), Tourism and Environmental Degradation in Himalayas, Seminar paper, presented in Indian Institute of Geographers, X Conference, Delhi.

Pirazizy, A.A. (1989) "Impact of Construction activities on environment Changes in Kashmir Himalayas". In. I. Mohan's (Ed.) *Environmental Issues and Programmes*, Ashish Publishing House.

Shah, S.K. (1978), Facies Pattern of Kashmir within the Tectonic Framework of Himalaya, in Salklani, p.s. (eds.) *Tectonic Geology of Himalaya*, Delhi.

Singh, J.S. (Ed.) (1985), *Environmental Regeneration in Himalaya : Concept and Strategies*, The CHEA and Gyanodhya Parkashan, Nainital.

Stankey, G.H. & Lime, D.W. (1973) "Recreational Carrying Capacity: An Annotated Bibliography". INT-3, *Intermountain Forest and Range Experiment Station*, Ogden, Utah.

Stankey, G. H. (1974) , "Patterns of Wilderness Use as Related to Congestion and Solitute". *Intermountain Forest and Range Experiment Station*, Missoula, Montana.

Taylor, P.J. (1975), Distance Decay in Spatial Interactions Concepts and Techniques in Modern Geography, No. 2 Geography Abstracts, Norwich.

Tocalis,T.R.(1978), Changing Theoretical Foundations of the Gravity Concept of

Human Interaction in B.J.L. Berry (eds.) *Perspectives in Geography : The Nature of Change in Geographical Ideas,* N. Illinois, UP.

Wadia, D.N. (1934), *Geology of India,* Tata Mc-Grow Hill Publishing Co., New Delhi.

Wagner, P.L. (1960). *The human use of the earth.* Free Press, New York.

Wall, G. and Wright, G. (1977), *The Environmental Impact of Outdoor Recreation,* Department of Geography, Publication, Series, No. 11. University of Waterloo, Waterloo.

Weber, P.J. (Eds.) 1979) *High Altitude Geo-ecology,* Boulder, West View Press.

White, H.P. and Senior, M.L. (1983) *Transport Geography,* Longman Group Limited, London.

ROAD TRANSPORT IN MAGADH PLAIN BIHAR STATE

G. D. Singhal, Awadh Tiwary and *Abha Pathak*

Introduction

Magadh plain occupies the core part of central Bihar joining north and south Bihar road transport. The prominent urban centres of this plain is well-connected with roadways. But the conditions of roads and the transport system is infested with several problems. The prevailing conditions should be ameliorated at the earliest in view of the region fast gaining international tourist status. Magadh plain is, structurally, a segment of the great Indo-Gangetic trough of India. The plain spans over an area of 17,872 square kilometres consisting of six districts viz. Patna, Nalanda, Gaya, Jehanabad, Nawada and Aurangabad and seventy-four Anchals of the south Ganga Plain (Central Bihar) in the state of Bihar. It extends forms 24.°17' North Latitudes to 25.°44' and from 84.°30' to 86.°33' East Longitudes. The large human agglomerations of the Magadh plain are namely : Patna, Gaya and Biharsharif, the state capital of Bihar, Divisional and District headquarters respectively. Patna is situated on 25°36' North Latitudes and 85°15' East Longitudes Gaya 24°37' North Latitude and 85°1' East Longitude, Biharsharif 25°11' North Latitudes and 85°31' East Longitudes.

Magadh region seems to have been occupied by human settlements deep down during the prehistoric times. The word "Magadh" appeared in some of the pages of *Atharva Ved* before it took a definite shape during the Mahabharat (Singh, and Kumar, 1970). Before the advent of the Aryans, the region appears to have developed an organised agricultural society which was acquainted with how to form well-ordered villages how to make forts and building and to prepare reservoirs and tanks for irrigation" (Beden Powell, 1898). Resourceful Magadh which was not fully organised even upto Buddha and which became one of the earliest homes of Buddhism and remained a centre of faith until the Muslim conquest (Singh, 1971). The limits of Magadh empire waxed and waned in its history. This, with some alternating vicissitudes rose several times as the greatest regional and national power by annexing the neighbouring territories in the north, east and west and super-imposing on others, as in the regime of Bimbisar, Ajatshatru, Mohapadmananda, the Mauryas and the Guptas, etc. (Singh, 1971). In Bimbisara's time Magadh comprised an area corresponding to the old districts of Patna and Gaya. (presently the said six districts). The glory of Magadh was screened for a long time and the story of the region consists of many hyphens. The cities of Patna, Gaya and Biharsharif are endowed with rich historical background (Pandey, 1985).

Road Transport

Road, a symbol of motion, may be defined as a "convenient way over which vehicles, cyclists, pedestrians etc. may lawfully pass for going from one place to another" Priyani, V. B. Magadh plain is the region of central Bihar linking both the industrially rich Chhotanagpur and agriculturally most developed north Bihar plain.

Historical Perspective

Roads, as one of the initial and vital means of communication have been universally recognised everywhere and their development can be traced back to the very beginning of civilization. The excavation at Mohenjodaro have revealed that the art of road construction was known to India as far back 5,000 years. At the time to Aryan period, the major roads have been referred to as *Maha-Patha*, which in other words of a highway engineers may be called "National Highways" Road development was attached a special attention by Indian engineer during the times of Chander Gupta Maurya who has established a separate communication department to look after it. The various amenities like shady trees, wells and rest houses along the roadside were provided at the times of Ashoka. (Sehgal and Bhanot, 1966). During the regime of the

Mughal emperors, there were much improvement in road conditions.

The conditions of roads in India was pretty bad during the times of the arrival of the Britishers. It was only somewhere in the middle of the 17th century that a few trunk roads were constructed on the remains of the roads of the Mughal period and his practice continued upto the middle of the 18th century. It was realised that during the Governor-General of Lord Dalhousic in 1855 that the administration of the highways under the military engineer was not a very satisfactory arrangement. Consequently a department of Public Works, known as the Central Public Work Department (C.P.W.D.) was established. Its main works were the construction and supervision of the major roads in India. Thereupon various committees, acts and plans were framed for the development of roads e.g., appointment of Jayakar Committee and creation of Central Road Fund, establishment of Indian Road Congress 1934, Nagpur Conference, 1943, National Highway Act, 1956, 20 Year Road Development Plan, 1961-81 etc.

Road Condition in 19th Century

The account of the roads presented by Francies Buchanan who toured the region in 1811-12 presents a vivid picture of the deficiency of communication at that time. According to him, "During the rainy season, all internal commerce is at a complete standstill, as the roads are then so bad, as not to admit of even cattle, travelling with back loads. The cross roads from market to market are those which are chiefly wanted, and not one who has not seen the condition of these could believe that a country so extremely populous and rich, and having such occasion for land conveyance, could be so ill provided. The object in such roads is not to enable gentlemen to drive their curricles, but to enable cattle carrying back loads to pass at all season from one market to another, and in the fair season to enable carts to do the same" (Kumar, 1970). The actual development of roads had taken place in these conditions of the 19th century. Actually the period from 1851 to 1947 was the era of transport revolution (Singh, Jagdish, 1964). In other words of O' Malley (1906), at present time, the region is intersected by a number of excellent roads which place every part of it whitin easy reach of the markets" (Roy Choudhary, 1957). After this period there had been a tremendous development of road in and the around the major urban centres including the three cities. Presently these centres are interlinked with metalled roads.

Establishment of Bihar State Road Transport Depot

On the 26th January, 1953 a nationalised road passenger service under the State Government came into existence, by the name of "Bihar Rajya Transport" upto 30th April, 1959, when it was renamed as "Bihar State Road Transport Corporation" (B.R.S.T.C.). The Organisation is entirely owned and operated by the State Government. A State Transport Depot was established at Gaya on 1st April, 1954. It has also been established at Biharsharif. Within the region, only Patna has local (within the city) bus service. The traffic flow within the city is maximum on Bankipur Chowk, Bankipur-Malsalami, Patna Junction-Danapur road and Patna-Hajipur (on Gandhi Setu bridge) . The traffic flow on the bridge is also very heavy and buses run at short intervals (Fig. 23.1). Besides, the state buses, a number of private buses including the luxury coaches, video coaches are on the road to different part of state and even outside to West Bengal, Uttar Pradesh and Orissa. The major urban centres including the three cities of the region are well-connected with all the industrial mining and commercial centres of the state with good roads. These are served with cow-web patterns of roadways. Besides buses, the urban centres are facilitated with other mechanically propelled and power driven vehicles. The existing road pattern of the cities and other urban centres is the mixture of old and new planned roads layout. In the old city areas we may see unplanned organic pattern of street pattern while the newer settled parts have definite planned roads. Now, the roads system of Patna is typical of a linear type. The Ashoka Raj Path running on a ridge along the river Ganga has to a very greater extent, influenced this linear pattern of growth. The roads of western Patna was laid out according to a plan prepared by J.F. Munnings in 1911-12 when Patna was made the capital of the newly created province of Bihar and Orissa.

Table 23.1 : *Bus Routes, Trip and Parsenger*

Sr. No.	Name of Routes	Total number of Trips per day both ways	Total number of passenger on journey Per day as per bus capacity
1	Patna-North Bihar	105	5250
2.	Patna-Dinapur	50	2500
3.	Patna-Biharsharif	15	750
4.	Patna-Gaya	4	200
5.	Patna-Ranchi	6	300
6.	Patna-Tata	4	200
7.	Gaya-Ranchi	5	750
8.	Gaya-Dhanbad	2	100
9.	Gaya-Hazaribagh	8	400
10.	Gaya-Chatra	4	200
11.	Gaya-Lakhaipur	2	100
12.	Gaya-Sherghati	8	400
13.	Gaya-Aurangabad	8	400
14.	Gaya-Daltonganj	3	150
15.	Gaya-Daudnagar	2	100
16.	Gaya-Tikari	10	500
17.	Gaya-Mirabigha	2	100
18.	Gaya-Mau	2	100
19.	Gaya-Goh	2	100
20.	Gaya-Deoghar	2	100
21.	Gaya-Wazirganj	4	200
22.	Gaya-Rajauli	2	100
23.	Gaya-Rajgir	2	100
24.	Gaya-Saotor	6	300
25.	Gaya-Jethian	2	100
26.	Gaya-Fatehpur	2	100
27.	Gaya-Islampur	4	200
28.	Gaya-Khizersarai	4	200
29.	Gaya-Jehanabad	14	700
30.	Gaya-Nawada	6	400
31.	Gaya Bodh-Gaya	12	600

Source : Field Works.

The study of road can be examined on the following items

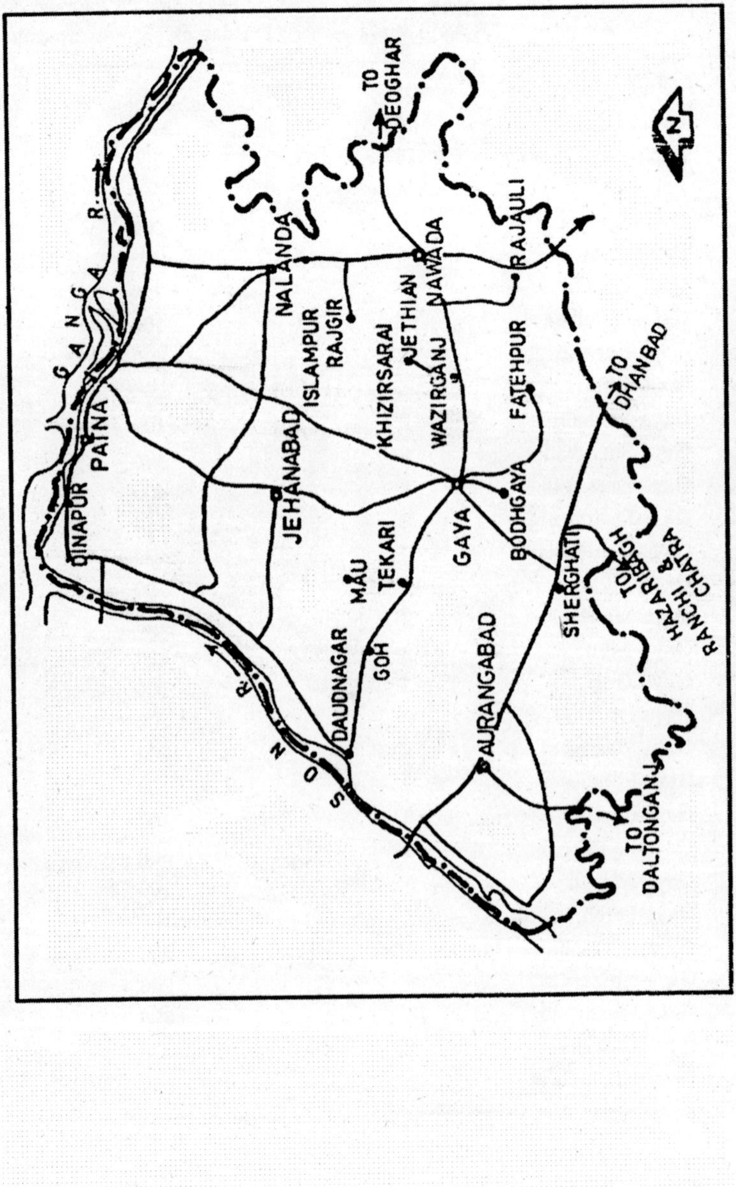

Fig. 23.1 : Roads of Magadh Plain (1998)

providing vast horizon to further investigation. These items can be listed as :

(i) Traffic Studies
(ii) Rural Traffic Studies
(iii) Motor Vehicle Volume Studies
(iv) Origin and Destination Studies
(v) Loadometer Studies
(vi) Accident Studies
(vii) Speed Studies
(viii) Urban Traffic Studies
(ix) Economic Studies
(x) Roadway Fund Allocation Studies

Trips and Volume of Bus Traffic Flow

Fig.23.2 shows the traffic flow of Magadh plain. Gaya, Bodhgaya, Patna, and Nalanda found heavy traffic of tourist from native and foreign as well. The Table 23.1 depicts the number of buses both state and private buses from Patna and Gaya to different places of the Magadh plain and within state and even beyond.

Bypass road in Magadh plain are, namely, Patna bypass road, Gaya bypass road, Biharsharif bypass road, Rajgir bypass road and Bodh Gaya bypass road. Aurangabad bypass road, Berh by pass road, Rajgir bypass road and Bodh Gaya bypass road. Presently the bypass roads have gone a long way in easing traffic congestion, traffic-jams at the "traffic-bottlenecks" of existing narrow, out-dated roads of urban centres.

Sugestions for Better Transport

Problems of road transportation and accidents exits in all cities of the world. Though on a varying magnitude. Every city provides for the movement and circulation of different types of a vehicles on roads. The most famous road of the country, the grand trunk road is so dilapidated that it makes travelling very uncomfortanble and dangerous. Dobhi-Barhi road is in very bad condition and is full of cracks and potholes and difficult to commute. In general, travelling on Bihar roads is a nightmarish experice for one and all, whether they are car-borne or bus-driven. For fulfilling the functions of road properly, it must provide by way of width and layout, adequate

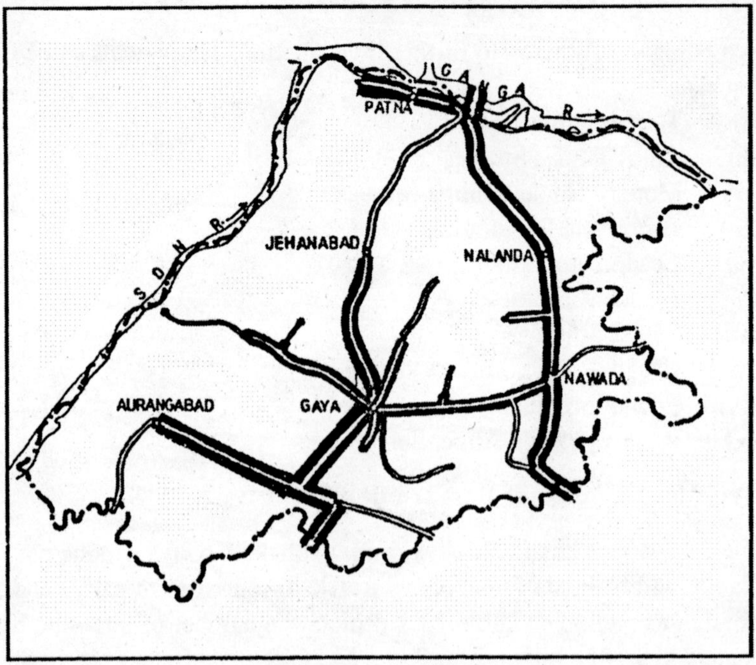

Fig. 23.2 : *Flow Line Map*

accommodation for the different types of traffic both slow and speedy using the roads. Safety is primarily a matter of traffic signs, traffic signals, street lighting, proper road junction, separation of different types of traffic streams etc. Road expansion projects not kept up, and indisciplined traffic can drive even the Mercedes round the Benz. Now the question arises, will urban India become an easy rider, or will the ultimate symbol of having arrived being stuck in a traffic logjam, fuming with the fumes ? Now, the fast increasing volume of mechanically propelled vehicles, combined with slow moving traffic, have outgrown the width of roads and lanes in almost the whole of the built up are of the cities. So, the roads accidents are not only for reasons of human elments, but also the character of roads, its design and layout.

Concluding Remarks

Magadh plain is endowed with a number of religio-historic centres of international tourist fame. Patna, Gaya, Bodhgaya, Rajgir,

Nalanda, Pawapuri, Biharsharif, Maner etc. occupy very significant places on the tourist map of India. These centes are provided with good road communication and transport facilities and fall on the Buddhist circuit, still more transport ease is desirable keeping in view of the lakhs of foreign and native tourist arrivals.

References

Beden-Powell, S.H. (1898) : *The Indian Village Community*, Pongman Green and Co. London

Chaudhary, P.C. Roy (1957) : *Bihar District Gazetteers-Gaya*, Patna Secretariat Press.

Kumar, N. (1970) : *Bihar District Gazetters-Patna*, Patna Secretariat Press.

Priyani, V.B : "Road Engineering".

Singh, R.P. and Kumar, A.(1970) : *A Monograph of Bihar*, Bharat Bhawan.

Sehgal, S.B. and Bhanot, K.L. (1966) : *A Texbook on Highway Engineering and Airports*, S.Chand and Sons, New Delhi.

Singh, Jagdish, (1964) : *Transport Geography of South Bihar*, B.H.U., Varanasi.

Singh, R.L.(1971): *India-A Region Geography*, National Geographical Society of India, Varanasi 185.

24

ANALYSIS AND PLANNING OF RURAL TRANSPORT SYSTEM IN GORAKHPUR DISTRICT UTTAR PRADESH

R. C. Tiwari and *Shankreshwar Tripathi*

Introduction

Transport system is a mirror of economic development and material prosperity. Division of labour, areal specialization and commercialisation which are integral part of modern civilization, are the off springs of modern means of transport and communication (Tiwari and Tripathi,1987). The greater the centralization and specialization of man's activities, the higher is the need for efficient transport (Janella, 1969). As a matter of fact, transport provides the basic infrastructure for the growth and development of agriculture and industry (Tiwari and Yadav, 1990). In modern society, transport has attained much more significant role not only in terms of economic development but removing scarce conditions, maintaining uniform price line or safeguarding the people from hunger, disease and natural calamities. It is with this view that an attempt has been made in this Chapter to evaluate the existing network of rural transport and suggest suitable guidelines for its scientific planning taking Gorakhpur district of Uttar Pradesh as a case study.

The Region

Gorakhpur District (26°13' to 27°29') North Latitude 83°05' to 83° 56' East Longitude) is a part of Saryupar region of the Ganga Plain. It is surrounded by the districts of Deoria in the east, Azamgarh in south, Basti in the west and international frontier along Nepal in the north. The total geographical area (some part of Gorakhpur district have been utilised in curving out a new districts named as Maharajganja) 6,272 km. (2.2 per cent of Uttar Pradesh) is divided into 4 tehsils, 31 community development block and 291 Nyaya Panchayats. The region is a part of the vast plain of northern India, which has been built up by the riverine deposits in a fore deep during the Pleistocene period. The slope is gentle (average gradient is less than 1°). The average height ranges from 75 metre along the Ghaghara (in the south) to 110 metres (in the north). The total population of 3,764,473 (1981 Census; 1991 estimate being 4,347,000) resides in 4,110 villages and 11 towns which are connected through *kachha/pucka* roads many of become inoperative during rainy season due to frequent floods and water-logging.

Existing Transport System

Like other districts of Uttar Pradeh "The road transport plays the most important role in the spatial interaction of Gorakhpur district (Singh, J., 1979). These Roads may be grouped into National Highway, State Highways and District roads. The national highway passes through the central and southern part of the district (Fig. 24.1). It includes (a) Varanasi-Gorakhpur (52.17km) and (b) Lucknow-Gorakhpur-Kasia (42.17 km.) roads. The construction of bridges over the Ghaghara at two places (near Ayodhya and Doharighat had made significant changes in the mobility of goods and passenger traffic to and from the district.

Besides, National Highways, there are 300 km. long state highways within the study region. Again the blocks of Gorakhpur tahsil enjoy the maximum density of state highways. These include Sonauli-Nautanwa, Gorakhpur-Ballia (96 km.), Gorakhpur-Golabazar (62 km.), Gorakhpur-Maharajganj (54 km.) and Gorakhpur -Deoria (17 km). Besides, there are 325 km. long district and other roads maintained by the state public work department. These are mainly confined to Pharenda and Maharajganj tahsils. The forest department maintains another 150 km. of roads of the Tarai tract of the districts. Similarly, Zilla Parishad is another agency which maintains nearly 400 km.of roads of the district. The construction of link roads is a recent major step in transport development which links the central villages with important towns and cities of the region and help in accelerating the process of rural development.

The density of railways has been confined to the northern and central parts of the district (Fig. 24.1) and only 42 per cent of the blocks of the district are enjoying this facility. The southern region occupied by Bansgon tahsil is devoid of railway transport and is considered as the most backward track of the district. With the conversion of the metre gauge line into the broad gauge, the region has been provided direct rail line to Lucknow in the west, north-western Bihar, West Bengal, Assam and eastern India in the east and central and southern India in the south. Prior to the introduction of railways in 1885 and subsequent construction of metalled roads, Ghaghara and Rapti river were extensively used for transporting goods and people. There was regular boat service up and down in the Ghaghara river even during the British period. But after the construction of railways and metalled roads, the importance of waterways has been considerably minimised and today, these are used only for crossing the river course and for floating the logs down the Rapti from the Himalayan foothills to Deoria, Ballia and Bihar.

Density of Roads

The average density of metalled roads in Gorakhpur district comes to 18.17 km. per 100 km. and 29 km. per lakh population (Uttar Pradesh has 18 km. and 60 km. respectively). The lowest density (less than 10 sq. km. per 100 sq. km is recorded in the blocks susceptible to frequent floods and in Tarai tract. In order to attain a more realistic picture of the density of roads, the region has been divided into grids of one square km. and length of metalled roads has been measured in each grid. Here again Gorakhpur city and its neighbouring areas have emerged on the top (Fig. 24.2). Likewise, Khajani in the south and Maharajganj in north record moderate road density. The lowest road density is observed in the forested area of the Tarai zone and in the flood prone areas of the Rapti, Rohini and Ghaghara rivers.

Figure 24.2 shows the density of railway line on the basis of grid square which exhibits that nearly 70 per cent of the region's area is devoid of railways facilities. On an average, there is 3 km.

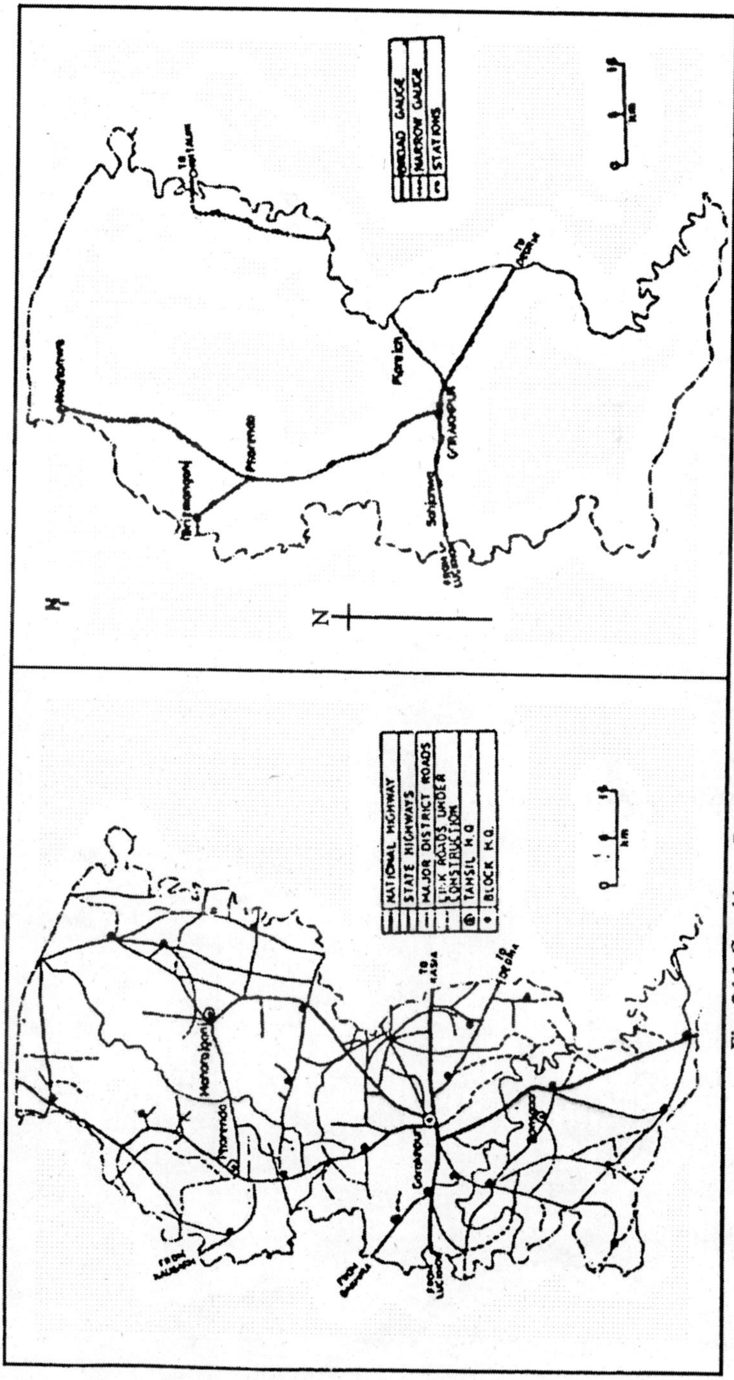

Fig. 24.1: *Gorakhpur District Transport Network Road and Railways (Existing)*

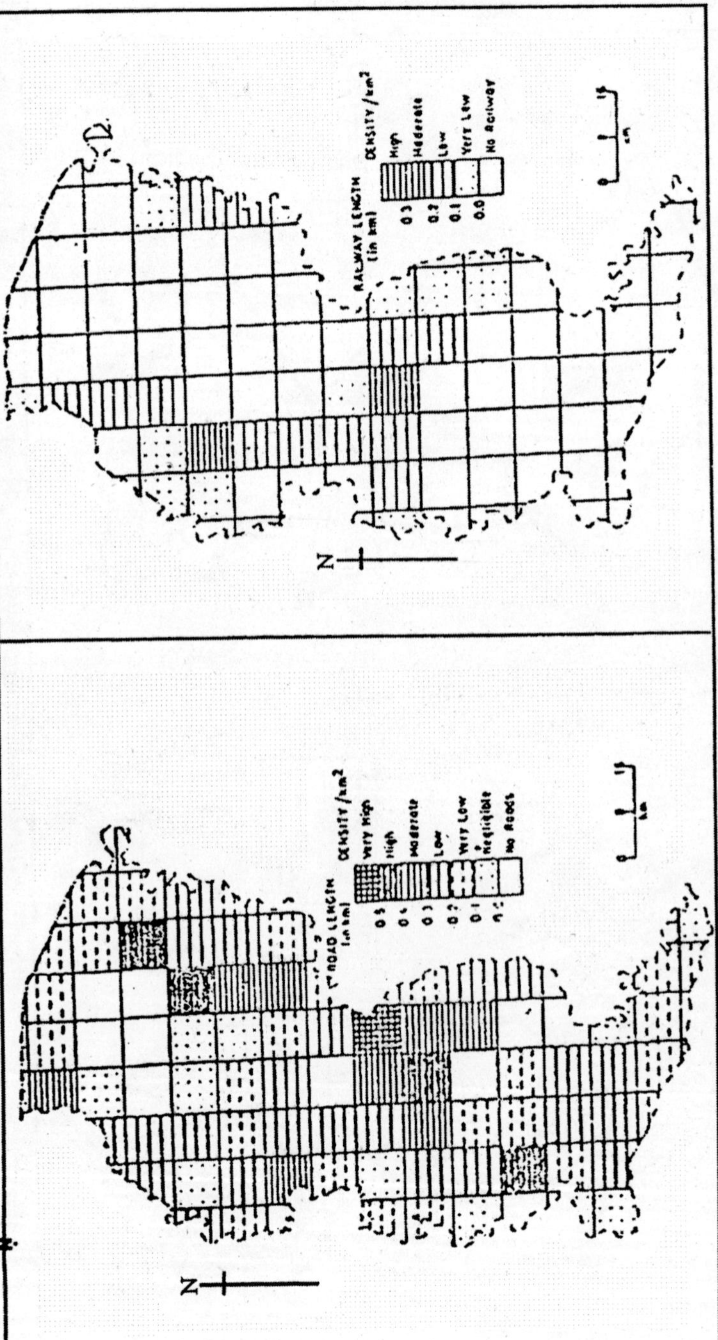

Fig. 24.2 : Density of Railway Lines and Metalled Roads of Gorakhpur District

of rail length for every 100 sq. km of the district's area against the state average of 29 km. Pharenda and Maharajganj tahsils have only 2.8 km. of railway line for every 100 square km. of the area. In the central part of the region this figure rises to 5.1 km. Here again the highest density of the railways has been recorded around Gorakhpur city and moderate density (0.20 to 0.30) near Pharenda town.

Accessibility

In fact no place is really inaccessible on the earth in strict sense of the term. But here the area lying within 5 km. from main roads and 3 km. from sub-roads has been considered highly accessible. Similarly, the area lying within 10 km. distance from the road enjoy medium accessibility beyond which lies inaccessible area. In the present analyses following four aspects of accessibility, namely, (a) accessibility by Roads, (b) accessibility by Railways, (c) Net accessibility, and (d) travel-time accessibility.

Accessibility by Roads

Road accessibility (Fig.24.3) has been measured in terms of distances of 5,10 and 15km. from the road. The 10 km distance from the road, on an average, has been taken as accessibility area. As is evident from Fig. 24.3 the majority of the inaccessible area are found in Nautanawa. Laxamipur and Nichloul blocks of the Tarai region due to its difficult terrain and dense forest cover. Similarly Bramhpur, Khajani and Belghat blocks and a small chunk in Sahjanawa block fall under the inaccessible areas. On an average 30 per cent of the villages of the district enjoy very high accessibility, 26 per cent high accessibility, 18.89 per cent medium accessibility, and remaining 25.11 per cent poor accessibility, Pharenda tahsil occupies the first place in the district whose 65 per cent of the villages are found within three kilometres distance from the metalled or unmetalled road. Maharajganj tahsil stands second whose 57 per cent of the villages are situated within 3 km limit from the road. In remaining two tahsils (Gorakhpur and Bansgaon) more than 50 per cent of the villages lie within 3 km. distance from the roads. It is worth mentioning that large number of villages in Pharenda and Maharajganj tahsils are located along the roadside because these tahsils belong to the newly settled area of the district. On the contrary, villages of the old settled area (Bansgaon and Gorakhpur tahsils) do not align with the metalled roads which are mainly the development of the present century. Their sites and situations have been much influenced by the fertility of soils, the availability of drinking and irrigation water and security of life and property rather than the transport accessibility.

Accessibility by Railways

As mentioned above the railways in the district are mainly confined to the central and northern parts of the region. That is why the accessibility by railways is maximum in Gorakhpur tahsil closely followed by Pharenda tahsil (Fig. 24.3). The conditions are not so favourable in remaining two tahsils. It is evident from Figure 16.3 that a belt of inaccessible area by railways runs from Bhathat and Paniyara to the international boundary with Nepal. Another small chunk of an inaccessible area is noticed in Pali block.

Composite Accessibility

A composite picture of accessibility may be obtained by superimposing the two maps of railways and roads over each other. Thus Mithoura block of Maharajganj tahsil record the maximum accessibility while the entire central part of the district enjoy good accessibility. Other areas of poor accessibility include Khajani, Urwa and Barhalganj blocks of Bansgaon tahsil.

Accessibility by Time

Accessibility may also be analysed with the help of the isochronic diagram (Singh, J. 1964) called relative accessibility and is largely determined by the speed of travel, degree of traffic congestion, frequency of traffic Jams, directness of routes, etc. (Brown, 1959) Figure 24.4 has been prepared on the basis of isochronic diagram taking into account the travel time distance from the district headquarters. It has been found that almost all the block headquarters of Gorakhpur tahsil are approachable within an hour from the district headquaters. This distance is the highest in the west, south and east along national highways and state ; highways. But its limit decreases on the roads of Sonouli, Maharajganj and Pipraich. Similarly, the

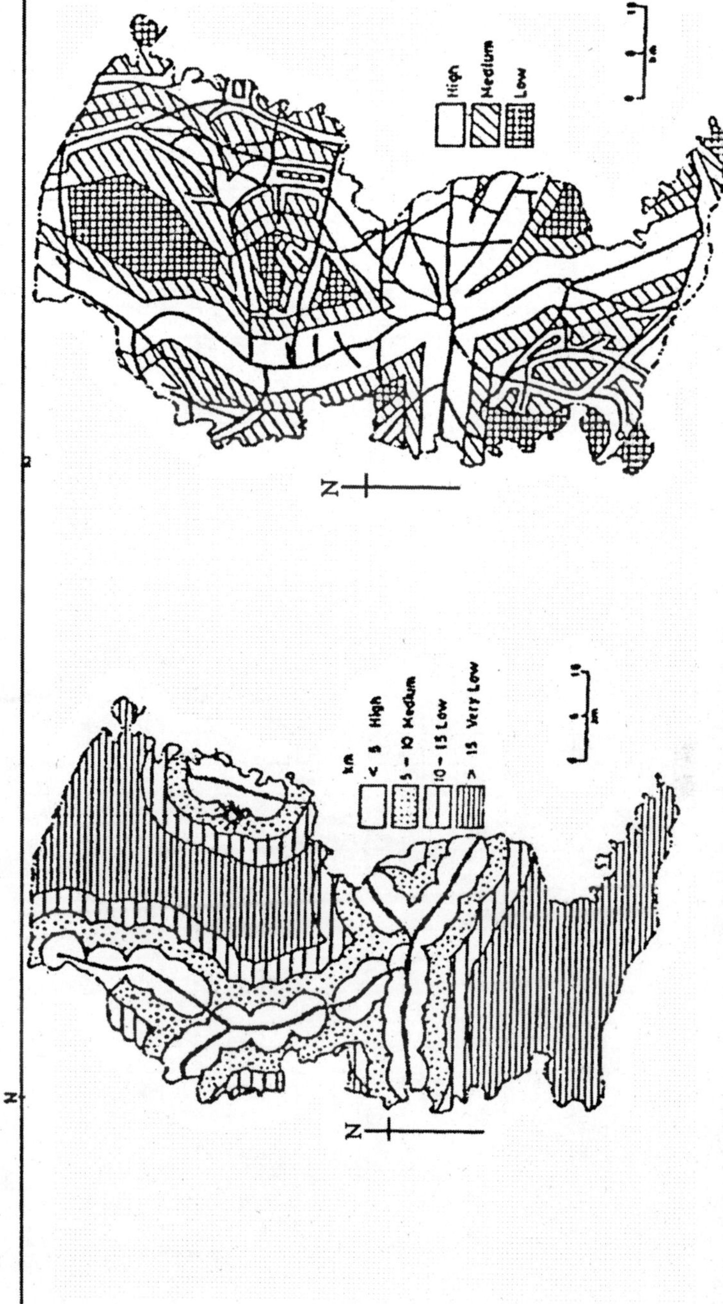

Fig. 24.3 : *Accessibility by Roads and by Railways in Gorakhpur District*

Table 24.1 : *Connectivity Matrix*

S.No.	Name of Blocks	1	2	3	4	5	6	7	8	9	10	11	12	13	14	15	16	17	18	19	20	21	22	23	24	25	26	27	28	29	30	31	
1.	Jungle Koutia	0	0	0	0	0	0	0	1	0	0	1	1	0	0	1	1	0	0	0	0	0	0	0	0	0	0	0	0	0	0	0	5
2.	Chargaon	0	0	1	0	0	0	0	0	0	0	0	0	0	0	0	0	0	0	0	0	0	0	0	0	0	1	1	0	1	0	1	5
3.	Bhathal	0	1	0	1	1	0	0	0	0	0	0	0	0	0	0	0	0	0	0	0	0	0	0	0	1	1	1	0	1	0	1	0
4.	Pipraich	0	0	1	0	1	1	0	0	0	0	0	0	0	0	0	0	0	0	0	0	0	0	0	0	0	0	0	0	0	0	0	3
5.	Sardar Nagar	0	0	1	1	0	1	0	0	0	0	0	0	0	0	0	0	0	0	0	0	0	0	0	0	0	0	0	0	0	0	0	3
6.	Kharobar	0	0	0	1	1	0	0	1	0	0	0	0	0	0	0	0	0	0	0	0	0	0	0	0	0	0	0	0	0	0	0	3
7.	Bramhpur	0	0	0	0	0	0	0	0	0	0	0	0	0	0	0	0	0	0	0	0	0	0	0	0	0	0	0	0	0	0	0	0
8.	Sahjanawa	1	0	0	0	0	1	0	0	0	1	1	0	0	0	0	0	0	0	0	0	0	0	0	0	0	0	0	0	0	0	0	4
9.	Pali	0	0	0	0	0	0	0	0	1	0	0	0	0	0	0	0	0	0	0	0	0	0	0	0	0	0	0	0	0	0	0	1
10	Piprouli	0	0	0	0	0	0	0	0	1	0	0	0	0	0	0	0	0	0	1	0	0	0	1	0	0	0	0	0	0	0	0	3
11.	Nautanwa	1	0	0	0	0	0	0	0	0	0	0	1	1	0	1	1	0	0	0	0	0	0	0	0	0	0	0	0	0	0	0	5
12.	Laxamipur	1	0	0	0	0	0	0	0	0	0	1	0	0	0	1	1	0	0	0	0	0	0	0	0	0	0	0	0	0	0	0	4
13.	Brizmanganj	0	0	0	0	0	0	0	0	0	0	1	0	0	1	1	0	0	0	0	0	0	0	0	0	0	0	0	0	0	0	0	3
14.	Dhani	0	0	0	0	0	0	0	0	0	0	0	0	1	0	0	0	0	0	0	0	0	0	0	0	0	0	0	0	0	0	0	1
15.	Pharendo	1	0	0	0	0	0	0	0	0	0	1	1	1	0	0	1	0	0	0	0	0	0	0	0	0	0	1	0	0	0	0	6
16.	Comperganj	1	0	0	0	0	0	0	0	0	0	1	1	0	0	1	0	0	0	0	0	0	0	0	0	1	1	0	0	0	0	0	6

(Contd.)

Table 24.1 : *(Contd)*

17.	Barhalganj	0	0	0	0	0	0	0	0	0	0	0	0	0	0	0	0	0	1	0	0	1	1	0	1	0	0	0	0	0	0	0	4
18.	Gagha	0	0	0	0	0	0	0	0	0	0	0	0	0	0	0	0	1	0	0	0	1	0	0	0	0	0	0	0	0	0	0	2
19.	Khajani	0	0	0	0	0	0	0	0	0	1	0	0	0	0	0	0	0	0	0	1	1	0	1	0	0	0	0	0	0	0	0	4
20.	Bansgaon	0	0	0	0	0	0	0	0	0	0	0	0	0	0	0	0	0	0	1	0	1	0	0	0	0	0	0	0	0	0	0	2
21.	Kouri Ram	0	0	0	0	0	0	0	0	0	0	0	0	0	0	0	0	1	1	1	1	0	0	0	0	0	0	0	0	0	0	0	4
22.	Urwa	0	0	0	0	0	0	0	0	0	0	0	0	0	0	0	0	1	0	0	0	0	0	0	1	0	0	0	0	0	0	0	2
23.	Belghat	0	0	0	0	0	0	0	0	0	0	1	0	0	0	0	0	0	0	1	0	0	0	0	0	0	0	0	0	0	0	0	2
24.	Golo Bazar	0	0	0	0	0	0	0	0	0	0	0	0	0	0	0	0	1	0	0	0	0	1	0	0	0	0	0	0	0	0	0	2
25.	Paniyara	0	0	1	0	0	0	0	0	0	0	0	0	0	0	0	1	0	0	0	0	0	0	0	0	0	1	0	0	0	0	0	3
26.	Partawal	0	1	1	0	0	0	0	0	0	0	0	0	0	0	0	1	0	0	0	0	0	0	0	0	0	0	1	0	1	0	1	7
27.	Maharajganj	0	1	1	0	0	0	0	0	0	0	0	0	0	0	1	0	0	0	0	0	0	0	0	0	0	1	0	1	1	0	1	7
28.	Ghughuli	0	0	0	0	0	0	0	0	0	0	0	0	0	0	0	0	0	0	0	0	0	0	0	0	0	0	1	0	0	0	1	2
29.	Mithura	0	1	1	0	0	0	0	0	0	0	0	0	0	0	0	0	0	0	0	0	0	0	0	0	0	1	1	0	0	0	1	5
30.	Siswa	0	0	0	0	0	0	0	0	0	0	0	0	0	0	0	0	0	0	0	0	0	0	0	0	0	0	0	1	0	0	1	2
31.	Nichloul	0	1	1	0	0	0	0	0	0	0	0	0	0	0	0	0	0	0	0	0	0	0	0	0	0	1	1	1	1	1	0	7
		5	5	6	3	3	3	0	4	1	3	6	5	3	1	6	6	4	2	4	2	4	2	2	2	3	7	7	3	5	2	7	7

two hour limit is the highest from national highways and lowest for other roads. It is also clear from Fig. 24.4 that the marginal tahsils and Khadar areas of Urwa block require more than 6 hours of journey by bus. Here topography seems to be more dominant for influencing the relative accessibility than other factors. Similarly, the gravitational pull of regional and state capitals (Varanasi-Lucknow) and the nearness of international boundary have also influenced the relative accessibility of the district.

Traffic flow

Transport flow involves the volume and direction of the movement of goods, people and messages. This is a focus of many geographical investigation, from the diffusion of innovations of the movement of consumers (Hurst, 1974.). This study region is predominantly an agricultural area, hence, the traffic by and large, is confined to agriculture goods. It is more obvious at micro level where foodgrains, oilseeds, sugarcane and vegetables are principal commodities of movement from one place to another. Besides, consumer goods of daily use also constitute an integral part of the traffic flow. Gorakhpur city being the centre of district administration together with its four tahsils, 31 development blocks and a number of regulated markets, forms the nucleus for bus traffic flow in the region (Fig. 24.5). Maximum flow is marked on the routes of Gorakhpur-Sahjanawa-Lucknow and Gorakhpur-Barhalganj-Varanasi which join two principal cities of the state. Similarly, heavy traffic is also observed on Gorakhpur-Deoria, Gorakhpur-Kasia and Gorakhpur-Pharenda routes. Each block headquarters, local town and important mandi centre is connected by buses originating from Gorakhpur city whose frequency declines away from the city centre. As far as the intra-regional flow of traffic is concerned, it shows seasonal variations maximum after the harvesting of rabi crops and maximum traffic flow is recorded during the rainy season. But in sugarcane producing areas, maximum traffic flow is recorded during winter season.

Conncetivity Matrix

The effectiveness of transport network may be analysed through the connectivity matrix for which each block headquarter has been given the status of vertex and the matrix chart (Table 24.1) has been prepared by counting the number of edges. It is apparent that Bhathat, Nichloul, Maharajganj, Compeirganj, connectivity (each block having more than five vertices to its credit). On the contrary, Pali, Dhani, Gagha, Bansgaon, Urwa, Belghat, Gola Ghughuli and Siswa blocks have poor connectivity (less than 3 vertices). Others like Pipraich, Sardar Nagar, Khorabar, Sahjanawa, Piprouli, Laxamipur, Brizmanganj, Barhalganj, Khajani, Kouri Ram and Paniyara block represent moderatae connectivity.

On an average connectivity is high in the neighbourhood of urban centres and in fertile tracts. On the other hand, low connectivity is observed in those blocks which are either located in flood prone areas or on the periphery of the district. It is worth mentioning that these blocks are not even well connected with each other and hence, lack mutual exchange of expertise and ideas.

Alpha, Gama and Beta Indices

Table 24.2 : *Topological Indices*

Sl. No.	Tahsils	Alpha	Beta	Gamme
1.	Pharenda	0.08	1.18	0.46
2.	Maharajganj	0.14	1.15	0.41
3.	Gorakhpur	0.19	1.19	0.42
4.	Bansgaon	0.13	1.09	0.40
	District	0.17	1.18	0.40

Source : Author

The efficiency of the road network may be better evaluated with the help of topological indices (Kansky, 1963). Amongst these the value of the Alpha index in the region is only 0.17 (17 per cent of the optimum condition) ranging from the maximum of 0.19 in Gorakhpur tahsil to the minimum of 0.08 in Pharenda tahsil. It has been observed that the number of alternate paths between vertices is very poor. Similarly, the higher value in Gorakhpur tahsil may be attributed to the maximum concentration of development programmes there. The Gamma index representing the ratio of the

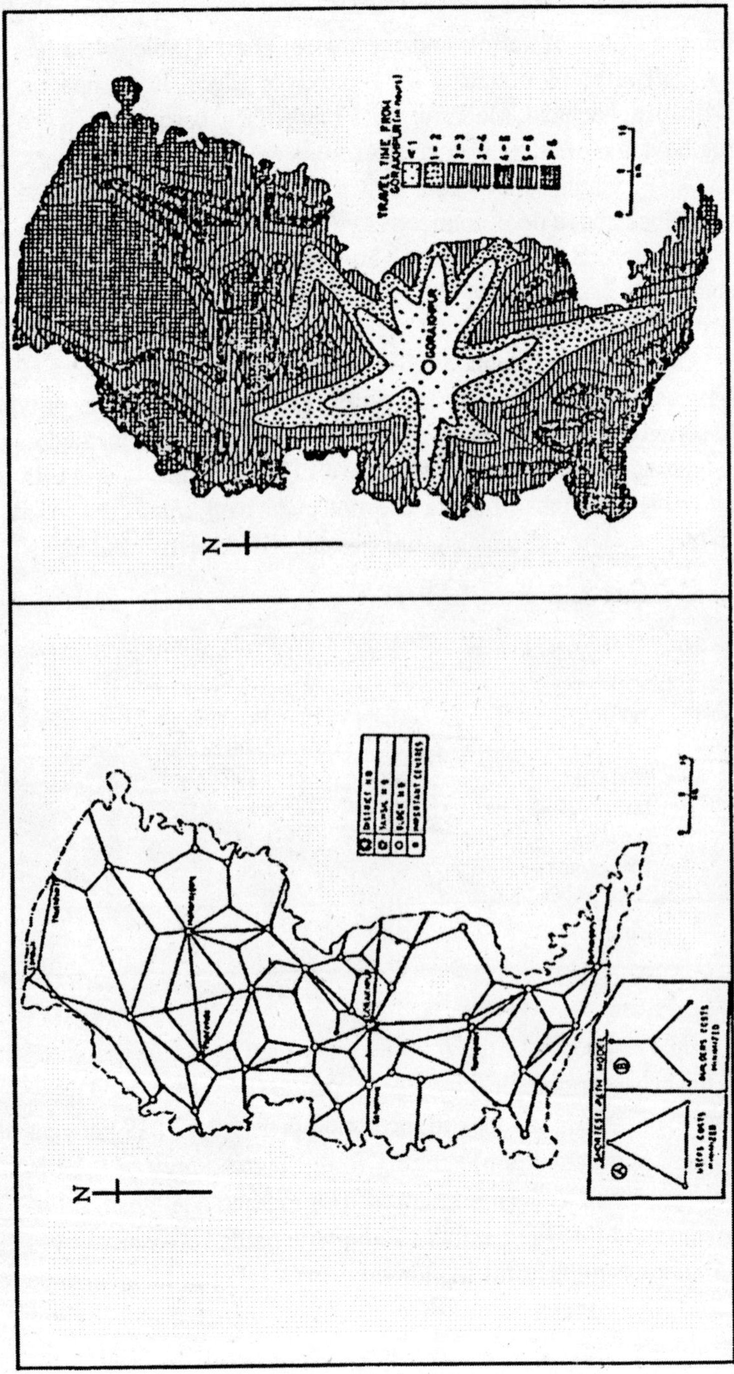

Fig. 24.4 : Gorakhpur District Bus Accessibility From Gorakhpur

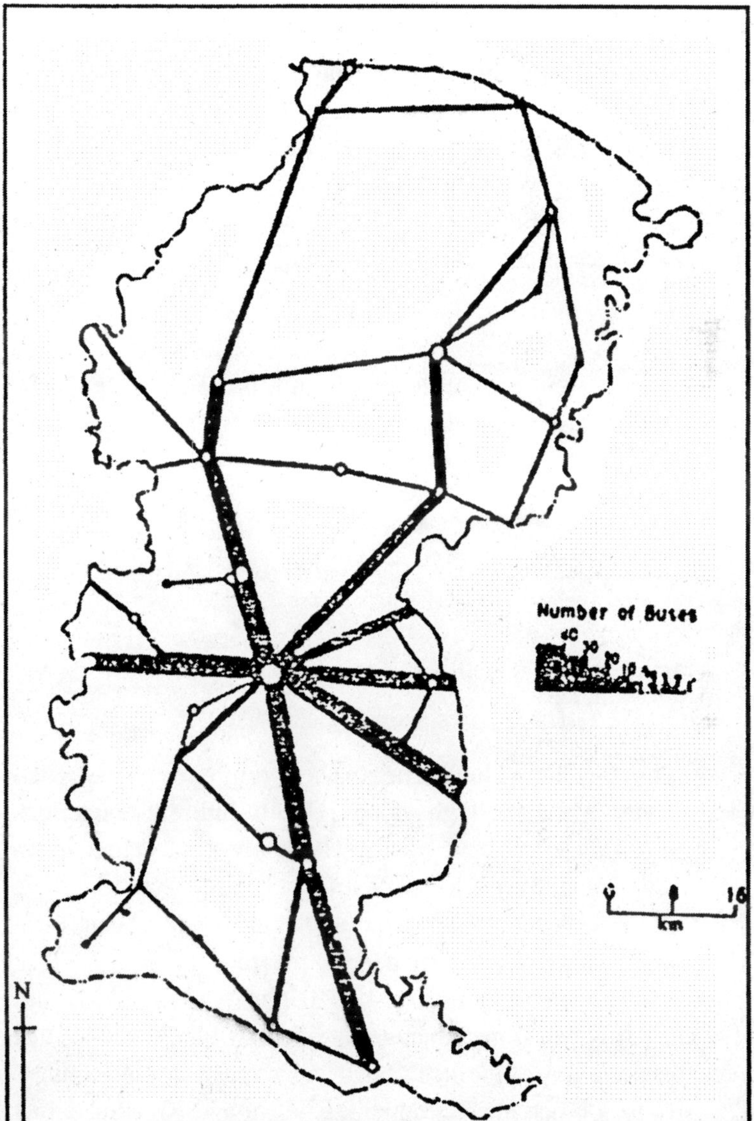

Fig. 24.5 : *Gorakhpur District BusTraffic Flow*

observed number of edges to the maximum possible number of edges, exhibits an average value of 0.40 for the district. Here Pharenda tahsil tops the list (0.46) while Bansgaon tahsil (0,40) occupies the lowest place. Intensity of transport linkages may also be analysed through Beta index which exhibits the number of linkages per node. A Beta index value of more than 1 indicates increasing number of edges. The study region has an average Beta index value 1.18 exhibiting its highest (1.19) and lowest(1.09) values in Gorakhpur and Bansgaon tahsils respectively. This suggests a very low degree of connectivity prevailing in the district.

Detour Index

Detour index, an important dimension of transport network is the ratio of direct and actual distances between two centres and may be obtained by using the following formula :

$$Di = \frac{Ad}{Dd} \times 100$$

where, Ad = Actual shortest distance on the network, and

Dd = Direct (topological) distance.

The low D*i* value indicates a straight or direct road connectivity while the high value exhibits indirect conntions between different transport nodes. The analysis of detour index values reveals that there is a lack of straight line connection between the block headquarter of the district (detour index value being 127). The highest value of 157 per cent is recorded by Gagha block followed by Siswa (151). Gola (148), Belghat (143) and Mithoura (142) etc. It is worth mentioning that all these blocks are located along the periphery of the district and lack direct road connectivity. Contrary to it block like Barhilganj, Khajani and Nichloul depict low detour index indicating direct road connections (Table 24.3).

Table 24.3 : *Detour Index*

Blocks	Detour Index	Blocks	Detour Index
Jungle Kouri	129	Barhalganj	103
Chargaon	102	Gagha	157
Bhathat	131	Khajani	110
Pipraich	120	Bansgaon	137
Sardar Nagar	124	Kouri Ram	125
Khorabar	112	Urwa	139
Bramhpur	125	Belghat	143
Sahjanawa	113	Gola	148
Pali	121	Paniyara	133
Piprouli	127	Partawal	120
Nautanawa	116	Maharajganj	134
Laxamipur	120	Ghughuli	122
Brizamanganj	137	Mithoura	142
Dhani	122	Siswa	151
Pharenda	125	Nichloul	110
Campeirganj	113		

Source : Author

Problems and Prospects of Transport System

Preceding discussions reveal that Gorakhpur district lacks an efficient transport system so as to promote development activities. Here transport system is encountered with a number of problems of whom following are worth mentioning :

1. Majority of the roads of the district are not being utilized to their full capacity while those near Gorakhpur city and tahsil headquarters generally remain over congested.
2. The road-length is inadequate indicating very poor accessibility in Mithoura, Nautanawa, Belghat, Urwa and Khajani blocks. The situation is almost alarming in Brahampur, Gagha, Belghat and Mithoura blocks which have less than 10 kms. of metalled road 1100 square km. of area.
3. Intra-block density of roads is not uniform throughout the district nearly 25 per cent villages of the district are located

beyond 5 km. of distance from the roads. The situation is alarming in Sahjanawa block where 63.57 per cent of the villages are located beyond 5 km. of distance from the roads. Similar conditions are also noticed in Brahampur, Laxamipur and Belghat blocks. Accessibility is also poor in respect of travel time in the region.

4. As evident from connectivity matrix, topological and detour indices, the connectivity of roads is poor in the region. Several block headquaters like Pali and Dhani are connected with only one centre in the region. Analyses of the road network shows an alpha index of 0.17 and the Gamma index of 0.40 displaying a low interaction between different parts of the district.
5. The district also lacks complementary nature of railway and road network. Concentration of both these means of transport is found in the central and north-western parts of the district. On the other hand, entire Bansgaon tahsil is not only deprived of the railway facilities, but is also lagging behind in road network except the places located along the National Highway from Gorakhpur to Varanasi.
6. There are structural problems in the development and maintenance of road network due to their varying width, depth of solling, construction material, management and ownership. Several important roads lack bridges and metalled surface. This problem is more pronounced in Pali, Sahjanawa, Bramhpur and Belghat blocks.
7. Heavy bus traffic, on some roads, creates traffic congestion and rapid deterioration of the road surface.
8. The nature of traffic problem varies from season to season. Generally the traffic almost remains suspended during rainy season due to floods and water logging (Khadar and Tarai region). At times roads come under one metre deep water causing immense damage to their surface.
9. Traditional mode of transport also uses (bullock cart, tonga, rickshaw, cart, thela, cycle etc.). Same road retarding its efficiency in terms of passenger traffic and good movement.
10. Regular bus services are hardly seen on sub-roads. Also there is no co-ordination between Utter Predesh State Road Transport Corporation and private buses resulting into huge

loss to the state exchequer.

11. Rural transport system also suffers from a number of maladies. Besides lack of good all-weather roads it lacks better transport planning and management. Though a scheme has been launched to connect all big villages (with more than 1,500 inhabitants) through link roads but due to faulty construction and poor maintenance these roads become unserviceable during rainy season. Even the consolidation scheme has not been able to improve village transport system due to poor planning of farm roads, cart tracks and footpaths.
12. There is resource inadequacy and lack of civic sense amongst people due to which roads are not properly used and frequent damages are caused while laying down pipelines, erecting electric and telephone poles, constructing drains, regulating rain water, or damaging and obstructing pavements for personal purposes.

Planning for Transport System

Since, the degree of efficiency in transport network is positively correlated with the levels of economic development, the transport planning is a prerequisite for rural development. Since the study region largely consists of a plains surface, gentle slope and uniform distribution of settlements, rail and roads are the best suitable means of transport. Due to rapid increase of road transport in recent years, the railway transport has been relegated to secondary position. In fact, "roads and road development are increasingly making first claim on nations budget allotments for road transportation because of its versatility and flexibility, represents uniquely useful instrument for economic and social progress" (United Nations Report, 1963.

In fact the concept of isotropic plane (Alber, R., 1971) provides the best solutions for the problems of (i) line set, and (ii) area set. This concept is largely based on the principle of shortest route network between three centres. This shortest route of straight lines equals or exceeds 120 degree angle from the point to which all the three paths are joined. Here a model plan for transport development with the help of critical path analysis based on isotropie plane concept has been proposed.

Following new roads been proposed joining important centres

of the region. The include (i) Thuthibari-Laxamipur road, (ii) Laxamipur-Maharajganj road (iii) Kouria Jungle-Pali road, (iv) Panjyara-Laxamipur road (v) Kouria Jungle-Bhaghat Gangha road. (vi) Brahmpur-Kouri Ram road, and (vii) Khajani-Belghat Gangha road. Besides these, major roads, some of the important feeder and approach roads should also be developed to ease out transport problems in the region (Table 24.4).

Table 24.4 : *Proposed Link Roads*

Sl. No.	From	To	Main Road to the linked
1.	Maharajganj	Chowk Bazar	Maharajganj
2.	Jhangaha	Bishunpur	Jhangaha
3.	Bhathat	Bailo	Bhathat-Maharajganj
4.	Laxamipur	Mahua-Adda	Laxamipur
5.	Maharajganj	Bagapar	Maharajganj
6.	Mithoura	Khajuria	Mithoura-Nichloul
7.	Mithoura	Sekhui	Mithoura-Nichloul
8.	Maharajganj	Khutaha	Maharajganj-Pharenda
9.	Khajani	Rampur Pandey	Khajani-Gola Bazar
10.	Nichloual	Baithwalia	Nichloul
11.	Malhanpur	Gaur Khas	Urwa-Gola Bazar Road
12.	Gopalpur	Gaur Khas	–do–
13.	Gola Bazar	Pakari	Gola Bazar
14.	Jungle Kouria	Bhiti	jungle Kouria-Pharenda
15.	Compeirganj	Ragauli	Compeirganj Karmaine Road

Source : Author

The district planning unit has also suggested the linking of all service centres and central villages by metalled roads. All block headquarters with atleast one regulated market by metalled roads and all service centres with surrounding villages by unmetalled roads. It is worth mentioning that mere provision for optimum transport network is not going to provide guarantee for the efficient transport system. In order to improve the traffic flow, some more vehicles should be brought on the roads. Brizmanganj, Dhani, Panjyara, Siswa, Khajani and Belghat blocks should be granted license for plying on more private transport and Uttar Predesh State Road

Corporation buses. This should also be supplemented with more and more request bus stops to improve transport accessibility of the region. The opening of a new bus Depot at Nichoul is a similar step to improve transport conditions in the tarai region of the district. Similarly the construction of new road bridges repairing and remodelling of old bridges and construction of by-passes are other suggestions to streamline the existing transport system. Similarly, construction of roads along the canal embankments or protective bunds/walls system, minimise its construction cost, but also increase durability of the canals and the bunds.

References

Alber, R. *et al.* (1971): Spatail Organisation, *The Geographers' View of the World*, New Jersey, Prentice Hall.

Brown, P.A. (1959) : The Local Accessibility of Nottingham, *The East Midland Geographer*, No. 11

Colley, C. H. (1974) : "The Theory of Transportation" in M. E. E. Hurst (Ed) *Transportation Geography : Comments and Reading*, New York : Mc-Graw Hill Book Company.

Hurst, M. E. E. (1974): *Network Analysis*, in M.E.E. Hurst (Ed), *Transportation Geography*, London : Mc-Graw Hill Book Company.

Janelle, D.G. (1969): Spatial Reorganisation: A Model and Concept, *A.A.A.G.*, Vol. 59, No. 2 .

Kanasky, K. J. (1963) : Structure of Transport Network, Research Paper No. 84 Chicago : Department of Geography, University of Chicago.

Martin, B.V. and Warden, C. B. (1965) : Transport Planning in Developing Countries, *Traffic Quarterly*, Vol. 19, No.1.

Singh, J. (1964): *Transport Geography of South Bihar*, Varanasi : National Geographical Society of India, B. H. Varanasi.

Singh, J.(1979) : *Central Places and Spatial Organization in a Backward Economy: Gorakhpur Region – A Study in Integrated Regional Development Gorakhpur* : Uttar Bharat Bhoogol Parishad Gorakhpur.

Tiwari, R C. and Yadav, H.S.(1990) : Spatial Characteristics of Transport Network is Allahabad District Uttar, Pradesh, *National Geographer*, Vol. 25, No. 1.

Tiwari, R.C. and Tiwari, S. (1987) : Role of Transport in Development, *Avadh Journal of Social Sciences*, Vol. 1 (Kalakankar : Uttar Predesh).

United Nations Report on Science and Technology for Development, (1963) : Vol. 1, New York : U. N. O.

SPATIAL CHARACTERISTICS OF TRANSPORT IN ALLAHABAD DISTRICT, UTTAR PRADESH

R.C. Tiwari and *H.S. Yadav*

Introduction

Transport plays a significant role in all development processes. It not only removes scarcity in the areas of deficit production but maintains uniform price line throughout the vast territory. (Tiwari and Tripathi, 1987). In the modern times transport has assumed much more importance and no country or region can think of its agricultural, industrial and socio-cultural development unless it has an efficient and quick transport system. Transport is the focal point of all economic activities and busy transport nodes become sites for large cities and service centres. A well developed transport encourages not only cheap production but plays a significant role in the expansion of its market area. With the help of advanced transport system transport distributors do not bother to stock a hug quantity of commodities. Therefore, an increase in transport speed reduces the average volume of goods in transit and makes possible smaller inventories, thereby raising the rate of turnover of capital invested in stocks saving space and lessening rehandling. Instead of building up large stocks of goods, dealers can practice, hand to mouth buying (Truman and Merrill, 1952). The transport system encourages competition in market and price stability. It not only distributes consumer goods to the doorsteps of the consumers but enables them to cover long goods distances to purchase goods of their choice. According to Milne (1955) "transport is not only a means of production, it is itself an object of consumption. Travel is undertaken not only for the purpose of reaching destination but also for the enjoyment". Besides these, economic gains the transports system also provides social advantages like high standard of living, attitudinal changes, dissemination of knowledge and new ideas, diffusion of innovations, development programmes, social welfare etc. "Transport makes for homogeneity of people breaking the barriers of language, customs and environment (Daggett, S. 1934). It political and administrative fields the transport and communication provide a dynamic support to the government and international bodies maintaining national and international unity and integration, defence, administration of justice, law and order etc. In order to assess the multi-functional role of transport in rural development an attempt has been made in this chapter to examine the existing pattern of transport in Allahabad District and suggested suitable guidelines for its improvement and development planning.

The Region

Allahabad district (40° 47' to 45° 47' North Latitude and 81°19' to 82°.21' East Longitude) with a total area of 7,261 square km occupies a part of the fertile Ganga plain. The district has a total population of 3,797,033 (1981) of which 20.37 per cent resides in 16 urban centres and remaining in 3,531 villages of different size. Physiographically, the region is divided into three distinct units: the Ganga-par, the Yamuna-par and the Doab. It is well-linked with two principal cities of north India i.e. Kolkata and Delhi and is well served by means of transport and communication lines.

Existing Network of Transport System

The present transport system of Allahabad district mainly consists of roads and railways. Although the district enjoyed navigation facilities through the Ganga and Yamuna rivers during ancient time but same have almost lost their importance due to the introduction of faster means of transport. More recently the city has been connected through Indian Airlines service operating on Delhi-Kolkata route.

Railways

The district has a total rail length of 327.6 km which falls under three railway zones, namely, Northern Railway, Central Railway and North eastern Railway. Of these the Northern Railway alone covers a length of 250 km and includes three sections. Mughalsarai-

Kanpur section, forming part of the double track electrified busiest trunk route between Kolkata and Delhi, has a total length of 124 km in the district. It passes through the stations of Manda Road, Unchaatih, Meja Road, Bheerpur, Karchhana, Cheoki, Naini, Allahabad Junction, Subedarganj, Bamrauli, Manauri, Saiyed Sarawan, Muratganj, Bharwari, Bidanpur, Shujatpur, Sirathu, Athsarai, and Kanwar. Allahabad-Faizabad section which covers about 32 km. in the district passes through the stations of Prayag, Phaphamau, Siwaith, Dayalpur, and Mauaima. Allahabad-Jaunpur and Allahabad-Rail-Bareli-Lucknow sections branch off from Phaphamau Junction serving the eastern north and western north parts of district.

The Central Railway (broad gauge) through its Jabalpur-Allahabad section, which covers 38.25 km. of distance in the study region, passes through the stations of Shankergarh Jasra, Iradataganj, and Naini and serves the southern part of the Yamunapar area. Similarly Allahabad City-Varanasi section of North Eastern Railway (metre-gauge) has total length of 49.34 km. in the study region and connects Allahabad city with Daraganj, Jhusi, Ramnathpur, Saidabad, Handia Khas, and Bhiti etc.

Roads

The roads of the district are generally classified as national highways, state highways, major district roads and other district and village roads. The Central Government maintains the national highways, the State Govt. looks for the state highways and the Zilla Parishad, Nagar Mahapalika, Allahabad and Allahabad Cantonment Board, the other district and village roads.

National Highways

There are two nationat highways passing through the district (i) Kanpur-Varanasi section of N.H. -2 (130 km.) (Fig.25.1) and (ii) Allahabad-Rewa section of N.H.-27 (41.8 km.). The first enters the district from the north-west and running almost parallel to the course of the Ganga river traverses the study region from the west to the east. Its breadth is about 5 metres but in Allahabad city, it is about 9 metres broad. Constituting the busiest route of the country it is wide enough to facilitate two-lane traffic. Prior to the construction of Shastri Bridge N. H.-2 was linked through a bypass (22 km.) via Phaphamau. This route has become important with the completion of the new bridge across the Ganga river near Phaphamau.

Table 25.1 : *Allahabad District : Road Transport System*

Sl. No.	Blocks	Total Lengths of Roads (kms)	Length of Roads	
			per 1000 pop (km.)	per 100 Sq. km of area
1.	Dhanupur	73	6.39	32.75
2.	Handia	92	9.07	52.48
3.	Pratappur	95	8.26	51.51
4.	Saidabad	83	6.46	42.78
5.	Bahadurpur	84	5.22	30.21
6.	Baharia	47	3.40	19.15
7.	Phulpur	100	9.00	44.74
8.	Holagarh	38	3.83	23.77
9.	Kaurihar	71	6.43	32.15
10.	Mauaima	41	4.51	26.24
11.	Soraon	67	6.35	47.85
12.	Chail	93	7.35	51.04
13	Newada	65	5.46	24.19
14.	Muratgani	66	6.47	26.69
15.	Kanaili	76	8.40	43.03
16.	Manjhanpur	65	7.89	31.04
17.	Sarsawan	71	7.19	25.63
18.	Kara	88	8.03	32.75
19.	Sirathu	85	6.02	26.27
20.	Chaka	52	5.71	34.12
21.	Karchhana	92	7.50	35.85
22.	Kaundhiara	65	8.54	27.51
23.	Jasra	85	8.94	32.80
24.	Sankergarh	116	13.99	25.95
25.	Koraon	120	9.07	17.80
26.	Manda	69	6.92	16.49
27.	Meja	108	12.52	23.96
28.	Uruwa	93	9.04	54.10
Region (Rural)		2200	7.27	30.92

Source : Zilla Sankhikiya Patrika, Janpad, Allahabad, 1982-83

The Allahabad-Rewa section of N.H.-27 runs south-west from Allahabad city and after traversing a length of 41.8 km. in the district goes on to Rewa (in Madhya Pradesh). There are some important link roads connecting railway station treated as national roads which cover a total distance of 27.06 km.

State Highways

The total kilometerage of state highways traversing through the district is about 216. These highways connect the region with the district of Pratapgarh, Sultanpur., Faizabad (S H -9), Unnao, Rae Bareli, Lucknow (SH-38) Banda (S H-44) and Mirzapur (N H-27). Besides there are railway feeder roads at the railway station of Phaphamau, Manuari, Kanwar, Mauaima, Karchhana and Naini. Their total Length being about 12.34 km. Such roads connect Bharatganj and Manda (3.21 km.), Sirsa, Meeja and Koraon (32.18 km.) and Mooratganj and Rajapur (44.25 km.) The south Ganga Grand Trunk portion has a length of 57.6 km. in the district and connects Allahabad city with Mirzapur. The Allahabad-Rewa road (provincial) is confined to the municipal limits of Allahabad city and is 3.29 km in length. The average breadth of these State highways is usually about 3.66 metres.

Other Roads

The district is also served by a number of other metalled and unmetalled roads each being under the charge of eighter the Public Work Department and Zilla Parished, the Mahapaliaka, Allahabad and Allahabad Cantonment Board. There are about 2,200 km. of roads in the rural areas of the district (Table 25.1 and Fig. 25.1) which help in the continuous flow of the traffic and goods to and from in the district.

The total length of urban road in the district is about 244 km. The Public Works Department maintains about 1,692 km. of roads in the district while rest is under the charge of local bodies like Zilla Parishad (190 km.) Nagar Mahapalika and Cantonment Board (44 km.) and other organization (518 km.). Besides, these metalled roads there is a good network of unmetalled roads in the district which are the main linkages to the rural settlements during the dry season. But in rainy season these unmetalled roads become inoperative for vehicular traffic and goods movement. The city of Allahabad is situated more or less midway between Kolkata and Mumbai and also between Delhi and Kolkata and is connected with these places with major taken up the programme of massive road construction in rural areas so as to connect all villages with and above 1,500 population with all-weather metalled roads. These link roads are providing new linkages to the rural areas of the district and expanding them to new socio-economic changes.

Rail-Road Connectivity

The economic development of a region very much depends upon transport route connectivity and availability of the means of transport and communication. The connectivity also helps in the study of traffic flow and transport accessibility. In the study region the rail-road connectivity for three physiographic regions i.e. the Doab, Gangapar and Yamunapar has been calculated by using the following formula" ß = e/v where, 'e' is the connectivity index, e. is the number of routes, and v is number of points (service centres), railway station and terminal points. The ß index with a value lower than '1' shows unconnected network of route, index of '1' indicates many above '1' shows many alternative routes between central points of route network. The study areas shows an average ß index of 1.34 indicating the occurrence of alternative routes. The connectivity index has the higher value (1.40) in the Yamunapar followed by the Doad (1.30) and the Gangapar (1.29) areas. The Gangapar area records the lowest index value because here routes indicate less branching. That is why the northern part of the study region has comparatively poor accessibility.

The roads and railways are the main modes of transport in the study region. In order to assess their importance in regional development road densities have been calculated both in terms of area (road length per 100 km of area) and population (road length per 10,000 population) and the results have been shown through Fig. 25.3. An analysis of these figures and Table 25.1 indicate that road density in respect of area is the minimum in the south-eastern and southern parts of the region, moderate in the western and northern parts and maximum in the eastern part of the Doab and study region (Fig. 25.3). On the contrary the road length of the

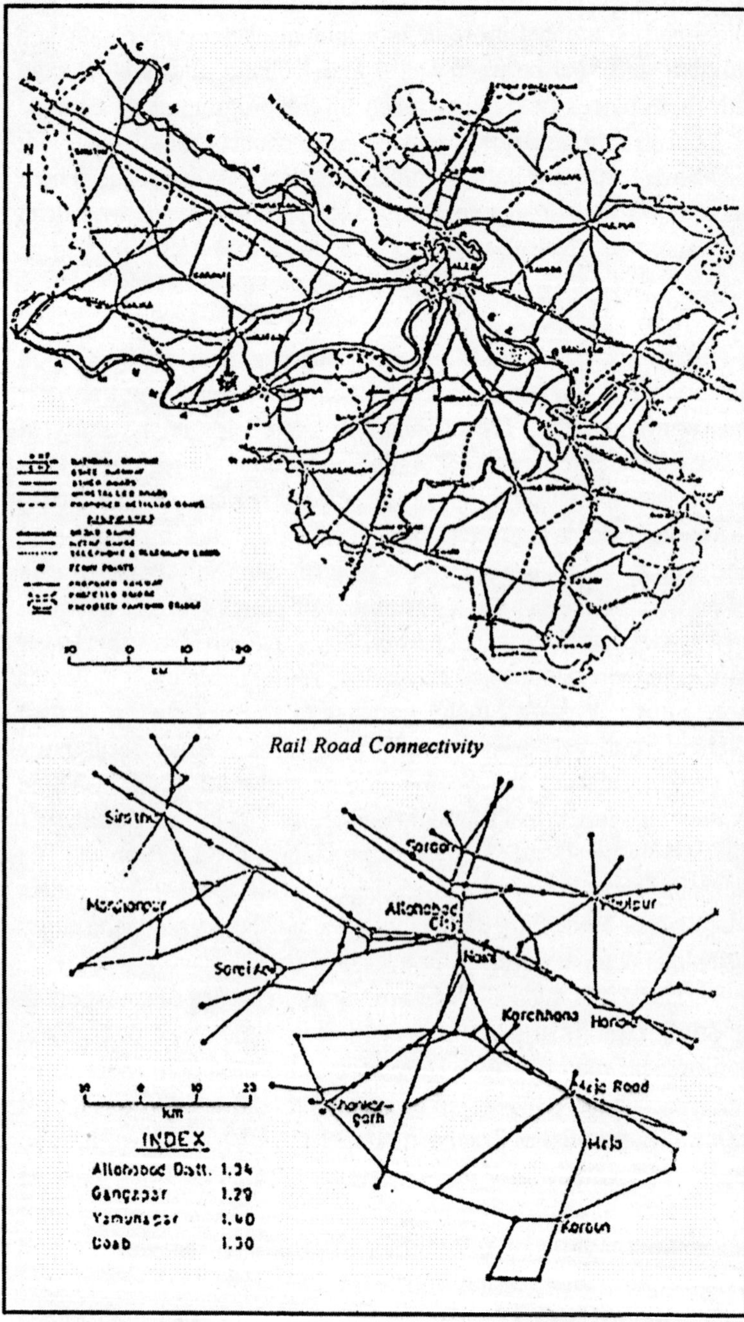

Fig. 25.1 : *Allahabad District : Means of Transport and Communications*

Doab area and minimum in the northern parts of the Gangapar area and in the vicinity of Allahabad city which have (Fig. 25.3) This very well confirms that plain area of the district requires more roads due to population pressure while plateau area needs their development to plug the area gap in road network.

Rail and Road Accessibility

The concept of accessibility by railways and road is based on the minimum distance (i.e. 5 km) from the transport slightly different from the railways which only connect major centres and trains stop only at railways station. On the contrary roads provide linkage to even the smallest settlement unit located along its sides irrespective of their distance from the bus station and bus stop. But in India bullock carts and lights vehicles use railway side lanes to negotiate distances between important settlements and service centres. Hence rail routes become an additional source of cheap transport and their accessibility becomes as important as the roads. Fig. 25.2 shows the accessibility by railways and roads in the study area. High accessibility (<5 km.) by railways is found in the Gangapar and northern part of the Doab areas. Similarly moderate accessibility (5-10 km.) is noticed in the eastern part and in a narrow area along the Ganga. Contrary to it there is a large tract in the southern part of the Doab and the Yamunapar areas which have low accessibility by railways (<10 km. from the railways line) (Fig. 25.4) This is because the areas is traversed by four rail lines towards Varanasi, Jaunpur, Pratapgarh and Lucknow while the Yamunapar area is traversed by two line towards Mirzapur and Jabalpur, and the Doab area by one rail line to Kanpur

The accessibility by roads (Fig.25.4) has altogether a different story to tell major parts of the Gangapar and Doab and central part of the Yamunapar areas exhibit high accessibility (< 5 km.) by roads. Moderately accessible areas (5-10 km.) are located along the Ganga, in central part of the Doab, western part of Karchhana tahsil and extreme north werstern part of the Yamunapar areas (Fig. 25.4). Poorly accessible areas are scattered all over the region mostly along the periphery of the district. Such low accessible (>10 km. from road) pockets are located in Mauaima, Baharia, Holagarh, Dhanupur, Koraon, Shankergarh, Newada, Kanaili, Sirathu and

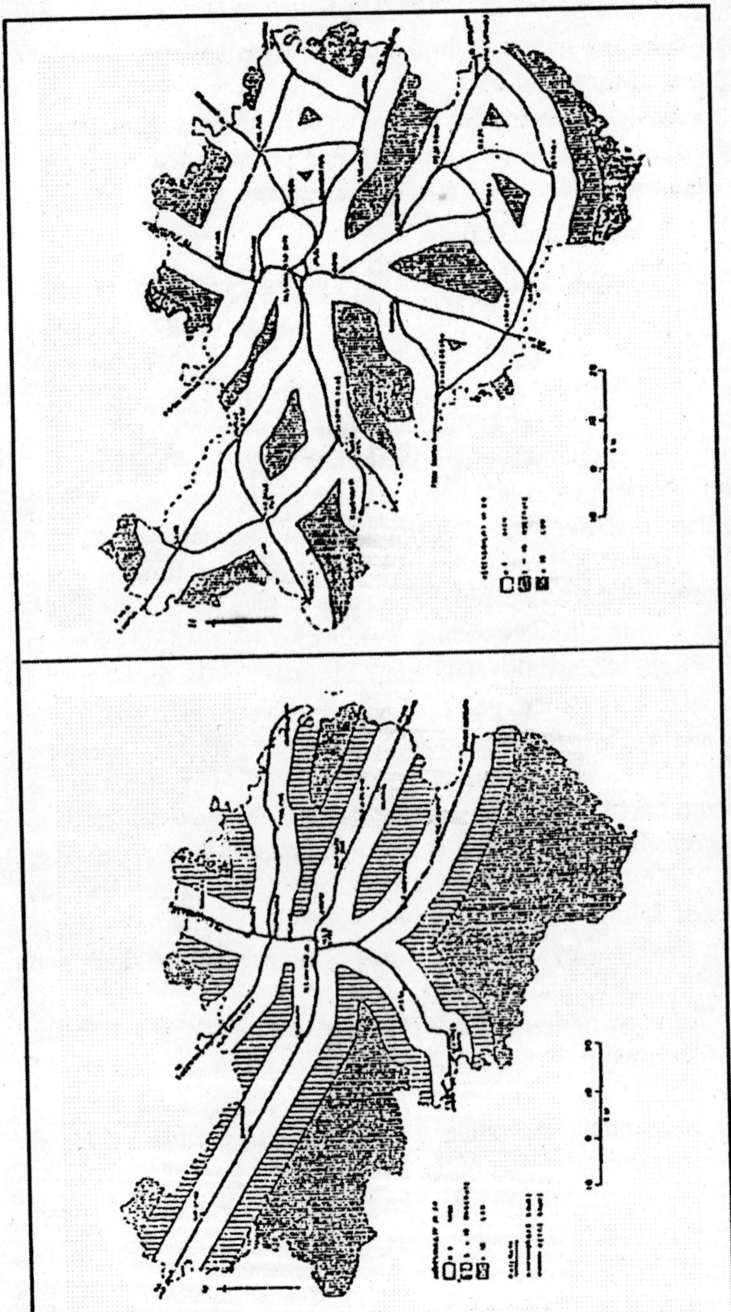

Fig. 25.2 : Accessibility by Railways in Allahabad District

Kara blocks of the district where transport deficiency has been the major constraint in development activities.

Another assessment of the transport network may be made by measuring the flow of traffic on the transport routes. A look of the figure shows that maximum number of buses ply on Allahabad-Phulpur-Jaunpur route, followed by Allahabad-Mauaima-Pratapgarh, Allahabad-Handia Varanasi, Allahabad- Karchhana-Mehja-Mirzapur, Allahabad-Naibari-Rewa-Jabalpur and Allahabad-Maratganj-Fatehpur-Kanpur routes. At present there is heavy load of goods and passanger traffic on Shastri bridge (across the Ganga) and the Yamuna bridge (across the Yamuna) and there is an immediate need to construct one more bridge over the Yamuna (near the exising Yamuna bridge) to avoid transport bottleneck.

Schematic Linkage Plan

On the bases of aforesaid discussion it may be asserted that the existing transport system of the region required scientific and careful planning to accelerate the process of development. This amounts to the construction of new roads, railways, bridges and culverts, repairing and widening the existing roads and railways, and converting the *kachha* roads into metalled ones and metre gauge rail track into broad gauge one. Following are some of the suggestions to realise these goals.

New Roads

Since an all-weather approach road for a village is a prerequisite for subsidiary activities like poultry farming, dairying etc. a sound rural road transport policy will help to achieve national objectives like social integration, generation of employment, removal of poverty and rural development in general and removal of regional imbalances in particular. Thus the development of rural linkages has great significance from the point of efficient administration, integrated rural development, political cohesion, trade and commerce, social justice and economic upliftment of rural poor (Ramanujam, 1984). In view of the scarce resources of the study area a care has been taken to propose small link roads to provide all-weather roads facilities to the maximum number of villages of the study area rather than going for construction of new long metaled roads (Fig. 25.1).

Efforts should be made to convert existing kachha roads into all-weather metalled roads to improve the transport facility in remote rural areas of the district especially during rainy season. Also the existing village roads and chak roads etc. should be made motorable to improve inter and intra-village transport system. Apart from these some important metalled roads have also been proposed to fill up the wide gap in the existing road net work of the region. For example, in the Gangapar area of the district one ring type metalled road starts from Allahabad city via new bridge (near Curzon bridge) over the Ganga towards west in Kaurihara block and turns towards north before Lalgopalganj, and reaches as far as up to Mauaima and Baharia block headquarters. The other important proposed metalled road runs almost parallel to the railway line in the Doab area. Apart from these some new bridges (including Pontoon bridges) have also been proposed to increase the flow of traffic and goods on the existing roads and remove transport bottleneck.

Fig. 25.2 exhibiting railway accessibility clearly demonstrates that southern part of the Doab and Yamunapar areas have poor accessibility by railways. Hence, two new railway lines have been proposed to improve the situation (Fig.25.1) One of them starts from Allahabad Junction and runs south-westward up to Sarira via Sarai Aquil. Another railway line starts from Meja road railway station and running in the southern direction it goes up to Deoghat near the border of Madhya Pradesh via Meja and Koraon.

Upgradation and Repairing of Old Roads

The widening and repairing of old roads increase their accessibility and volume of traffic. The S. H-38, SH-9 and Allahabad- Jaunpur roads require their upgrading to the status of national highway to accommodate increasing traffic load. Similarly Kaushambi-Karchhana road in the Doab and Meja-Koraon-Badokhar road in Meja tahsil need proper soling and widening to improve their traffic conditions. There are a number of important district roads which are metalled only on paper. Their surface condition is pitiable. Such roads need repairing and widening keeping in mind their importance in view of the passenger and goods traffic. Many of them may be used as by-pass for important roads of the district.

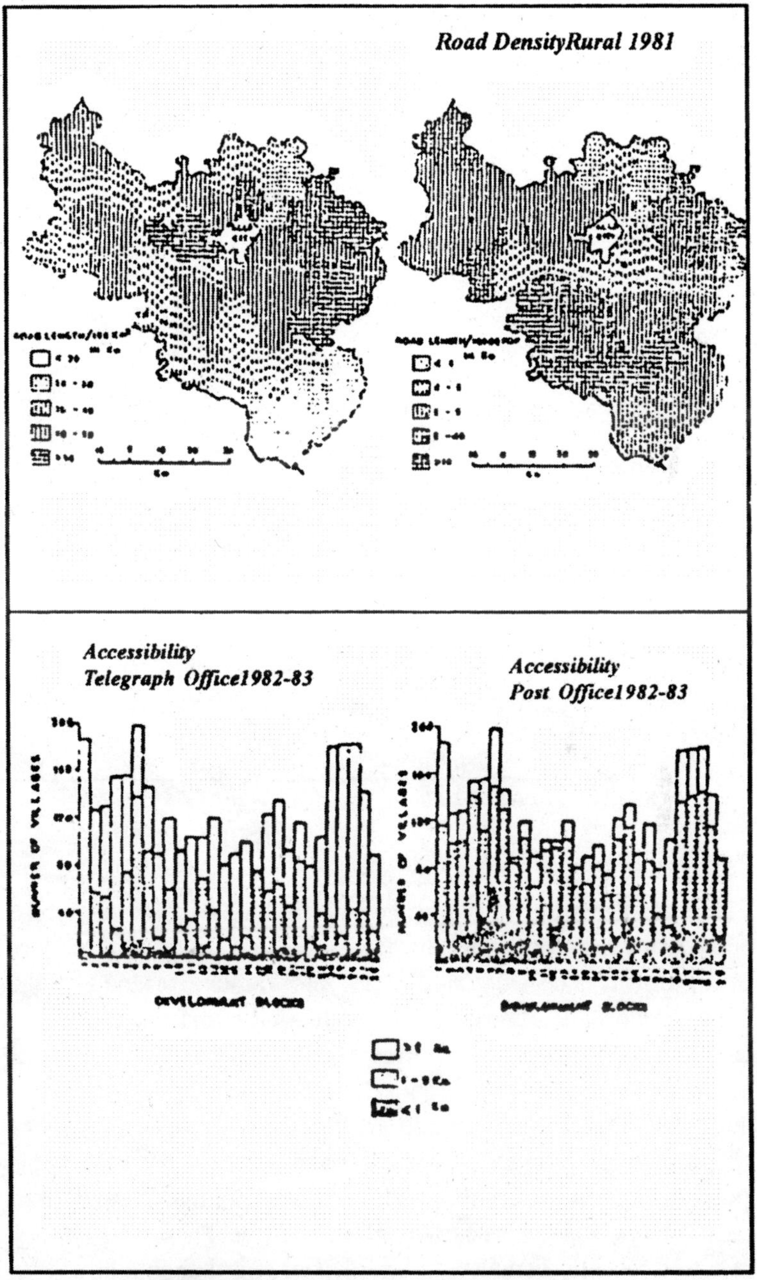

Fig. 25.3 : *Rural Road Density and Accessibility by Telegraph Office and Post Office*

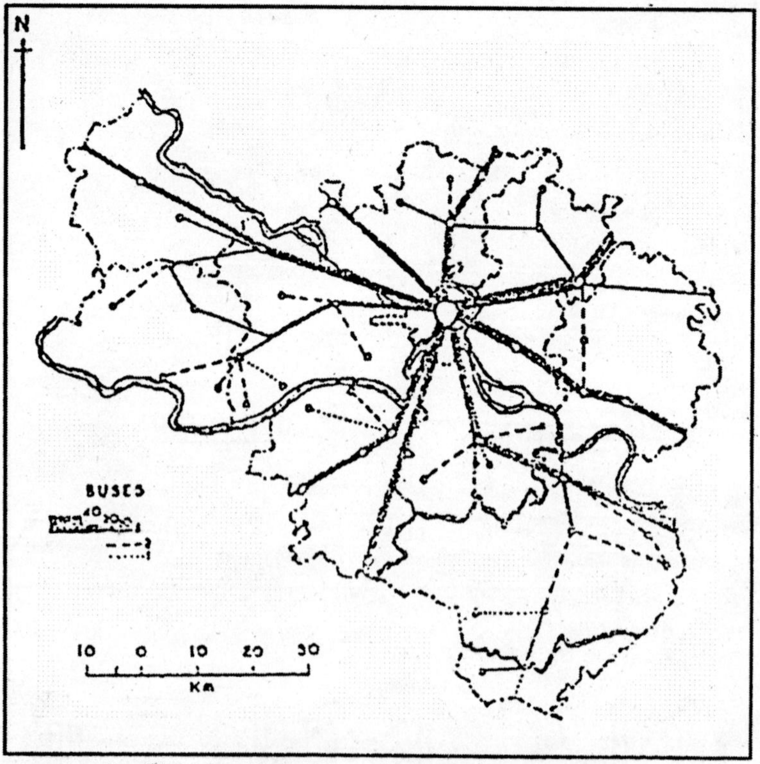

Fig. 25.4 : *Bus Traffic Flow of Allahabad District (1983-84)*

Bridges and Culverts

There are a number of old bridges on the important roads which become weak and outdated to accommodate free flow of modern heavy traffic. The Curzon Bridges, the Yamuna bridge and the Tons bridge have almost completed their life and are too old to bear the load of fast moving present day heavy traffic and goods movement. Accordingly proposal have been made to construct multilane new bridges (Fig.25.1). Two such bridges have been proposed near Allahabad city, one over Ganga river west of the existing Curzon bridge and other over the Yamuna river south of the existing bridge. It is heartening that one lane. Phaphamau bridge has been recently completed. A further proposal should be sent to add one more lane to accommodate more traffic through this bridge.

One new bridge has been proposed near Sirsa town across the Ganga river to connect Sirsa and Meja road with Saidabad and

Handia through proposed new roads to accelerate development processes in the Yamunapar area of the District. One pontoon-bridge across the Ganga needs to be built up to improve transport linkages between the north-eastern part of Uruwa block and south-eastern part of Handia block. Another bridge over the Tons river near Naya Kohdar in Meja block will connect Karchhana with Koraon and its surrounding, thereby helping in the economic development of Koraon and Meja blocks. Besides these bridges, a number of small bridges and culverts should also be built up to imporve road conditions and drainage which causes immense damage to roads and highways during rainy season.

There are 68 ferry points in the district which are locked on the Yamuna, Ganga, Belan, Tons and Lapri rivers. Some of these ferry points need to be replaced by pontoon bridges while others need the development of steamers and large boats to allow quick and safe transport facilities. Zilla Parishad and Public Works Department while auctioning these Ghats should ensure safe travel facilities to the public.

Improvement in Communication System

Along with the improvement in surface transport system a detailed proposal should be finalised to improve existing rural communication system to facilitate fast flow of news and information. In this connection new post office should be opened to locate one post office in each Nyaya Panchayat, one telegraph office in each block headquarters and one public telephone call booth in each Nyaya Pancheat center.The All India Redio, Allahabad should upgrade its transmitter and there should be at least four-hour programme for rural masses promoting agricultre, industry, education, health-hygiene, environment, etc. low cost cheap community, T.V. sets should be provided to the villages which should get uninterrupted supply of electric power for its agricultural and industrial activities. Besides national programme the local telecast should devote at least two hours time for socio-economic development of the rural areas. There should be frequent screening of films in schools, hospitals and community centres to apprise the village folk about the new development trends in the field of agriculture, industry), transport and communication etc.

References

Bigham, C. Traman and Merril, J. Robert 152 : *Transportation : Principles and Problems.*

Bonovia, M.R. (1958) : *The Economics of Transport.*

Choudhary, M. Subbarayulu (1983) : Film as a Medium of Communication : Its Potential and its Achievement, *Yojana*, Vol.21 No. 11.

Envyclopedia of Social Science, Vol. 15.

Joshi, E. B. (Ed) (1968) : Uttar Pradesh District Gazetteers, Allahabad (Lucknow: Govt of Uttar Predesh Deptt. of District Gazetteers.

Milne, A.M. (1955) : *The Economics of Inland Transport*, p 28.

Nayak, U.V. (1984) : Satellite Communication for Rural Development, *Kurukshetra*, Vol 32 No. 5 Feb. 1984.

Nayak, K. N. (1984) : Roads for Prosperity, *Kurukshetra*, Vol. 32, No. 9, June 1984.

Strart Daggett (1934) : *Principles of Inland Transportation.*

Tiwari, R. C. and Tripathi, S. (1987) : Role of Transport in Development, *Avadh Journal of Social Sciences*, Vol. 1.

TRANSPORT NETWORK IN GANJAM DISTRICT, ORISSA

Patrali Chatterjee

Introduction

A transport network is usually established in a region with a view to facilitate economic and social interaction in space–both at the intra-and inter-regional levels. But once it is established, it plays an important role in shaping the space economy. Again, it is obvious that the intra-and inter-regional interactions between different nodes are operationalised throught the efficient links or routes. This transport networks represent the determining factors for the functional association between the dominant and subordinate centres in a region. However, once a transport network comes to be established in an area, it is considered as a neutral agency providing equal accessibility to all the nodes connected by it. The differentials in accessibility are brought out by the relative location of nodes with reference —

(i) Firstly, to each other, by geometrical centrality and aggregate travel distance; and

(ii) Secondly, to the network as a junction–an intermediary point or a terminal.

So, in order to assess the importance of transport links in connection with accessibility and dispersion more meaningfully through the interpretation of logically aligned numbers than a mere quantitative projections, attention often is given to network transformation. A recent approach to the explanation of certain observed features of network has its roots in the significant advances in the quantitative analysis of watershed geometry over the last twenty years from Horton's (1945) recognition of stream order and its modification to a simple combinational system by Strahler (1952). The success of the Horton-Strahler[2] approach in the recognition of fundamental regularities in a drainage system suggests that the search for comparable ordering systems for more complex network, notably transport circuits, would be worth pursuing (1966). Attempts to order or categorise transport networks containing circuits have not so far been conspicuously successful. To be sure, useful functional classifications of railroad networks in terms of their flow characteristics have been attempted by Wallace (1958). But the recognition of the classes 'internal', 'originating', 'terminating', 'bridge' and balanced traffic lines demands rarely available data. For this consideration, circuit hierarchies have been suggested by Haggett in a systematic way. Very simple transport system may consist merely of single lines or sets of lines arranged in a tree-like fashion. In these cases, the transport net is simple enough for Strahler type ordering systems to be directly applied. But simple networks are however in the minority as the road systems unlike the stream systems, are more or less noticed to consist of a highly complex interlacing circuits.

Direct ordering of 'natural circuits' is thus inherently more difficult that that of natural 'trees' studied by Horton. So circuits or a higher order topologic form are decomposed in the trees of a lower order on the basis of accessibility and dispersion of the network for measure of the efficiency of the road network. An interesting simple quantitative analysis of the networks of road is successfully done in considering the mean dispersion index calculated from matrix of shortest paths (in links) between different nodes expressed by a notation D.I.

$$\text{Dispersion Index D.I.} = \sum_{i=1}^{n} \sum_{j=1}^{n} d_{ij}$$

where, d_{ij} denotes shortest paths (in number of links between i^{th} and j^{th} nodes.

The dispersion index is extended from the case of double summation from accessibility matrices.

Study Area

The author is tempted to work out on the study of the above aspect of transportation taking undivided Ganjam (undivided district of Orissa before 1992) district in Orissa as areal laboratory since she is well acquinted with this region on the one hand, and on the other this district has the unique geographical location and ecological set-up in Orissa with greater degree of physical diversity which in turn is expected to cast a great spell on the existing transportation network. This undivided district plateau and at the same time it has the home of a large number of tribal population providing ample socio-economic diversity. And, all these must have influenced the structure and efficiency of transport which in turn must have the influenced on micro-regional development. The district is also acting as a transition socio-culturally being situated as cultural border between the Oriya and the Telugu speaking people of Orissa and Andhra respectively where the inter-state, inter-district and intra-district transport routes are expected to have impact of spatio-cultural variations in the levels of development.

Moreover, this district is part of such a State like Orissa, whose transport system has long been neglected. The district has also attracted the attention of the author's doctoral research supervisor, Professor D.K. Singh of Utkal University, Orissa who being deeply involved in different types of micro-level study on various geographical aspects, rendered some valuable and intersting suggestions about the uniqueness and specific criteria of transport of the district. All these points indirectly or directly touched the mind of the author in making Ganjam such a specific regional choice for a spatial study of the dispersion matrix and decompositon indices in transport geographic analysis.

Fig.26.1, 26.2 and 26.3 show the administrative, physiographic, and road accessibility situations of the district. Considering the old 13 districts of Orissa, Ganjam is the third largest in the state and it extends over an area of 12,5000 sq. km. lying between 18°46'N and 20°17'N Latitudes and 83°48'E and 85°11'E Longitudes. With 100 km. of coastal lines of the Bay of Bengal, it now ranks third in respect of population size. It is divided, for administrative convenience, into 4 sub-divisions, 10 talukas, 22 police stations, and 29 community development blocks. Of the 4 sub-divisions, the Paralakhemundi sub-division has been made a new district, the Gajapati district.

Dispersion of Road Network

The dispersion index as expressed earlier, has been calculated for the 29 blocks of Ganjam district and given in Table 26.1. To have the idea of gross accesibility of roads in the blocks the matrix has been worked out taking each community development block as a micro-unit. The matrix itself is an indicator of the number of interception in communication links between any pair of nodes.

The mean dispersion index is calculated from the summation of *dij* matrix for each block and then averaging it out by *n(n-1)* where is the number of nodes in the region. The mean dispersion index is shown in the choropleth Fig. 26.4. The higher the mean dispersion index, the poorer is the transport efficiency in the region. In terms of the inter-node connectivities the Fig.26.1 shows that Gumma (1.5), Chatrapur (1.8), Sheragada (1.67), R. Udayagiri (1.42) and Kasinagar (1.67) have good accessibility as the mean dispersion index is well below average while Ganjam (2.3), Purushottampur (2.47), Kabisuryanagar (2.40), Aska (2.80), Patrapur (2.00), Dharakote (2.00), Belaguntha (2.40), Parlakhemundi (1.20), Nuagada (2), Sorada (2.61), have moderate accessibility. On the other hand, Mohana (3.13), Jagannathprasad (3.71), Polasara (7.00), Chikiti (6.00), Sankhemundi (3.21), Rangeilunda (3.21) and Hinjilikatu (5.33) are having poorer accessibility and road efficiency.

Some blocks like Gumma, Nuagada, R.Udayagiri do not confirm with physiographic situation and ther connectivity indices, showing unexpectedly good accessibility condition. But, dispersion in road network is calculated only within the block area limited by its political boundary. So at this micro-level calculation, accessibility is not much influenced by physiography.

The poor transport condition in the western part of the district and fair situation in the east as depicted by the above transport-space analysis should draw the attention of the authorities of transport development in the district in particular and in the state as a whole, because block level transport development is vital for future economic development. The present analysis is significant as it considered all connections right from as low as the village level roads in all the 29 blocks of the district. The advantage of the eastern blocks over the west is, however, more due to the favourable geographical environment than any other human and political considerations.

Based on the ideas of accessibility and dispersion Haggett (1966) has suggested a five stages algorithm for the breaking down of circuit type networks into trees through the stages of pole identification, polar network isolation, network hierarchies and lastly decomposition of circuit and path ordering. This method is an important tool in assessing transport efficiency around regional nucleus but as our present study is micro-regional so identification of such important transport centres which influence greatly the surrounding zone is very difficult to mark out and rare in number also. This basic idea of decomposition is applied by D.K.Singh, J.K.Routray and V.L.Singh in other way tc suit the decomposition of road network for micro-regional analysis. In this chapter the same method is followed to calculate the decomposition index of the road transportation of Ganjam district.

The concept of decomposition of transport network around the primary and secondary distributing centres has been based on the proposition that the flow of goods and ideas of development measures, in geographical space have been affected by the undesirable spatial differentiation in effectiveness of transportation and communication links.

Hence, Ganjam district is considered as a regional unit and the district headquarters as the principal controlling centre of distinguishing measures (like the pole identification of Haggett's method) and the block headquarter as the secondary level of distributing points so that development could filter into the villages at grassroot level in an equitable manner in economic development.

But the hierarchy in transport flows and the lack of balance in existing transport links cause the disparity in development at the micro-regional level. The dispersion of transport network the decomposition of the roadways with increasing distance from the district headquarters and the poor road condition and non-availability of good roads between block headquarters and the villages cause

Table 26.1 : *Dispresion Index of Road Network in Ganjam District*

Name of the Block	Dispersion Index d*ij*	Mean Dispersion Index d*ij*/n(n-1)
Chatrapur	36	1.80
Ganjam	46	2.30
Khalikote	88	2.93
Kodala	432	4.80
Purushottampur	74	2.47
Hinjilikatu	166	5.53
Kabisuryanagar	48	2.40
Rangeilunda	180	3.21
Kukudahandi	107	2.55
Digapahandi	85	2.83
Sankhemundi	160	3.81
Chikiti	36	6.00
Patrapur	40	2.00
Polasara	14	7.00
Aska	56	2.80
Sheragada	20	1.67
Dharakote	12	2.00
Belaguntha	48	2.40
Bhanjanagar	152	2.71
Jagannathprasad	156	3.71
Parlakhemundi	44	2.20
Buguda	40	2.00
Sorada	146	2.61
Gumma	2	1.50
Nuagada	2	2.00
R. Udayagiri	17	1.42
Kasinagar	20	1.67
Rayagada	147	2.63
Mohana	282	3.13

much hindrance to the development. According to K. Singh, Routray and V.L. Singh (1980) :

(a) The character for flow and transfer of goods, men and ideas which are considered as the indicator and determinants of economic development at regional level are influenced by

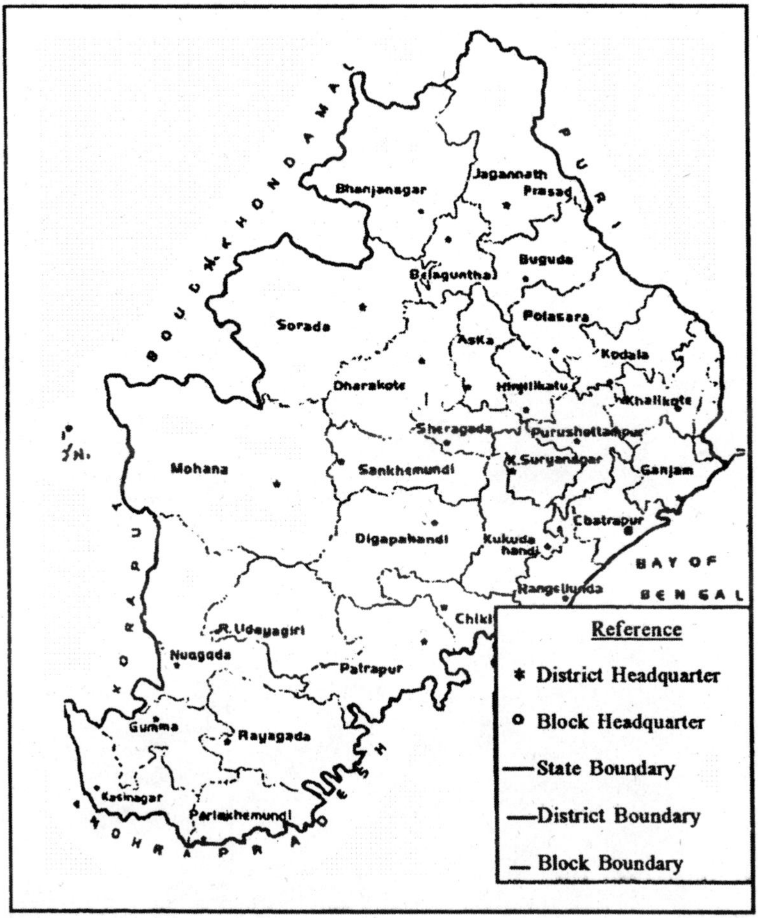

Fig. 26.1 : *Ganjam District of Orissa Administrative*

the intensity of decomposition of transport facilities in any regional unit.

(b) The decomposition of transport facilities considered as the district level (a convenient regional unit of administrative development programmes within the state) influences in two ways : (i) Between the district headquarters and the block headquarters; (ii) Between the block headquarters and the village. Hence in a spatio geographical context, one has to see in the first place the nature of decomposition in transport facilities around the district headquarters as the focus

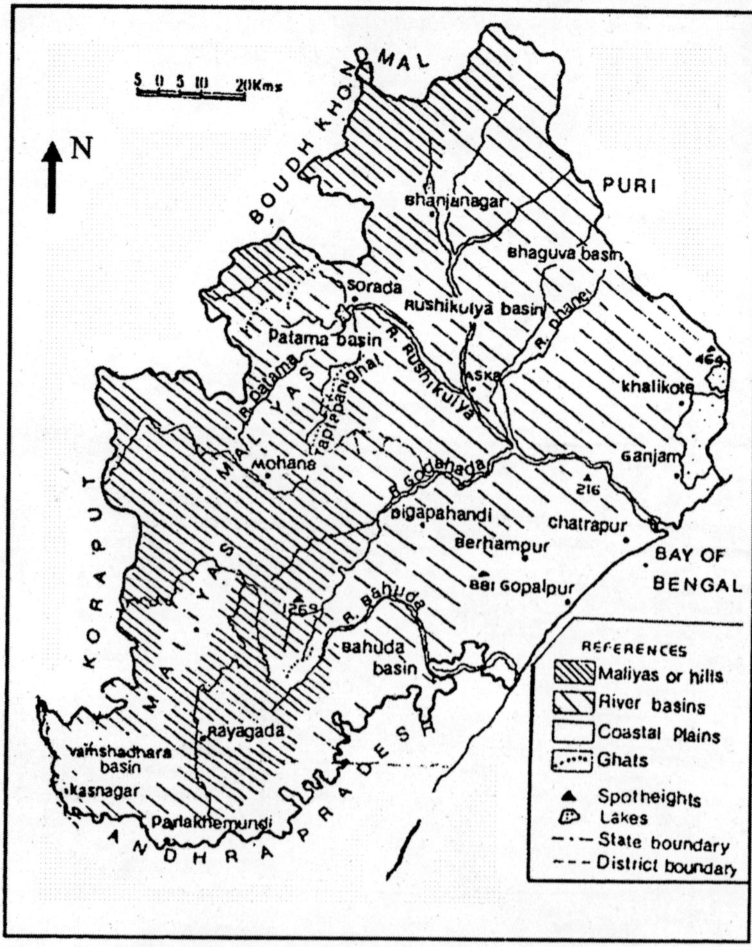

Fig. 26.2 : *Ganjam District of Orissa Physiography*

radiating to the various block headquarters as the initial points of receiving nodes through the transport routes working at the arteries of transfer, and in the second place, the nature of decomposition around each block headquarters directed along the links from them towards their controlled villages which are the final points of receipt at micro-regional spatial level or grass roots level.

(c) The intensity of decomposition between district headquarter centres and the block headquarters may be considered to be higher if their inter-distance is greater and that their distance has to be the weighted distance, considered by the

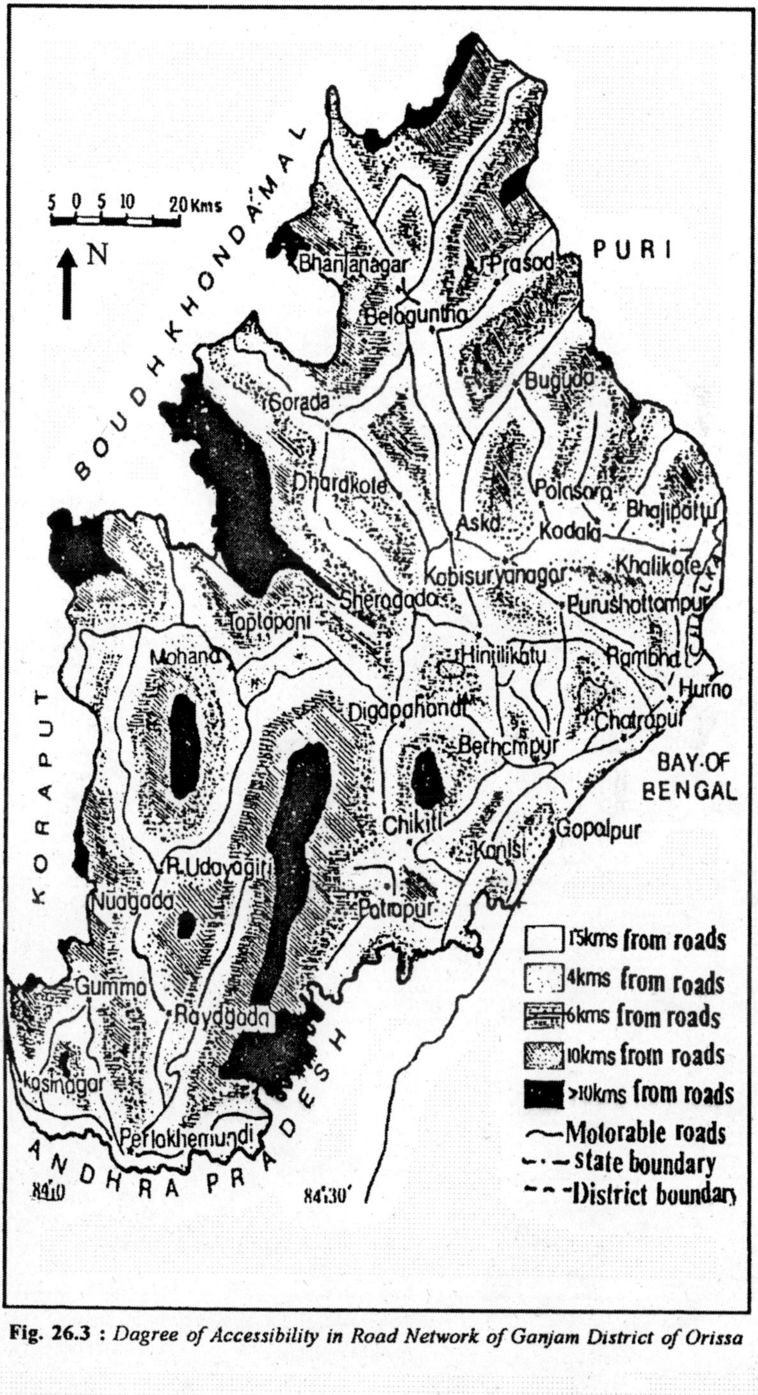

Fig. 26.3 : *Dagree of Accessibility in Road Network of Ganjam District of Orissa*

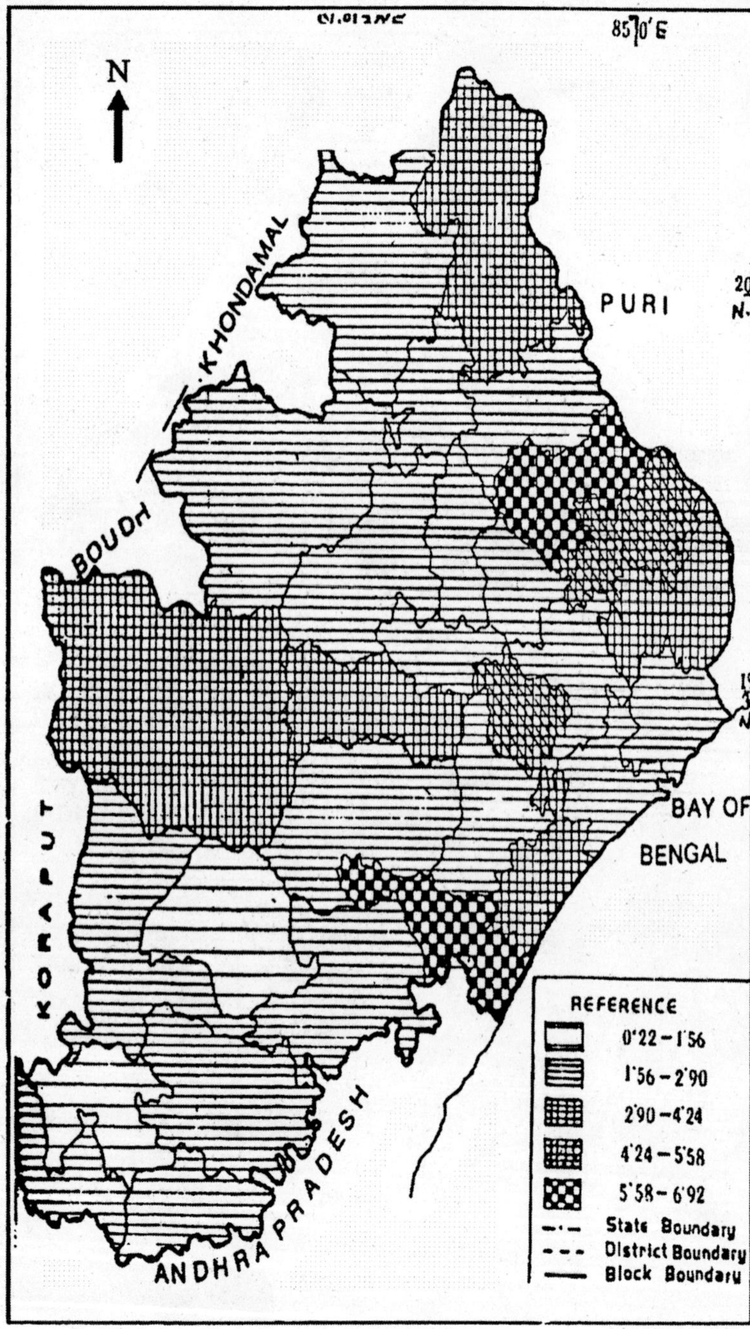

Fig. 26.4 : *Dispersion Index of Road Network of Ganjam District of Orissa*

appropriate path order of the respective segments of the connecting link as in case of stream order in drainage system.

(d) The density of decomposition within each block may be regarded to be higher if there is a poorer village level link between the settlement and the connected block headquarters. These links to be weighted by multiplier of their respective path orders and higher the total length of links within a block, lower is the intensity of decomposition.

(e) The above two assumptions of decomposition of transport facilities could be combined for each individual block to arrive at composite decomposition indices of transport facilities in order to find out the nature of the correlation between the composite decomposition indices and the composite indices of development of individual blocks keeping in view that the key of control of transfer lies at the primary focus i.e. the centre of district headquarters.

Taking into view all the above observations the decomposition index of transport of Ganjam district has been prepared.

Composite Micro Regional Decomposition Index

The effective distance (Ed_{ij}) between the district headquarters centre (i)and any centre of block headquarters (j) is calculated by considering the true distance (D_{ij}) as the summation of its segments (S_{ij}) in their true lengths and dividing each of the segment by its path order and then finally summing them up. Because the effective distance would be essentially perceived shorter if it has more efficiency due to higher path ordering and shorter the effective distance, less is decomposition effect.

So, $$Ed_{ij} = \sum_{i=1}^{n} (S_{ij}/K_i)$$

where, n = number of segments,

K_i = variable indicating the path order.

The Ed_{ij} of the 29 blocks of Ganjam district has been calculated and tabulated in Table 26.2. Only a modification in path ordering was adopted as by V.L. Singh, D.K. Singh, J.K. Routray. Though the general principle of team order in drainage system was followed, an additional weighted was given to the various categories of links or paths, beginning initially from the peripheries with 1 for village road, 2 for other district road (ODR), 3 for Major District Road (MDR), 4 for State Highway (SH), and 5 for National Highway (N.H.).

Then the total linkage *(Tli)* is derived with slightly different ideology. Each and every intra-block link *(L)* has been measured and then multiplying each of them by their respective path orders *(K)* and finally these are summed up. Because the total linkage would be enhanced due to higher path ordering and greater the effective linkage, less is the decomposition effect.

$$\text{So,} \quad Tl_i = \sum_{i=1}^{m} (l_i \times K_i)$$

where m = Number of road links,
L_i = Length of the ith link, and
K_i = Variable indicating path order

The path ordering is done with above mentioned modification. The values of Tl_i for each block are calculated and tabulated in Table 26.2.

To have the composite Decomposition Index, the Ed_{ij} and Tl_i are combined by the following formula,

$$\text{Composite Decomposition} = \frac{Ed_{ij}}{Tl_i}$$

because decomposition varies inversely with Tl_i, but directly with Ed_{ij}. These are given in Table 26.2.

A choropleth Fig. 26.5 is drawn to represent the degree of decomposition in transport network in different blocks of the Ganjam district from the district headquarters. It is seen that mostly in the eastern and northeastern blocks, the decomposition of transport is less, which means that effective distance from the district headquarters to those block headquarters like Gunjam, Khalikote,

Table 26.2 : *Computed Values of Effective Distance, Total Linkage and Decomposition Index of Ganjam District*

Name of the blocks	Ed*ij*	TL*i*	Decomposition Index Value
Chatrapur	0.00	225.00	0.000
Ganjam	1.25	298.00	0.0042
Khlikote	7.08	323.50	0.022
Kodala	9.58	205.00	0.047
Purushottampur	7.03	105.00	0.073
Hinjilikatu	8.08	240.00	0.034
Polasara	13.96	107.50	0.130
Kabisuryanagar	10.58	280.00	0.094
Ranbgeilunda	5.83	292.50	0.020
Kukudahandi	10.58	280.00	0.038
Digapahandii	9. 58	152.50	0.063
Sankhemundi	16.08	208.75	0.050
Chikiti	9.79	195.00	0.076
Patrapur	11.04	145.00	0.076
Bhanjanagar	19.08	326.25	0.059
Belaguntha	21.42	285.00	0.075
Jagannathprasad	26.42	312.50	0.085
Buguda	17.96	187.50	0.096
Aska	11.08	275.00	0.040
Sheragada	11.83	197.50	0.060
Dharakote	14.21	245.00	0.058
Sorada	19.83	230.00	0.086
Parlakhemundi	35.25	245.00	0.160
Kasinagar	39.25	245.00	0.160
Rayagada	28.58	167.50	0.171
Gumma	41.88	90.00	0.465
Nuagada	27.34	100.00	0.273
R.Uayagiri	21.71	170.00	0.128
Mohana	17.08	342.50	0.050

Hinjilikatu, Polasara, Rangeilunda, Kukudahandi are less, providing inhabitates more or less short journey.

On the other hand, western hilly tribal blocks and the blocks like Chikiti and Patrapur in the south of the district are having great degree of decomposition in transport network.

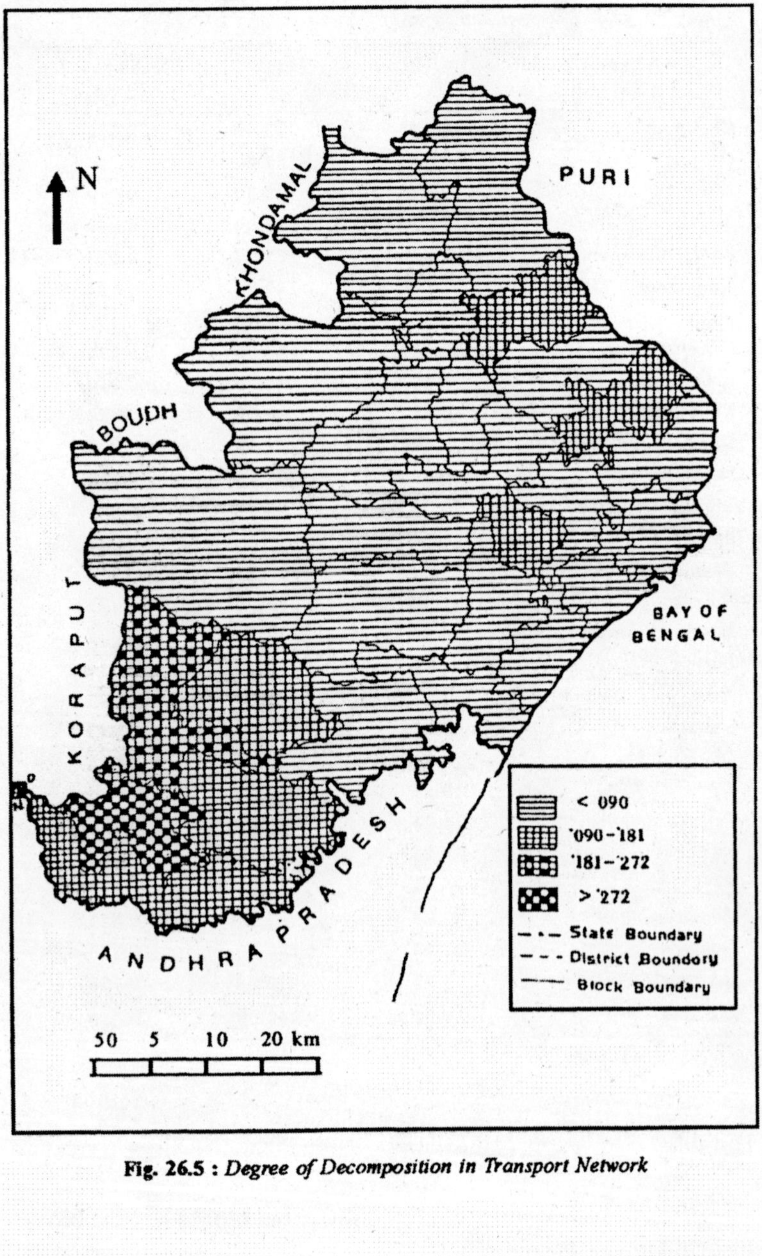

Fig. 26.5 : *Degree of Decomposition in Transport Network*

Concluding Remarks

This idea of dispersion and micro-regional decomposition index is an attempt to have precise idea of the condition of road network and transportation facilities in the case study area and in this way a bit different from the conventional cartographic representation of the theme. It is hoped that the decomposition indices of the transport and road network will give a guideline for the planners to chalk out viable plans for harmonious development of the regional transport which will further initiate the development of the macro-regional economy, so that there would be a healthy growth of living environment.

References

Haggett, P. : "Network Models in Geography", Chapter XV in *Integrated Models in Geography*, (ed.), R.J. Chorley and P.Haggett, Methuen, London, 1966.

Horton, R.E. (1945): Erosional Development Streams and their Drainage Basins, "Hydrophysical Approach to Qualitative Morphology, *Bulletin of the Geological Society of America*, Vol. 56.

Singh, D.K., Routray, J.K., Singh, V.L.(1980) : Intra-regional decomposition indices in space for transport links and levels of micro-regional development : An inverse correlation-Phulbani district- A Case study in an approach to micro-level planning, *Recent Trends* and *Concepts in Geography*, Vol. III, edited by P.B. Mandal V.N.P. Sinha, Concept Publishing Company, New Delhi.

Strahler, H.N. (1952) : "Hypsometric (Area Altitude) Analysis of Erosional Topography", *Bulletin of the Geological Society of America*, Vol. 63.

Wallace, W.H. (1953) : "Railroad Traffic Densities and Patterns", *Annals of the Association of American Geographers*, Vol. 48.

TRANSPORT NETWORK AND RURAL DEVELOPMENT IN BURDWAN DISTRICT, WEST BANGAL

Gopa Samanta and *Kuntala Lahiri Dutta*

Introduction

Transportation has been described as the lifeblood of civilization constituting an important item of infrastructure for socio-economic and cultural development (Misra, 1996). In the present day, no economy can flourish in isolation. Transport network is a prime physical linkage creating a host of other linkages like economic, social, technological, administrative etc., to open up an economy and to ensure greater circulation within it (Rondinelli, 1976). Therefore, transport network is a vital key in the development of an economy. The objective of the present study is to identify the nature of transport network and rural development as well as the nature of relationship between these two. The study area considered here is constitued by eleven rural development blocks located in the eastern part of Burdwan district, West Bengal.

Burdwan district as a whole occupies an important place among the district of West Bengal with regard to rural development. It was one of the sixteen districts of India during the mid-1960. So agriculture-cum-rural development received early thrust in this district, specially in the eastern part of it. With regard to crop productivity, the district's achievements have remained quite high till date. Our study region has Burdwan town in its central part acting as a focus of development. The region, therefore, is essentially the rural hinterland of Burdwan town. Our hypothesis is that nearness of an urban market promotes rural development if improved rural-urban linkage is established early.

Transport Network

A transport network may be of several types like rail, road, water etc.each of which has its own characteristics. Road network is the most significant mode of transport among them, though the region has had early development of water transport both during Mughal rule and during the British Raj (Paterson, 1910). The Damodar river has long lost its navigability and, therefore, has acted more as a transport barrier in recent past. The river flows through the middle of the region and its canals benefit the agricultural tracts on both the banks. Its contribution as a mode of water transport, however, has declined to a great extent today. Only ferry services are provided by local panchayats at a few places along the river. Even these are neither very effective nor reliable due to the highly seasonal nature of the river. Railways too had an early beginning in this part of India (Munsi, 1980). One branch of the Eastern Railway main line crosses the region through its middle part in an east-west direction connecting the region with Kolkata in the east and Durgapur-Asansol industrial belt in the west. The role of this railway line in the region's development as a whole has been immense. However, for the general rural-urban integration, the railway network is nowhere operative at a local level in comparison to the road network.

The road network reaching into the rural hinterland of the region is comparatively a recent phenomenon, mainly a result of last two decade's rural development programmes and administrative efforts, enabling the panchayats to undertake road construction work. The region is well-integrated by road netwok. The roads have a large number of bus services on them connecting the smaller market centres and other nodes of the region. Burdwan, the central urban market centre is well connected by buses with the surrounding rural areas.

A very rapid improvement has also taken place in the provision of buses reaching into the remote villages of the region. Many of these recently built roads are yet to be metalled. Buses ply not only along the metalled roads but along morum roads as well. Due to the dominant nature of road network among the different modes of transport our analysis heavily hinges upon it.

Road Network

The road network of the study area is constituted by one national highway (G.T. Road), several P.W.D. roads and panchayat roads. The national highway (NH-3) connects the region with Kolkata in the east and Delhi in the west especially facilitating the goods transport of the area with the whole of northern India. Burdwan town being located at the nodel point i.e. at the junction of all roads is the most accessible point of the region. All kinds of bus routes (inter-state, inter-district and intra-district) serving the region originate from and terminate at Burdwan town. The vast rural tract around it is well-connected by intra-district routes of but service with the central town.

Besides these, intra-district routes, a new dimension in the transport network of the region has been added by the town bus service. The routes ranging between 10 to 40 kilometres into the remotest villages integrate the rural areas with Burdwan. These areas were previously unserved by intra-district services and had much difficulty connecting with the central town.

This town bus service, though in operation only since 1991, has opened immense opportunities for rural development. Getting an easy access to the nearest large urban market, rural economic development in the form of vegetable gardening, dairy farming, development of rural market centres etc. have been speeded up to a great extent in recent years.

Dichotomy in the Transport Network

Looking closely into the transport network of the region, it is found that there is a dichotomy in the degree of accessibility between the northern and southern parts of the Damodar. The southern part, commonly known as the Trans-Damodar region, is relatively poorly connected by transport with the central urban focus. The number of existing P.W.D. roads are fewer in number in this region (Fig.27.1). Even the town service network in a recent addition to the transport scenario of the region, too has not been able to remove the dichotomy the between the north and the south.

The river Damodar has played a vital role since the historic past in creating dichotomy of transpot network of this region. The northern part of the region was always well-connected to the rest of Bengal since the construction of what is now called the G.T. Road in Mughal period. The introduction of railways during the Brithish Raj further increased the connectivity of the northern part (Munsi 1980) whereas the southern part continud to remain less accessible due to the existence of Damodar as a major physical barier. A bridge has been constructed in 1978 on the Damodar but it is highly inadaquate in serving the needs of the entire region.

Rural Development

Rural economy of the region is mainly based on agriculture, therefore, the development of agriculture in the post-independence period has highly benefited the rural areas. Several factors have bought about agricultural development—canals of the Damodar Valley Corporation, early establishment of Panchayati Raj, introduction of technology and the proximity of the region to the Kolkata-Hooghly and Asansol-Durgapur industrial belts. Agricultural development, however, has remained concentrated in the eastern part of the district. The villages of the western part and the rural economy as a whole is in fact traumatized by the severe environmental impact of mining (Lahiri-Dutta, 1994).

The physicial conditions of the region are also conducive to agricultural development. The nearly flat land and productive soils of the region are due to its being part o the floodplain of the Damodar river. The fertile alluvial land has also facilitated the adoption of modern technology (canal irrigation, deep tube wells, use of tractors etc.) in agriculture. The installation of Damodar Valley Corporation during the early 1950s as multipurpose river valley project expanded the scope of irrigation for double and multiple cropping in the region. With this additional water, the subsistence level agricultre soon changed into a commerical one.

The success of agriculture in Punjab-Haryana become a model of agricultural development for the whole of India in the 1970. Burdwan, following that model (though not in all respects), also developed its agriculture by adopting modern machineries and technology like irrigation pumps (shallow and submersible), tractors, harvestors, cold storage etc.

The Panchayati Raj System in West Bengal in 1970 also facilitated the overall rural development of the region. The division

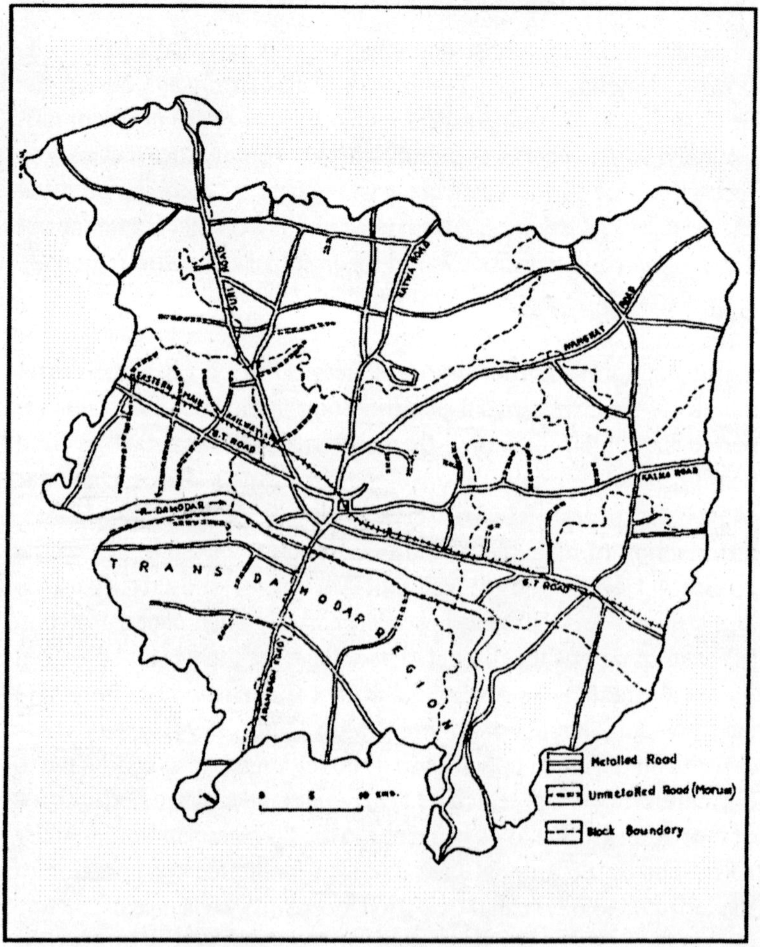

Fig. 27.1 : *Transport Region of Burdwon Region of West Bengal*

of rural areas into development blocks and Gram Panchayats and revitalizing the electoral perception process were positive steps in agriculture and other socio-economic development of rural areas. Availability of finance in the form of loan from branches of rural banks and co-operative credit societies for purchasing agricultural machineries and other monetary support from the block level administration to the marginal farmers have paved the way for agricultural development.

Infrastructural Improvements

Like, other economies, the development of rural areas is also closely related to the improvement of infrastructure. Much of the credit for the development of socio-economic infrastructure in rural areas goes to the Integrated Rural Development Programme. Development of rural areas depends not only upon agriculture but also upon several other socio-economic factors. Improvement of accessibility is one such important factor. Accessibilitiy by roads has been increased very rapidly in the region during the last two decades. Besides the national highway, P.W.D. roads, the panchayat roads, mostly made of morrum, have proliferated and reached into the rural areas breaking the remoteness of villages.

The development of road network (both metalled and unmetalled) has triggered off a number of infrastructural improvements in the rural areas. Amenities, like school level education, health care facilities, recreation and banking have developed along with the improved accessibility. Moreover, wider and better irrigation availability in the area and consequent multiple cropping have been able to create a considerable amount of marketable surplus. By marketing this agricultural surplus the cash income of the rural folk has somewhat increased and in turn has necessitated the development of better infrastructure. To keep pace with the overall changes in the rural features, socio-economic infrastructure, such as, electricity supply, piped water supply, telecommunications etc. have also improved at a rapid rate.

Development of Large Village and Rural Market Centres

An improved level of social and economic development has rapidly increased the business and service activities in the rural area. The villages located at favourable or more accessible points are increasing in size very rapidly. In the study area of eleven rural development blocks (total area - 2821.05 square kilometres) the number of large villges, defined as those settlements having population above 5,000, increased from only 8 in 1971 to 32 in 1991 (Samanta, 1996). Most of these villages have very high density of population and provide urban amenities like street lighting, pucca roads, water supply etc. They fall short of urban centres only with regard to their occupation structures. These large villages and their high growth rates are another indicator of rural development and rural prosperity.

Among those large villages some (such as Galsi, Seharabazar, Saktigarh etc.) are favoured by their nodal points location and high degree of accessibility. These centres have turned into rural market centres with a large number of retail shops, government offices, banks, co-operative credit and marketing societies, daily vegetable markets, schools, telephone exchanges etc. and shops of all kinds of consumer goods. These market centres are in the midway of the rural-urban continum. Another interesting feature is that all of these market centres are situated at an average distance of 20-25 kilometres away from the central town of Burdwan. These rural settlements are developed as satellites of the main town.

Dichotomy in the Level of Development

As a logical corollary to the dichotomous transport network, disparities have arisen in the level of rural development too. The rural landscape of the entire region is dominated by agriculture and hence, they are intricately connected. A distinct difference is found with respect to the level of agricultural development. The productivity index of crops of the trans-Damodar region is much lower (153) than the northern part (189). The backwardness of agriculture has led to a much lower level of socio-economic development of the region. There are several reasons behind this backwardness of agriculture : recurrence of floods and the scarce canal irrigation facilities are among them. However, the most important reason is the lower degree of accessibility of the trans-Damodar region.

Role of Transport in Rural Development

Clearly, the integrated road network of the region under study has played a vital role in the development of rural areas and *vice-versa*. An efficient and effective rural road transport system is imperative in order to bring the urban and rural areas functionally close to each other failing which the development of rural economy is adversely affected (Ojha, 1986). The developments in agriculture and improved accessibility of the market have initiated vegetable cropping and commercial cultivation in the region. Agricultural development would not have neen possible without opening up the rural areas to the urban market. This improved accessibility to the urban market is being provided by the integrated road network of the region.

The region's predominant crop, paddy, is the main raw material of the rice mills which form the backbone of the agro-processing industry of Burdwan town. The rice mills are the industrial base of the urban economy of Burdwan employing about 16.99 per cent of the total workers. The easy access of the agricultural produce to the urban processing units has further boosted the agricultural production of the rural hinterland. The integrated road network has encouraged the mobiliy of a large number of rural commuters to the Kolkata-Hooghly and Durgapur-Asansol industrial belts. The timing of buses (especially the town bus service) is to facilitate the commuters. Most of the buses halt at night in the villages providing an early trip to Burdwan town in the morning and a late trip to Burdwan town in the morning and a late trip to villages in the night. It also helps the vegetable and milk producers to sell their products in the urban market of Burdwan town in the early morning.

The increased accessibility also enabled the rural residents to avail themselves of the better health and educational facilities of the town. The existing facility of higher education (University, Medical Collage, Degree Colleges and Technical Colleges) is being obtained by the rural people with the help of the improved transport network. A greater degree of mobility provided by rapid means of transportation, therefore, leads to social development of the rural areas and break their isolation by increasing the rural-urban linkage. The important role of transport in the development of rural areas is also clear from the dichotomous nature of transport network and its associated imbalances in the level of rural development between the north-Damodar and south-Damodar parts of the region. The commercialization of agriculture, development of vegetable cropping and infrastructural improvements have speeded up the rural development. This has been possible with the help of high degree of connectivity. The construction of metalled and unmetalled roads and the introduction of the town bus service have helped to break the rural isolation and have opened to them the urban market in Burdwan town. Most of these developments of transport are limited to the north-Damodar region. Consequently, this tract has attained a higher level of rural development.

On the other hand, the south or Trans-Damodar region, which had still remained flood prone (Chatterjee, 1969) even a couple of years after the construction of Damodar Valley Corporation, is less accessible due to the physical barrier of the river Damodar. Only one bridge (Krishak Setu) built in 1978 connects the entire south-Damodar region with Burdwan town. The town bus service network too is weak here discouraging vegatable cropping. Inspite of the fertile alluvial land, the Trans-Damodar region has remained comparatively backward due to the lower degree of the development of transport network. Therefore, on the basis of findings of our enquiry, here it concludes that the transport network is the key factor in the rural development of a region.

References

Chatterjee, S. P. (1969) Damodar Velley Region : Report of Joint Committee for a Diagnostic Survey, Damodar Valley Corporation, Kolkata.

Dutta Ray, S. (1994), Agricultural Growth in West Bengal, *Economic and Political Weekly*, July 16.

Lahiri-Dutta, K. (1994) 'Mining Development and Consequent Degeneration of the Urban Environment : A Planned Alternative for Raniganj Town, *Geographical Memoirs, NBU Journal of Geography*, Siliguri.

Misra, K. K. (1996) Rural Environment of Banda District in Uttar Pradesh, *Indian Journal of Landscape System*, Kolkata, Vol :19 No. 1.

Munsi, S. K. (1980) *Geography of Transportation in Eastern India Under the British Raj*, K. P. Bagchi and Campany, Kolkata.

Ojha, G. D. & Reddy, M. P. (1986) Rural Road Transport System in Andhra Pradesh', *Kurukshetra*, Vol. XXXIV, No. 6, London.

Paterson, J. C. K. (1910) *Bengal District Gazetteers : Burdwan*, Bengal Secretariat Book Depot, Kolkata.

Rondinelli, D.A. and Ruddle, K. (1976) *Urban Functions in Rural Development: An Analysis of Integrated Spatial Development Policy*, Agency for International Development, U.S. Department of State.

Samanta, G. (1996) An Analysis of the Spatial and Temporal Dimensions of Large Villages of Eastern Burdwan Districts, *Geographical Review of India*, Kolkata, Vol . 58, No. 1.

28

EVOLUTION AND COMMODITY TRANSACTING BEHAVIOUR OF RAILWAY TRANSPORT IN BRAHMAPUTRA VALLEY

A. P. Mahanta

Introduction

The railway transport in India was pioneered by the Britishers in order to serve their own colonial interest. Therefore, there had always been existing a functional gap between the transport facilities and transport needs during the British period. In many parts of British India, Railway lines were opened just to mobilise certain traffic in bulk from the remote areas to the sea ports. The railways in the Brahmaputra Valley too appeared to feed the inland waterways or to act as an alternative land route to mobilise the colonially exploited resources from the remote areas to their way to London. Defence need of mobility at the time of the Second World War had reinforced the slumbering railway lines in the valley for the first time. After independence, innovative measures have been adopted in order to cope the railway transport with the contemporary political, economic and strategic exigencies in the region and also to keep pace with the development of railway transport in the rest of the country.

The railway, while providing linkages to altogether 43 commodities moving to and from 127 goods traffic handling nodes interacts with the spatial units of resource transaction within the valley and abroad. It appears that more than 84 per cent of goods handled at these nodes are of inter-regional range and are belonging to (i) Non-manufactured products, (ii) Semi-manufactured items, (iii) Manufactured goods for consumption, (iv) Food-stuff (v) Infrastructure goods, (vi) Strategic goods, and (vii) Unclassifiable items. The railway transport, while mobilising these traffic items from node to node interacts among the spatial units of resource functioning placed at different sectoral status. Thus, the commodity flow develops some linkage gradient, the ratio of vertical and horizontal displacement between the link-heads. The strength of resource linkage in general follows the rule of Gravity Model. However, the linkage strength referring to a traffic to and from the Brahmaputra Valley through railways is susceptible to the principles of telescopic rating and the commoditywise discriminated freight charges adopted by the Indian Railways. The commodity transacting behaviour of railway transport what is observed today in the Brahmaputra Valley is an outcome of the legacy of colonial drainage of resources at the non-manufacturing or semi-manufacturing phases. It is the existing rate structure of the Indian Railways aimed at reducing the friction of distance and an easy mobilization of primary products even at the cost the mobility of intensively manufactured items for short distances which deprives the interest of the region in ironic way.

Historical Background

The railway transport indeed had revolutionised the inland transport by mobilising men and materials at reasonably high speed for long distance. The economy of scale factor was the sole cause of worldwise acceptance of railway transport for the purpose of transferring the factors as well as the outputs of exploitation and transformation through the wheels on permanently fixed iron bars to desired destinations. The world's first inception of railway transport was in England, an insular country, when it was controlling over the political and socio-economic affairs of the countries in different continents. The pioneers of railway transport in England as well as in India were also found accustomed to the ocean trade routes to enhance colonial power through the coastal strips. George Stephenson, the father of railway transport who was even directly associated with the construction of Mumbai-Igatpuri line in India, himself was in the opinion of installing railway lines into the inaccessible areas to the water lane. With this background, a series of dendritic railway network were aligned in India since 1853, all of which were rooted at the major sea ports extending their branches into some resourceful areas away from the navigable streams.

The Brahmaputra Valley part of the British India was connected with the Kolkata port through the Brahmaputra Steamer Lane providing a fair access to the places of British interest in this region. However, an extensive part of upper Assam, earmarked for tea, coal, timber and petroleum products of specific colonial interest, was beyond the access of water transport since the Burhi-Dihing rivers through this sub-region was not navigable. The situation compelled the traders to set up the first railway line in the Brahma-Putra Valley connecting Dibrugarhghat with Lekhapani coal-field as early as in 1882.

Besides network shape, traffic mobilised on the network throughout the pre-independent period of the Brahmaputra Valley were observed to be of purely colonial nature. First of all, the railway transport could hardly impinge upon the traditional economic sector of subsistance farming and upon the life-style of the indigenous population associated with this farming practice. Secondly, there had been a sufficient expansion of tea and jute practices under direct colonial sponsorships synchronising the alignment of different railway line through the leading jute and tea growing areas in the Valley. Finally, the railway lines could provide relatively a better access to the forest, mining and some of the semi-manufacturing sectors rathers rather than the thickly populated built-up areas in the valley. In this way, the railway transport, even at its very beginning stage, had established a dilemma of colonial movement in the region.

Evolution of Railway Network

The Assam Railway Trading Company had constructed the first railway line to connect Dibrugarh with Makum in 1882. This railway system popularly known as the Dibru-Sadiya Railway had its 151 km long Y-shaped network connecting the tea, timber, coal and petroleum producing Burhi-Dihing Valley subregion of upper Assam. The said railway system could serve the mobiliary needs of its prosperous service-field so efficiently that the traffic density on its network in 1894 stood first among the then existing meter guage systems throughtout the country. The profit per route-mile of this railway for the period 1896-1901 was also recorded one of the highest among the various railways in the country. In 1942, in view of an intensive use of the available asset for strategic purpose at the time of the Second World War, the Government of India had taken up this railway system and amalgamated the entire system to the Bengal Assam Railway in 1945.

Construction of the *Jorhat Provincial Railway,* 49 km long 2 feet gauge alined from Kakilamukh to Titabar and Mariani at a cost of Rs. 4,73,000, was started in 1883. This railway was built, owned and worked by the State Government to mobilise tea from the flourishing gardens south of Jorhat town. After the alignment of the Assam Bengal Railway (1904) through its southern heads, its Y-shaped network had been serving as a link between the steamer lane on the north and the metre guage trunk line on the south. This system was handed over to the Bengal Assam Railway (1943) who had remodelled its network as a metre guage line from Mariani to Neamatighat instead of former Kakilamukh and its other branch to Titabar was abandoned.

The Tezpur *Balipara Light Railway* of 2.5 feet gauge was opened in 1895 by a private company to afford an outlet to tea products of Darrang district through Tezpurghat. Its northernmost 10 km route was merged into the Eastern Bengal Railway meter guage network in 1932, while the resmaining 26 km route was converted into meter guage standard with the initiative of the North Eastern Railway in 1953. The Chaparmukh Silghat Railway line, 82 km in 2.5 feet standard, linking the Assam Bengal Railway and the Brahmaputra Steamer lane at its two heads was constructed in 1920 to facilitate the tea gardens in Nagaon district. A branch line from Senchoa to Moirabari (49km) was opened in 1930 to encourage jute cultivation at its closed vicinity. However, this Y-shaped network touching two ferry-points at Silghat and Dhingghat could encourage the inter-district traffics of central Assam across the Brahmaputra river.

It was observed even in the third quarter of nineteenth century that the opening of rail routes from Kolkata to Patna (1860) and to Kustiya (1862) had compelled the GSN and the RSN companies to push back their vessels from the Ganges lane to the Brahmaputra lane where their steamers would remain free from a competition with a rail route. Such a supermacy of railway transport over inland waterways encouraged the *Assam Bengal Railway*

Campany to construct some railway lines in Assam and in East Bengal as an alternative land route of the Brahmaputra Steamer Lane. The Assam-Bengal Railway network was entirely of meter guage standard and its 925 km long trunk line was aligned from Chittagong port to Tinsukia through the shortest possible route involving even no break-point on the way. The extension of its trunk line was in three phases. The first phase of construction, Chittagong to Silchar in Assam through the Mymensing jute tract, was by the end of 1898, while the most critical 186 km long Hill-Section on the composite basement of steep grades upto 1 in 37 for 9 km, 37 tunnels aggregating to a length of 4745 metres and of 574 bridges and viaducts was completed in 1903 during the second phase. The third phase of construction (333 km) through the Dhansiri-Kopili Saddle and the tea gardens in Sibsagar district to reach Tinsukia, a centre of tea, timber and mineral products, was completed in 1904. Meanwhile, the same company had constructed another route (1897) from Pandu to Jamunamukh (129 km) to feed the steamer lane with timber and jute from the Kopili valley. In 1910, Jamunamukh was interlinked with the trunk line at Lumding involving an extension of 61km. Apart from the completion of the trunk line alongwith the Pandu-Tinsukia route as two alternatives of the Brahmaputra Lane, the Assam Bengal Railway had constructed three and more traffic towards its network. Opening of 14 km long Naginimara branch in 1917 was aimed to collect tea, timber, stone and coal traffic. The other two branches viz. Furkating-jorhat (70 km) in 1929 and Simaluguri-Khowang (69 km) in 1931 were opened through the flourishing tea gardens.

The easternmost 22 km segment of the Parbotipur-Dhuburi Section constructed by the Eastern Bengal Railway in 1902 was the oldest meter guage route on the north bank of the Brahmaputra valley. It was constructed under the pressure of the Indian Tea Association seeking a rail connection from Dhuburi to Kolkata. Subsequently, its network was extended to Amingaon, the opposite ferry point of Pandu rail-head in 1907 and also to Balipara (1932) through the timber, jute and tea yielding areas. This network on the other hand, could serve the requirement of the Central Government to pour foodgrain into the famine districts in Bihar and the influx of labour-force to Assam. Further, the same network had brought a dramatic change in the traffic-flow picture of the Brahmaputra Valley and its environs. Immediately after the setting of this network, popularity of the Chittagong bound rail route and also of the Kolkata bound water route had reduced to nominal. Kolkata, provided with a better form of dock facility and a wider inland market, could capture most of the Chittagong bound traffics from the entire North East Region. At the same time, the all-rail-route from upper Assam to Sealdah through the ferry-crossing point at Pandu and gauge transshipment at Parbotipur could prove its efficiency over the slow-moving water route to Kolkata.

In order to bring the various railway system in Assam under a single system of operation, responsibility of the Parbotipur-Tinsukia route (895 km) and other 460 km long branch lines in Assam was handed over to the new system of the Bengal Assam Railway in 1942. Till that stage, the railway lines in Assam with their low operating standard, inadequacy of rolling stock and terminal facilities could hardly make the way to a steady outflow of selected commercial goods from the Brahmaputra Valley and its environs. The Second World War, while turning the entire North-East Region into an active battle-ground seldom hitched the railway transport sector as well as the general economy of the whole region. Throughout the war period the slumbering railway lines had to bear a great strain in lifting ali defence and civil traffics to and from the region where there was no road linking any part with this frontier area. To face with the war time transport crisis, operation of this railway system was handed over to the American Military Railway Service who had to perform an intensive movement with the existing endowments even at the cost of safety.

In consequence of the partition of India (1947), the Assam Railway was formed with the railway assets remained isolated within the north east region. Apart from its odd isolation, the Assam Railway had to suffer a lot from the non-existence of a major workshop to repair rolling stock. However, the problems were solved with the construction of the 342 km long *Assam link* from Kishanganj to Fakiragram through the Doars region in West Bengal. With effect of the reorganisation of the nationalised railways, the Assam Railway was merged into the *North Eastern Railway* (1958). In 1958, the railway network of the Mahananda plain, North

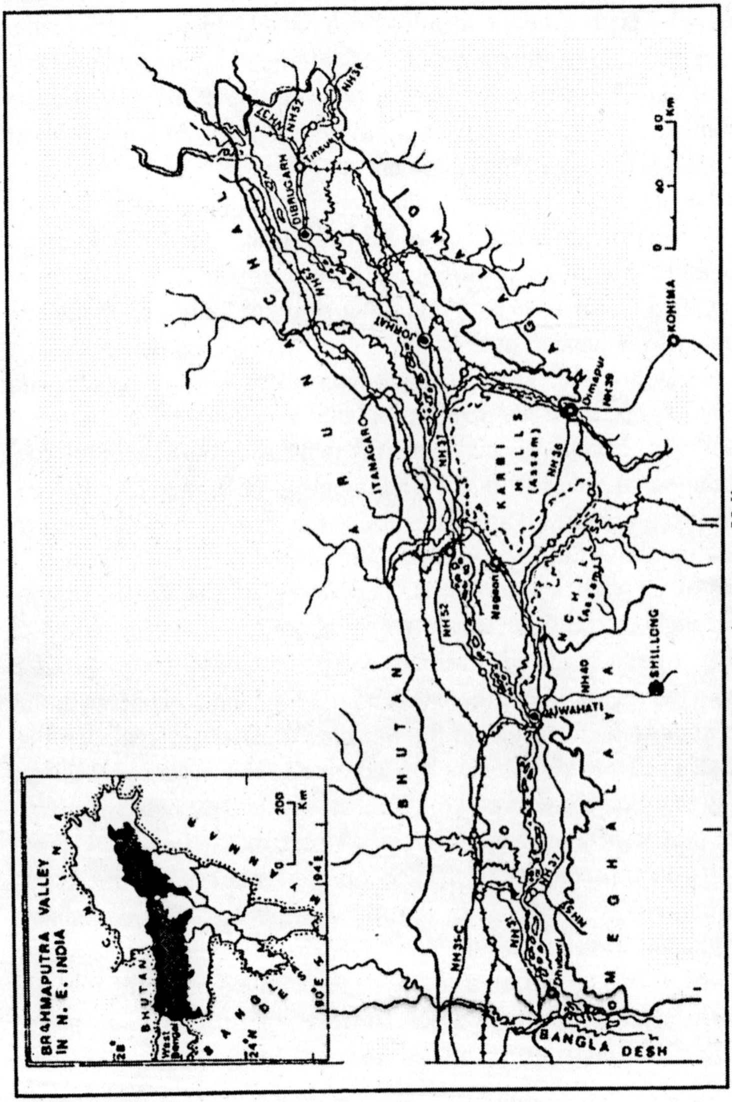

Fig. 28.1 : Brahmaputra Valley

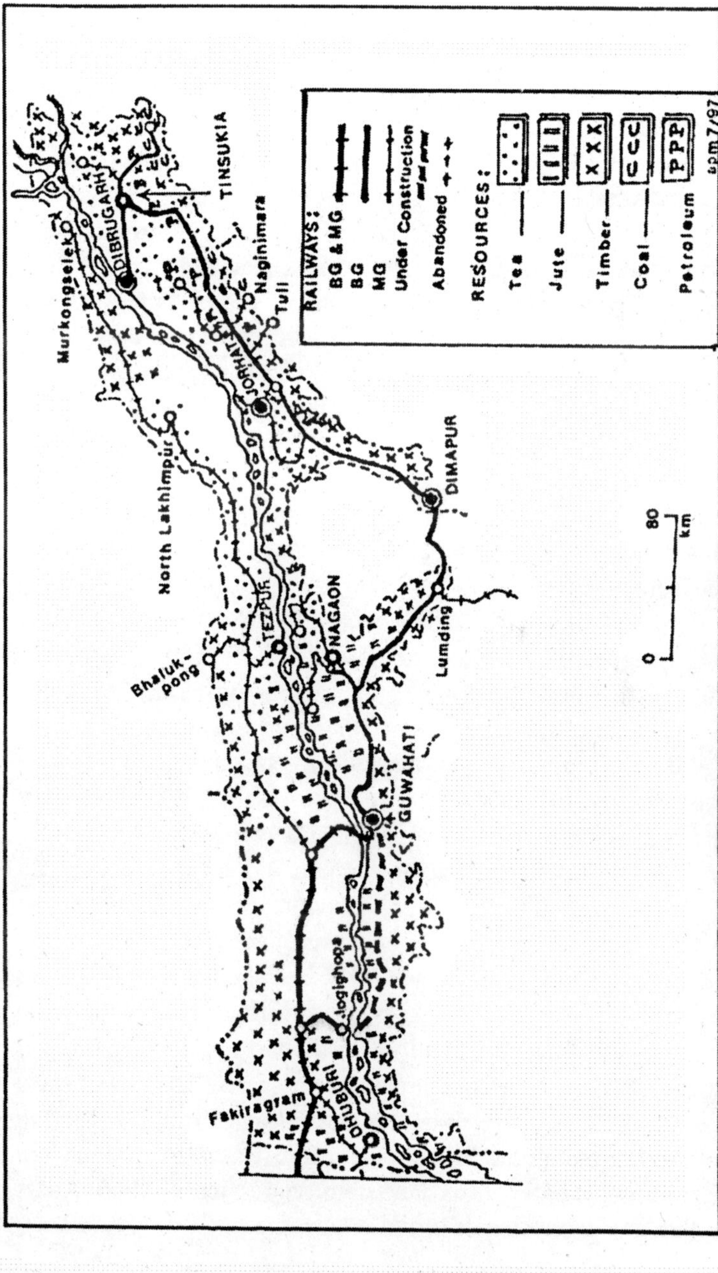

Fig. 28.2 : Brahmaputra Valley : Railway Network

Bengal and of the entire *North Eastern Region* was brought under a new zonal system of the *Northeast Frontier Railway.*

Immediately after independence, the railway system in this region, just a relic of the colonially aligned and operated railway system in the British period, could hardly able to keep pace with the nationwide process of development. First of all, the creation of the hostile state East Pakistan and delinked rail connections to Chittagong and Kolkata port from this region. Secondly, the meter guage network in the region could provide direct access only to Kandala port at a distance of 2812 km even from Guwahati and to the economically less prosperous North Bihar and Avadh-Rohila plain. Thirdly, the important urban, industrial and trade centres in south Bihar and West Bengal remained beyond the access of this network till .the construction of the Rajendra Pool on the Ganga and Garhara Yard near Barauni in Bihar. Further, the ferry break points at Pandu and at Maniharighat put hurdles on the free growth of traffics between this region and the rests. Over and above, its dendritic network, while aiming at the forest, mining and colonial plantation sectors hardly able to keep pace with the mobiliary needs of the leading townships in the Valley.

The Brahmaputra Valley being the heartland of the entire hilly frontiers in the northeast Region adjacent to China, Myanmar and Bangladesh has its acute demand of an improved rail transport facilities to cope with both defence and civil traffics at the time of war and peace. As an urgent measure to enhance the mobility on this network the capacity of the weakest Assam link section of the main line was increased by 93 per cent in 1962. Similarly, the ferry transshipment problem at Pandu-Amingaon point was solved with the construction of the *Saraighat Bridge* on the Brahmaputra. This first bridge on the Brahmaputra was opened for defence traffic in the time of Chinese aggression (1962) even at the stage before tis completion. The Chinese aggression, offering a bitter experience of inconvenient transport facilities in the North East Region, had compelled the Central Government to extend 327 km long meter guage line to Murkongselek on the North Bank Plain. Further, the Khejuriaghat-Haldibari broad guage line was also extended to Jogighopa immediately after Pakistani aggression in 1965. Apart from this, the expansion of broad guage line to Guwahati (1984), opening of meter guage branch lines to Bhalukpong (1990) and to Tuli (1995) and the ongoing process of constructing the second Rail-cum-Road Bridge alongwith a broad guage line from Jogighopa Valley are some of the milestones in the history of railway transport evolution in this region. Meanwhile, the construction of the Farakka Bridge-cum-Barrage (1971) on the Ganga and the broad guage conversion of the north east Railway main line (1984) had improved direct access of the region with the rest of the country. Over and above, the ongoing process of broad guage conversion of the existing meter guage network likely to bring about a new strategy of railway transport for the development of the valley in many ways.

Commodity Flow Structure

Out of 222 railway nodes in the valley, 127 nodes are opened for goods traffics. As many as 43 different commodities belonging to seven categories use to constitute a daily average bulk of 14948.39 tonnes of traffics handled at these nodes. Of the total freight volume, 15.22 per cent account for intra-regional transaction within the valley while other 3.9 per cent are exchanged between the valley and the other areas in the north east Region. The balance of 80.88 per cent, however, represents the transit quanta of the valley with the mainland of the nation through West Bengal territory (Table 28.1). Thus, the railway transport in the valley is found acting basically for interstate mobility of material for long distance.

A traffic, while moving from source to destination in horizonatal dimension use to cover a vertical distance, the difference of sectoral positions between the originating and terminating nodes in relation to the utilisation regime of the traffic item. Usually, the utilisation process of a materialistic resourse starts at primary sector and proceeds through the successive sectors of semi-manufacturing, intensive manufacturing, trade consumption and so on, but strictly at a distance apart from one sector to another. In this way, commodity movement from source to destination establishes a linkage gradient, the ratio of vertical to horizontal displacement involving the trip. The steepness of such a gradient, directly proportional to vertical and inversely proportional to the horizontal distance between the link heads, practically determines the intensity of flow on the link.

Table 28.1 : *Commodity linkages of the Brahmaputra Valley Through Railways*
tonne/day : 1985-95

Traffic Item	Intra-Regional	With N. E. Region		With Mainland	
		S	R	S	R
I. Non-manufactured					
Product of forest	3.47	—	21.93	149.92	37.68
„ of Mines	576.30	50.22	2.50	1925.08	75.37
„ of Agriculture	0.90	—	—	104.11	3.99
Total :	580.67	50.22	24.43	2179.11	117.04
II. Semi-manufactured:					
Product of forest	67.03	—	42.24	992.33	1.16
„ of Mines	219.85	28.62	—	1988.23	54.81
„ of Agirculture	—	—	—	3.05	—
Total :	286.88	28.62	42.24	2983.61	55.97
III. Manufactured : (For Consumption or trade)	10.77	122.86	0.43	61.50	573.95
IV. Food-Stuff: (from Primary Sector)	860.41	3.61	1.11	734.55	1874.09
V. Infrastructrue goods :					
for Construction	0.24	3.88	130.50	—	1153.72
„ Agriculture	8.52	6.78	—	330.80	1.53
„ Transport*	255.11	72.89	—	—	1415.27
Total :	263.87	83.55	130.50	330.80	2570.52
VI. Strategic goods :	25.57	6.22	3.78	13.14	20.62
VII. Non-Classifiable :	246.65	54.74	30.41	141.22	435.18
Grand Total	2274.82	349.82	232.90	6443.48	5647.37

*Includes materials for Broad guage Conversion work

S= Shipment R= Receipt

Source : Goods Earning Charts at railway nodes.

The vertical growth of economic activities throughout the northeast India is hindered by an overall scarcity of intensive manufacturing and trade functions in the Region. Predominance of the primary and the occasional semi-manufacturing activities, therefore, determines the vertical range of resource functioning in the region. Eventually, the limited vertical dimension remains incapable of establishing a steep gradient on the intraregional resource linkage in spite of the helping component of shorter horizontal distance involved in these local transactions. As such, the opposite heads of less than one fifth of the commodity linkages of valley through railways remain within the northeast Region.

The Brahmaputra Valley and its environs being one of the profound source regions of tea, jute, timber, plywood, coal and petroleum products use to establish a series of strong and permanent upward export linkage with several trade-cum-industrial sectors in the various parts of the country. On the opposite direction, the huge volume of inward bulk of foodgrain, salt, sugar, edible oil cement, steel goods and machineries from the various higher sectors in the country to the primary, semi-manufacturing and consumption sectors in this valley use to set up the import linkages overwhelmingly in downward direction. It appears that a distinctly steep gradient attributes to the strength and permanency of all outward and inward links joining the opposite heads at sufficient horizontal as well as vertical gaps.

The existing railway freight structure through its principle of telescopic rating has been greatly reducing the friction of distance on the intergional trade-links to and from the Valley with the various regions in the mainland of India. Owing to this telescopic rate structure, an overall monetary cost on the export links of wooden logs, cane, bamboo-cutting, coal, jute, railway sleeper, timber and planks, plywood petroleum products, tea, paper, fertilizer and oil-cakes has been reduced by 16 per cent over the proportionate mileage rate applicable for 50-100 km long trip. Similarly, the import bulk of foodgrain, potatoes, onion, vegetables, fruits, salt, sugar, edible oil and infrastructure goods to this region is availing of a 24 per cent reduction in transit cost indirectly gained out of this principle. Besides telescopic rating, the railway transport through its commoditywise discriminated tariff policy, based on the value of service principle, able to re-set the friction of distance and apparently the gradient of interregional linkages of different inward and outward commodities to and from the Brahmaputra valley in a critical manner. Through this principle of ability to pay, the lowest freight charge is imposed on the transit of the foodgrain, salt, coal and many other cheap commodities sources usually at primary sector. On the other hand, intensively manufactured items, machineries, electronic goods and costly chemicals are charged at the highest rate. Linkage gradient of the profound interregional commodity transits, therefore, undergoes its change susceptible to the discriminated tariff rates (Table 28.2).

Table 28.2 : *Discriminatory Effects on Selected Transits*

Commodity links	Trade Size (Tonne/Day)	Rate Index
Outwards :		
1. Coal, Fertilizers	2227	120
2. Log, Cane, Timber	896	130
3. Bamboo-Cutting	116	152
4. Tea	52	178
5. Jute, Paper	113	183
6. Plywood	131	235
7. Petroleum Products	1988	246
Inwards :		
8. Foodgrain, Salt, Potato, Onion	2114	100
9. Sugar, Cement	1028	141
10. Edible Oil	71	146
11. Steel Equipment	294	263

Source : 1. Table - 1
2. IRCA, Goods Tariff No. 42.

Commodity transit behaviour in the Bramaputra Valley through railway transport is relatively less sensitive of the factor of route distance than what is expressed in the normal Gravity Model. Rather this behaviour is susceptible to the factors like (i) Vertical range of the resoure functioning regime, (ii) Telescopic rating, and (iii) Commoditywise discriminated tariff structure of the Indian Railways. The mere route distance component, as has been worked out, determines 15.21 per cent and 29.16 per cent growth of the interregional commodity linkages to and from the Brahmaputra valley through railways. Growth of outward linkage, on the other hand, is determined by the vertical range of resource utilisation regime to the extent of 32.49 per cent, effect of telescopic rates - 8.41 per cent and by the discriminated tariff structure - 26.01 per cent. The same set of determinats effecting the growth of inward linkage is found to be 14.44 per cent, 12.25 per cent and 56.25 per cent respectively. Thus, it appears that behind each outward linkage from the valley there is the simultaneous considerations of (i) Where the consignment is awaited for, (ii) How far the destination is, and (iii) What is the transit charge. But, the growth of inward linkage is overwhelmed by the consideration of which one is receiveable at low transit charge.

Conclusion

The railway transport indeed has been providing a vital infrastructure of economic development, administrative efficiency and defence strength in the Brahmaputra valley and its environs. The histroy of railway transport evolution reveals that introduction, expansion and innovative measures of this facilitative sector are synchronizing with some sharp and immediate demand of mobility in the region where no alternative of those wheels on iron-surface could serve the purpose. Evolution of railway transport in the pre-independence decades had witnessed a river-port and sea-port oriented alignment of the network designed to drain out the exploited resources in the clonial manner. Alongwith the Independence, and overnight change in the geopolitical location of the Region had demanded an immediate re-design of its network and organisation as well. Besides, a gradual transformation of economic, cultural and defence regime makes a new headway of the city oriented railway linkages from this region to the rest of the country.

The entire process of resource functioning, a regime of resource exploitation, transfer, transformation structure of resource-links. Commodity flow occurs to provide linkage between the distinctly dichotomous sectors of resource functioning across the barrier of distance. Thus, the flow-frequency or the linkage strength is assumed correlating to the vertical range of the resource utilisation regime. Commodity linkages through railways to the from the various sectors in the valley too are susceptible to the vertical range of different economic sectors. A limited vertical range alongwith an imbalanced growth of the existing economic sectors in the valley boosts the strength of interregional commodity linkages at the cost of intra-regional links. Indeed the railway transport reinforces the outflow of commodities from the primary and the semi-manufacturing sectors in the region extending its peculier advantages of telescopism and commoditywise descriminated freight rates on transit. The telescopic rate decelerates the usual growth of intra-regional linkage by a general reduction in distance friction. The discriminated freight rates, on the other hand, deprive the growth of import linkage while mobilising the intensively manufactured infrastructure-cum-capital goods to this region at high rate. Besides this, the same rate principle is widening the way of drainage of raw materials while mobilising at a rate lower than that is applicable to the same bulk at manufactured stage.

Strategy of railway transport development for the commodity linkages in the Brahmaputra valley is of quite colonial manner. From this region, 80.12 per cent of the outward bulk use to move in non-manufactured or semi-manufactured stages. Similarly, 43.35 per cent of the receipt bulk of the region represent foodgrain and other edible items coming from the various sectors. This instance indeed enables to reflect a prolonged colonial phase in the regional economy of the valley where the nature creates, men exploit and the railway transport drains out the resources and finally where some foodgrains are poured to keep the egg-laying goose alive, but not to insist her to pluck directly from a paddy field.

An unrestricted outflow of the primary and the marginally manufactured products, a sluggish inflow of capital-cum-infrastructure goods and an easy entrance of edible-items are the basic involvement of railway transport in the valley. The first involvement use to hinder the growth of resource-based industries in the region to its optimum level, while the second one restricts the usual expansion of both resource and market-based industries in the region out of an inadequate supply of capital goods. The third involvement has been greatly discouraging the reginal economy to be self-sufficient in food-products and to put the agricultural lanscape into an intensive use. A slow process of industrialisation and agricultural development turns the regional economy to move round the intensive exploitation on the non-renewable primary sectors. This, a colonial pattern of exploitating the regional economy endures even long after the retreatment of colonial administration.

References

Govt. of Assam, (1967) Regional Transport Survey.

Mahanta, A.P. (1992) 'Railway Transport and Its Impact on Resource Utilisation in the Brahmaputra Valley', Ph. D. Thesis presented to Guwahati University

Medhi, S.B. (1978) *Transport System and Economic Development of Assam*, Publicaiton Board, Assam, Guwahati.

N.F. Railway, *Annual Report*, 1958-5950 1994-95.

North Eastern Geographer, Vol - 27.

National Academy of Sciences, (1963) *The Science of Geography*, Washington.

O.Dell, A. C. and P.S. Richard, (1971) *Railways and Geography*, Hutchinson & Co. London.

Raza, M. and Aggarwal, Y. (1986) *Transport Geography of India*, Concept Publishing Co., New Delhi.

Railway Board, (1902) *Report on the Company Managed Railways.*

Railway Board. (1978) Suppliment to the History of Indian Railways.

RITES, (1985) Regional Transport Survery of North East Region, Railway Board.

Singh, V. K. (1984) *Railway Freight Rates in India*, Verman & Co., Shaharanpur.

Thakur, P. (ed.) (1988) *Profile of A Development Strategy for India's North East*, Span Publications, Guwahati.

Werner, C. (1985) *Special Transportation Modeling*. Sage Publications, London.

APPENDIX - I

COMMODITYWISE BREAK-UP OF RAIL-BORNE GOODS

I. Products of Forest :

(a) Non-Manufactured : (1) Fire-wood (2) Cane (3) Brooms (4) Wooden log (5) Bamboo.

(b) Semi-Manufactured : (6) Rly. sleeper (7) Timber and Planks (8) Plywood and Tea-chest.

II. Products of Mine

(a) Non-Manufactured : (9) Coal (10) Stone, Ballast, Sand and Pebbles.

(b) Semi-Manufactured : (11) Petrol, oil and Lubricants (POL), (12) Bitumen, P. Wax, P. Coke and Naptha (13) Rock-Slabls.

III. Products of Agriculture

(a) Food-Item : (14) Grains and Pulses (15) Potato and Onion (16) Dry-fish.

(b) Cash Crops : (17) Tea (18) Oil-Seeds (19) Cotton (20) Jute (2.) Coconut. Areca-nut and dry-supari (22) Ginger, Tezpat, Chilly.

IV. Chemicals and Factory Inputs

(23) Sulphur (24) Iodized Salt and Sodium-sulphate for paper industry.

V. Infrastructure Goods

(25) Cement (26) Steel-goods (27) Non-Ferrous metalic goods (28) Non-metalic products (29) Tools and Machineries (30) Tea-garden equipments (31) Motor-vans and spare parts (32) ONGC/Oil consignment.

VI. Products of Manufacturing Sector

a) *Food Items :* (33) Salt (34) Sugar (35) Edible oil & Fats.

(*b*) *Productive Inputs* : (36) Chemical ftertilizer (37) Oil-cakes and Insecticides.

(*c*) *Consumer's goods* : (38) Paper (39) Gunny-bags (40) Kits (41) Soaps and Cosmetics.

VII. Redistributive Traffics

(42) Strategic goods (43) Equipments of the Rly. Dept. (RMC).

VIII. Unclassifiable Group

Includes Betel leaf, Tobacco, lime-stone, Bomboo-cane-wooden craft, Fruits and Vegetables, Raw-fishes and other occasional items in variable quantity.

29

THE GEOGRAPHICAL STUDY OF TRANSPORT NETWORK IN AMETHI TALUKA OF UTTAR PRADESH

S. C. Sharma, D. N. Tripathi and *O. P. Mishra*

Introduction

Transportation network of a region plays a pivotal role in reducing the spatial disparities and bringing about a balanced and integrated development. It is the basic infrastructure which helps in the proper exploitation of the available resources of a region and thus in establishing an economic equilibrium. Transportation network is, thus, a necessary element of the spatial reorganization of a region. It influences not only the economic life of a region but also affects the social, cultural and political life. Growth with equality and distributive justice is possible only with a proper development of transportation network. Thus, transportation planning must find an important place in integrated rural development planning of an area. This chapter tries to look into an optimum network of transportation system after analysing the existing position for the Amethi taluka of Sultanpur district in Uttar Pradesh in India. This taluka is an agricultural region characterised by inadequacy of social facilities and basic infrastructures.

Existing Transport Network

Total length of the roads in the study area is 318.27 km (145.20 km metalled and 173.07 km unmetalled). The density of metalled roads (per 100 sq. km is 24.76 km) and per lakh persons (59.45 km). If the unmetalled and other roads are also included in it, these figures rise to 54.29 km and 130.32 km, respectively. The data indicates that there are sub-regional disparities in the density of roads, for

Table 29.1 : *Length and Density of Roads*

	Metalled Roads			Unmetalled Roads			Total Roads		
Development Block	Length	Density per sq.km	Density per 100000	Length	Density per 100 sq.km	Density per 100000	Length	Density per 100 sq.km	Density per 100000
	persons			persons			persons		
Amethi	53.90	34.00	72.38	33.47	21.00	44.95	87.37	55.00	117.32
Sangrampur	20.50	19.99	42.62	37.60	36.57	48.18	58.10	56.05	120.81
Bhadar	34.50	21.60	51.27	50.50	31.63	46.96	85.00	47.00	126.33
Bhetua	36.30	21.97	66.75	51.50	30.91	56.83	87.80	52.70	161.32
Taluk Total	145.20	24.76	59.45	173.07	29.52	70.12	318.27	54.29	130.32

(*Sources : The Indian Geographical Journal*, 66(2) December 1991)

example, Sangrampur and Amethi blocks have higher density of roads than that of Bhadar and Bhatau. It appears that in the recent past road development has been more in the neighbourhood of Amethi town.

Railways

The study area is served by two rail-routes of Northern Railway, viz. Lucknow-Pratapgarh-Varanasi and Faizabad-Pratapgarh-Allahabad. The former passes through the eastern and the later through the central part. The total length of railways is 29.40 km the density being 5.02 km per hundred sq. km and 12.05 km per 1,00,000 persons.

Extent and Connectivity of Transport

The inter-connectivity oi the transport network of a region and its intensity change over time and with levels of socio-economic development. In the present study, for analysing the existing system of transportation network of metalled and semi-metalled roads. All the central places of the study area have been recognized as vertices. The central places which are completely unconnected with the edges have not been included in the study and the interaction points (crossings) have been considered as extra vertices. Thus, 29 vertices and 38 edges have been located. Extent and connectivity of the network are analysed by computing different indices. The Pie index is calculated by comparing the total length of the network with the network diameter. Larger the value of Pie, the greater is the extent of the network. This index for the region is 7.77 which indicates that there is a lesser extent of network in the region.

The connectivity of the network refers to the degree to which the vertices of a network are directly connected to each others greater degree of connectivity with a transport increases the efficiency of the system. Indices, like Cyclomatic number, Alpha index, Beta index and Gamma index are used to measure the connectivity. Cyclomatic number is the simplest and most useful way of measuring connectivity. This shows the number of circuits in the network. It increases with the completeness of the connections and the complexity of network.The Cyclometic number for the study is 8 which indicates that the region has moderate connectivity.

Alpha, Beta and Gamma indices may evaluate the connectivity in a better way. Alpha index is a ratio between actual number of fundamental circuits to the maximum number of circuits theoretically possible in network. The values of Alpha ranges between 0 and 1, 0 indicates a branching form of network and 1, a completely connected network. This can be expressed in per cent too. In the study area, the value is 0.17 or 16.98 per cent showing a weak connectivity. Gamma index relates the actual number of edges with the maximum possible number of edges. The value of Gamma index also varies from 0 to 1. It is 0.09 or 9.3 per cent for the study area, again indicating a weak connectivity between nodes. Beta index studies the linkage intensity i.e. the total number of linkages per node. It is calculated by dividing the total number of edges in a network by the total number of vertices. The value ranges from 0 to 3 for a planar graph and infinity for non-planar graph. The study area has a Beta index of 1.31 which shows the low degree of linkages. Thus, it is clear from the three indices that the connectivity in the existing transport network is quite weak due to the absence of proper inter-connections between the vertices.

The Koning index and connectivity matrix have also been analysed (Table 29.4). The Koning index for taluka, namely, Maharajganj, Kakwa, Punpur, Ramganj and Durgapur are the maximum indicating that these vertices are poorly connected, and the minimum index of Takar-mafi shows that it enjoys the most central location vertices, followed by Navgirwa, Vishesharganj and Bhadar.

Accessibility

Accessibility measures the effectiveness of transportation system. Physical accessibility is seen by drawing isodroms at an interval of 2.50 km. (Fig. 29.2). Table 29.2 shows that 88 per cent of the villages are within 2.50 km from the roads and railway station and the remaining 12 per cent from 2.5 to 5 km. There is no village beyond 5 km. 99 per cent of villages of block Amethi has higher accessibility (within 2.5 km.). Accessibility norm is defined for developed and agricultural areas as 2.4 km from any road and 6.4 km from main road, only about 11 per cent of villages seems to be inaccessible.

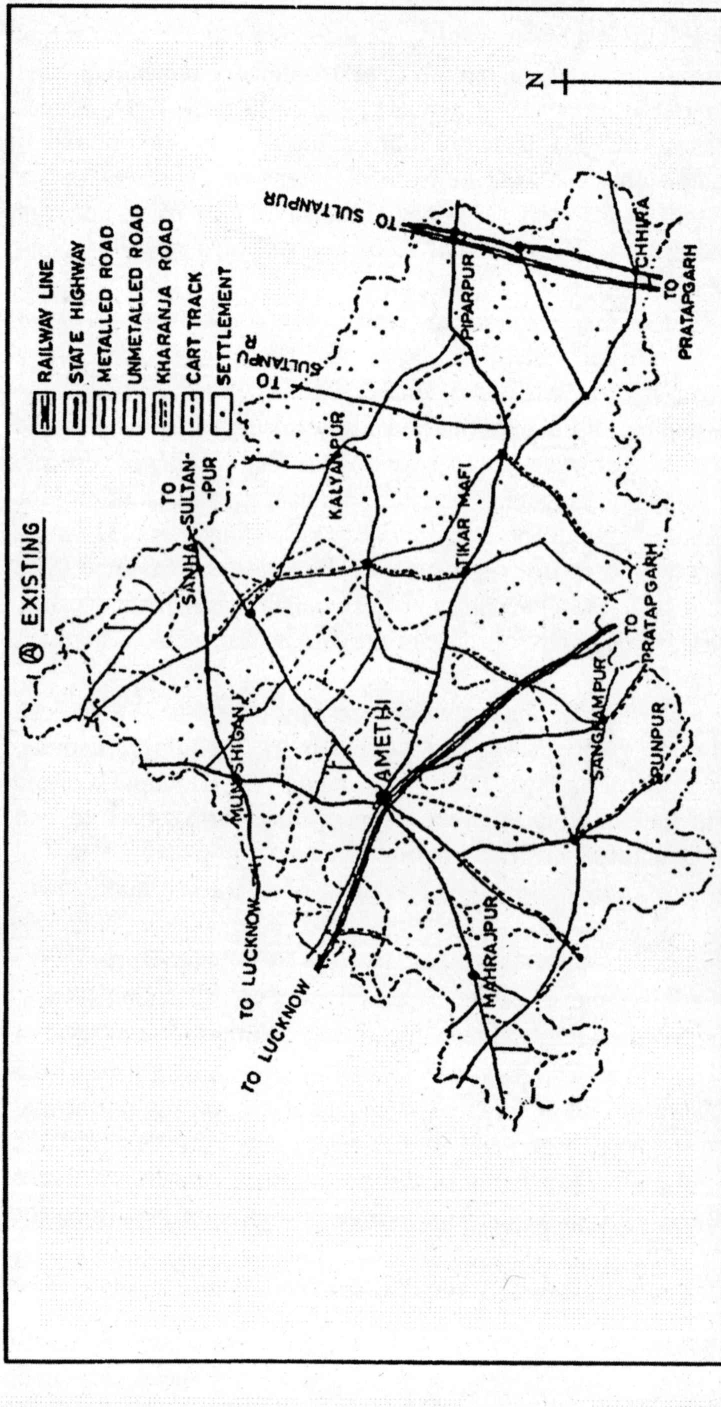

Fig. 29.1 : *Taluk Amethi : Road Network*

Table 29.2 : *Physicial Accessibility*

Development Block	Up to 2.50 km		2.51- 5.00 km	
	No. of Villages	Per cent of Villages	No. of Villages	Per cent of Villages
Amethi	85	99	1	1
Sangrampur	48	96	2	4
Bhadar	70	83	14	17
Bhetua	55	76	17	24
Total	258	88	34	46

Source : Fieldwork

Relative Accessibility is measured in terms of travel time distance between the highest order centre and independent places. In the present study, isocronic diagram (Fig.29.2) has been prepared by taking the travel time distance from Amethi, the taluka head-quarters, 48 per cent of villages is away from the taluka headquarters by a travel distance of 30 minutes and 39 per cent of 60 minutes. Again, Amethi block enjoys better accessibility, about 90 per cent of villages can be reached within 30 minutes (Table 29.3)

Table 29.3 : *Relative Accessiblity*

Development Block	30 Minutes		60 Minutes		90 Minutes	
	No. of Villages	No. of Per cent	No. of Villages	Per cent	Villages	Per cent
Amethi	27	90	9	10	---	---
Sangrampur	24	48	26	52	---	---
Bhadar	5	10	56	67	23	23
Bhetua	35	49	22	31	15	20
Total			17		38	

Source : Fieldwork

Nodal Accessibility is readability of one node from all nodes on the basis of presence node from all nodes on the basis of presence of absence of actual links and can be measured by the

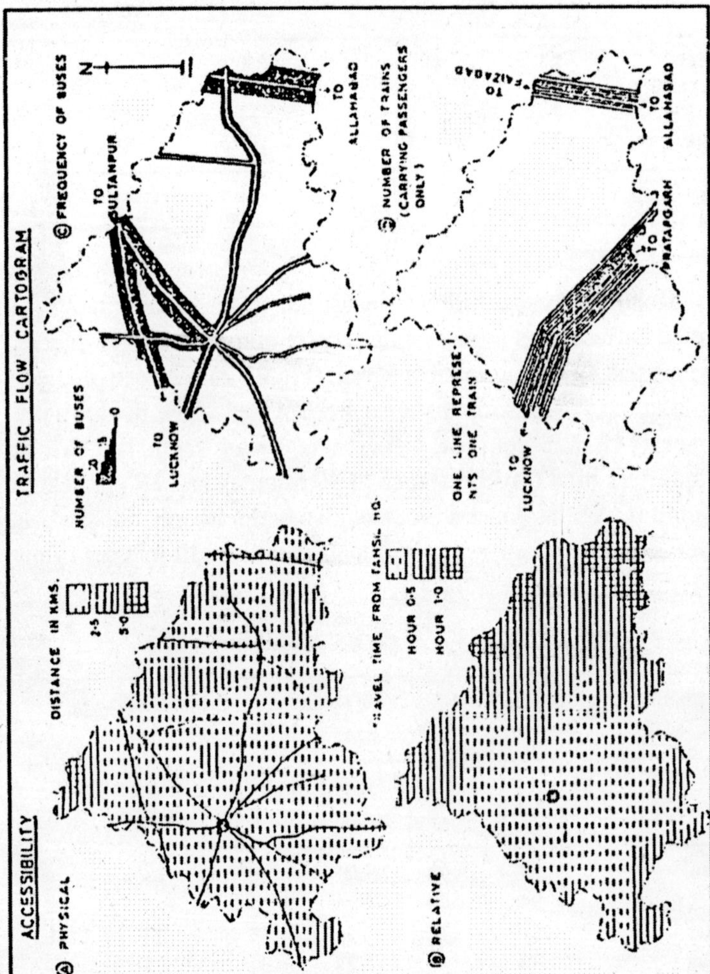

Fig. 29.2 : *Tahsil Amethi : Charactristics of Road Network*

shortest path matrix. Nodal accessibility indices of Amethi and Navgirwa are 66 each, the lowest in the region, showing that they enjoy the highest nodal accessibility. The highest value is that of Ramganj and Durgapur showing the lowest accessibility (Table 29.4).

Table 29.4 : *Identified Vertices and their Characteristics*

Identified Vertices	Koning Indes	Connectivity Matrix	Nodal Accessibility
Karari Hurshah	5	2	106
Munshiganj	7	3	99
Shivganj	6	3	87
Sanha	6	5	87
Navgirwa	5	5	66
Ram Nagar	7	2	90
Amethi	6	8	66
Tala	8	1	96
Karori-Girdharshah	7	2	92
Maharajpur	8	2	114
Kakwa	8	2	103
Kalyanpur	5	3	73
Bisheshraganj	7	5	87
Bargaon	7	1	112
Punpur	8	1	112
Sangrampur	7	3	90
Thengha	7	3	94
Hsyhkils	5	3	74
Tikar	4	3	77
Bhimi	5	4	79
Bhadar	5	5	82
May	6	2	103
Ayadhya	6	2	103
Nagar	6	1	108
Bhaurahara	6	2	103
Dhema	7	2	102
Trisundi	8	2	145
Ramgang	9	2	145
Duragapur	7	1	124
Piparpur			

Source : Author

Traffic Flow

Traffic flow is analysed by looking into frequency of bus trips. The frequency of bus trips is the highest on the Sultanpur-Pratapgarh road, where, on an average, 50 buses ply up and down in a day.

This is followed by Sultanpur-Raibareily and Amethi-Sultanpur road with 36 and 30 bus trips respectively. The concentration of bus traffic is very low on other rural roads (Fig. 29.2). There are 16 passenger-mail trains (up and down) on the Luckhnow-Amethi-Varanasi railwayline, while the number is 6 on the Faizabad-Allahabad routes.

Conclusion

The foregoing analysis of the transport network indicates spatial disparities and imbalances in their distribution pattern and connectivity. The density of roads and connectivity are higher in the block Amethi and Sangrampur than those of Bhadar and Bhatua. The link or feedar roads are indequate in the rural areas. The connectivity of all the central places except Amethi, Bhadar, Navgirwa and Visheshargani is low.

References

Bhatnagar, K. P. (1971) : *Transportation in Modern India*, Kishore Publication, Kanpur

Gillender, A. (1975) : Urban Transport Stuuies in Economic Policy", *Urban Studies Journal*, Vol.12, No. 2. June, 1975.

Patankar, P.G. (1978) : *Urban Mobility in Developing Countries*, Popular Publication, Bombay

Prakasa Rao, V.L.S. and V. K. Tewari (1979) : *The Structure of an Indian Metropolis: A Study of Bangalore*, Allied Publication, New Delhi.

Prakasa Rao , V.L.S. (1983) : *Urbanisation in India : Spatial Dimension*, Concept Publication, New Delhi.

Singh, J. (1964) : *Transport Geography of South Bihar*, Banarars Hindu University, Varanasi.

Singh, R.B. (1966) : Transport Geography of Uttar Pradesh, *National Geographical Journal of India*, Varanasi.

Sikdar, P.K. (1983) : Urbanisation and its Impact on Landuse - Transport Planning for Medium Size Cities in India, Paper submitted to I. R., April, 1983.

INDEX